Processing Series No. 5

Series Advisor: Professor B. L. Weiss

SILICIDE TECHNOLOGY FOR INTEGRATED CIRCUITS

Other volumes in this Series:

Volume 1　**Silicon wafer bonding technology for VLSI and MEMS applications** S. Iyer and A. Auberton-Herve (Editors)

Volume 2　**Process technology for silicon carbide devices** C.M. Zetterling (Editor)

Volume 3　**MEMS packaging** T.R Hsu (Editor)

Volume 4　**SIMOX** M. Anc (Editor)

SILICIDE TECHNOLOGY FOR INTEGRATED CIRCUITS

Edited by
Lih J. Chen

Department of Materials Science and Engineering
National Tsing Hua University

Published by: The Institution of Electrical Engineers, London, United Kingdom

© 2004: The Institution of Electrical Engineers

This publication is copyright under the Berne Convention and the Universal Copyright Convention. All rights reserved. Apart from any fair dealing for the purposes of research or private study, or criticism or review, as permitted under the Copyright, Designs and Patents Act, 1988, this publication may be reproduced, stored or transmitted, in any forms or by any means, only with the prior permission in writing of the publishers, or in the case of reprographic reproduction in accordance with the terms of licences issued by the Copyright Licensing Agency. Inquiries concerning reproduction outside those terms should be sent to the publishers at the undermentioned address:

The Institution of Electrical Engineers,
Michael Faraday House,
Six Hills Way, Stevenage
Herts., SG1 2AY, United Kingdom

While the authors and the publishers believe that the information and guidance given in this work are correct, all parties must rely upon their own skill and judgment when making use of them. Neither the authors nor the publishers assume any liability to anyone for any loss or damage caused by any error or omission in the work, whether such error or omission is the result of negligence or any other cause. Any and all such liability is disclaimed.

The moral rights of the authors to be identified as authors of this work have been asserted by them in accordance with the Copyright, Designs and Patents Act 1988.

British Library Cataloguing in Publication Data

Silicide technology for integrated circuits. – (Materials, devices and MEMS)
 1. Silicides 2. Integrated circuits – Materials
 I. Chen, Lih J. II. Institution of Electrical Engineers
 621 . 3'815

ISBN 0 86341 352 8

Typeset in India by Newgen Imaging Systems (P) Ltd., Chennai, India
Printed in the UK by MPG Books Limited, Bodmin, Cornwall

CONTENTS

	Editor	xi
	Authors	xiii
1	**Silicides – an introduction**	1
	J.W. Mayer and S.S. Lau	
	1.1 Origins and aluminum metallisation	1
	1.2 Pit formation and Al spikes	1
	1.3 Low temperature oxidation: the Au story	3
	1.4 Silicides and backscattering spectrometry	4
	1.5 Introduction to silicides of nanodimensions	4
2	**Silicide formation**	15
	L.J. Chen	
	2.1 Introduction	15
	2.2 Metal contacts	16
	2.2.1 Al/Si alloys	16
	2.2.2 Al/Si/Cu alloys	17
	2.2.3 Metal silicides	18
	2.3 Gate electrodes	32
	2.3.1 Polycrystalline WSi_x	32
	2.3.2 SALICIDE	33
	2.4 Interconnects	34
	2.4.1 Al/Cu alloys	34
	2.4.2 W interconnects	34
	2.4.3 Cu interconnects	35
	2.5 Plug	35
	2.5.1 Al-plug	35
	2.5.2 W-plug	37
	2.6 Diffusion barrier layer	39
	2.6.1 PVD-TiN	40
	2.6.2 CVD TiN and Ti	42
	2.6.3 Low-pressure chemical vapour deposition	42
	2.7 Adhesion and wetting layer	43
	2.8 Anti-reflection coating layer	44
	2.9 Conclusions	45
3	**Titanium silicide technology**	49
	Z. Ma and L.H. Allen	
	3.1 Introduction	49

	3.2	Formation of titanium silicides	49
	3.3	Fundamental aspects of Ti/Si thin film reaction	50
		3.3.1 Phase formation sequence	50
		3.3.2 Early stage of the Ti/Si thin film reaction	52
		3.3.3 Formation of metastable C49-TiSi$_2$	53
		3.3.4 C49-to-C54 TiSi$_2$ polymorphic transformation	54
		3.3.5 Agglomeration	59
	3.4	Integration concerns and scaling limits of titanium disilicide	61
		3.4.1 Effect of doping and dopant redistribution	61
		3.4.2 Reactions of Ti with SiO$_2$ and Si$_3$N$_4$	63
		3.4.3 Lateral silicide encroachment	63
		3.4.4 Thermal stability	64
		3.4.5 Chemical stability	65
		3.4.6 Contact resistivity	65
		3.4.7 Limitations of SALICIDE scaling	66
	3.5	Methods of enhancing the C54-TiSi$_2$ phase formation	68
		3.5.1 High temperature deposition	68
		3.5.2 Pre-amorphisation implant	69
		3.5.3 Metal impurity or interlayer	70
		3.5.4 Recessed spacer approach	72
	3.6	Conclusions	73
4	**Cobalt silicide technology** *T. Kikkawa, K. Inoue and K. Imai*		**77**
	4.1	Scope of the chapter	77
	4.2	Material properties	77
		4.2.1 Introduction	77
		4.2.2 Crystal structure	77
		4.2.3 Silicide phase diagram	78
	4.3	Fabrication technology	78
		4.3.1 Introduction	78
		4.3.2 Co salicide process	79
		4.3.3 Co single layer deposition	80
		4.3.4 TiN/Co bilayer deposition (TiN cap)	85
		4.3.5 Ti/Co bilayer deposition (Ti cap)	88
		4.3.6 Co/Ti bilayer deposition (Ti at interface)	89
	4.4	Electrical characteristics	90
		4.4.1 Introduction	90
		4.4.2 Effect of scaling	91
		4.4.3 Leakage current	92
		4.4.4 I–V characteristics	92
		4.4.5 Applicability	92
	4.5	Conclusions	93

5	**Nickel silicide technology**	95
	C. Lavoie, C. Detavernier and P. Besser	
	5.1 Scope of the chapter	95
	5.2 Introduction	95
	5.2.1 Limitations of $CoSi_2$	96
	5.2.2 Advantages and limitations of NiSi	98
	5.3 Characteristics and properties of Ni silicide phases	99
	5.3.1 Phase diagrams	99
	5.3.2 Crystal structures and volumetric changes	100
	5.3.3 Properties of Ni silicide phases	102
	5.3.4 Reduction in silicon consumption	104
	5.4 Phase formation	104
	5.4.1 Thermal budget	104
	5.4.2 Complex phase sequence	106
	5.4.3 Formation mechanism	114
	5.4.4 Disadvantages of low temperature diffusion	117
	5.5 Thermal expansion of NiSi and related stress effects	118
	5.5.1 Unit cell dimensions and CTE for bulk samples	119
	5.5.2 Unit cell dimensions and CTE for thin films	120
	5.5.3 Consequences of CTE anisotropy for stress in thin films of NiSi	122
	5.6 Texture of NiSi films on single crystal Si	124
	5.6.1 Pole figures and orientation distribution functions	124
	5.6.2 Standard classification of texture in thin films	125
	5.6.3 Pole figures for NiSi films on Si(001)	125
	5.6.4 Axiotaxy: a new type of texture	128
	5.7 High temperature limitations	129
	5.7.1 Formation of $NiSi_2$	129
	5.7.2 Morphological stability	130
	5.7.3 High temperature stabilisation of NiSi films	134
	5.8 Device characterisation	136
	5.8.1 Addressing the device roadmap	137
	5.8.2 Shallow junctions	139
	5.8.3 $NiSi_2$ and resistivity issues	141
	5.8.4 Dopant segregation	142
	5.9 Conclusions	143
6	**Light-emitting iron disilicide**	153
	L.J. Chou	
	6.1 Overview	153
	6.2 Growth methods	155

		6.2.1 Growing β-FeSi$_2$ by means of RDE	156
		6.2.2 Growing β-FeSi$_2$ through IBS	156
		6.2.3 Structure of the precipitate	158
	6.3	Effect of stress on the illuminating characteristics of β-FeSi$_2$	159
		6.3.1 Source of stress	159
		6.3.2 Effect of stress on β-FeSi$_2$	161
		6.3.3 Using Raman optical measurement to obtain the stress value	165
		6.3.4 Effect of stress on the feature of β-FeSi$_2$	168
		6.3.5 Optical properties	169

7 Silicide contacts for Si/Ge devices — 175
J.E. Burnette, M. Himmerlich and R.J. Nemanich

	7.1	Scope of the chapter	175
	7.2	Surface properties	176
		7.2.1 Si$_{1-x}$Ge$_x$/Si(001) surfaces	176
		7.2.2 Si$_{1-x}$Ge$_x$-Si(111) surfaces	178
		7.2.3 Stoichiometry of Si$_{1-x}$Ge$_x$ surfaces	179
	7.3	Formation and stability	181
		7.3.1 Interface thermodynamics	181
		7.3.2 Ti(Si/Ge) on Si/Ge	184
		7.3.3 CoSi$_2$ on Si/Ge	187
		7.3.4 NiSi on Si/Ge	190
		7.3.5 Other metals	192
	7.4	Electrical properties	194
	7.5	Summary	196

8 Silicide technology for SOI devices — 201
L.P. Ren and K.N. Tu

	8.1	Overview	201
	8.2	Source/drain engineering for SOI CMOS	202
		8.2.1 Introduction of SOI device structure	202
		8.2.2 Impact of series resistance	203
		8.2.3 Silicide design analysis	206
	8.3	Challenges to implement existing bulk silicide technology to SOI devices	208
		8.3.1 Titanium silicide process and voids	208
		8.3.2 Silicide thickness control	209
		8.3.3 Thin silicide thermal stability	209
		8.3.4 Alternative silicide choice	210
	8.4	Advanced silicide technology for SOI devices	211
		8.4.1 Ti silicide formation on pre-amorphised silicon	211
		8.4.2 Co silicide using Ti/Co or Co/Ti as source materials	217

		8.4.3 Ni silicide as a suitable candidate	220
		8.4.4 Low-barrier silicides of ErSi$_2$ and PtSi for ultra-thin devices	221
		8.4.5 Selective deposition of silicide on SOI	222
	8.5	Summary	225

9 Characterisation of metal silicides — 229
Y.F. Hsieh, S.L. Cheng and L.J. Chen

	9.1	Scope of the chapter	229
	9.2	Tools of materials characterisation	229
		9.2.1 Introduction	229
		9.2.2 Fundamental principle of beam–solid interaction	230
		9.2.3 Image resolution	232
		9.2.4 Probe size and detection limits	234
		9.2.5 Field of applications	235
	9.3	Morphology observation	237
	9.4	Crystal structure of metal silicides	240
		9.4.1 Epitaxial silicides	240
		9.4.2 Amorphous metal/Si alloy films	242
	9.5	Initial silicide formation	243
		9.5.1 Silicide formation in amorphous interlayer	243
	9.6	Phase formation and identification	245
		9.6.1 Silicide formation on patterned (001) and (111)Si wafers	245
		9.6.2 Effects of interposing layers on the formation of metal silicides	246
		9.6.3 Silicide formation by metal ion implantation into Si wafers	250
	9.7	Defect analysis	252
		9.7.1 Cracks (SEM, FIB, PTEM, XTEM, BF, DF, DP)	252
		9.7.2 Voids (XTEM)	254
		9.7.3 Pinholes (SEM, AFM, PTEM)	254
		9.7.4 Planar defects (PTEM, XTEM, 2B-DF, HRTEM)	254
		9.7.5 Vacancies (DP, simulation)	255
	9.8	Thermal stability (sheet resistance, PTEM, XTEM, BF, DF)	255

Appendix: Glossary — 263

Index — 271

EDITOR

L. J. Chen

Professor Chen received his Ph.D. in physics from UC Berkeley in 1994. He became a faculty member in the Department of Materials Science and Engineering (MSE), National Tsing Hua University (NTHU), Hsinchu, Taiwan in 1977. He is currently the Dean, College of Engineering and Chair Professor, MSE Department, NTHU. He has published more than 320 papers, mostly on metallisation and ion implantation in IC devices and more recently on nanomaterials in leading international journals. Professor Chen has won almost every major award and prize from the National Science Council and Ministry of Education in Taiwan. In 1998, he was elected the Outstanding Scholarship Foundation Chair (1998–2003) and Ministry of Education National Chair Professor (1998–2001) and a member of the Asian Pacific Academy of Advanced Materials. In 2001, he was awarded the Ministry of Education National Chair Professor on a permanent basis and elected to be a fellow by the American Vacuum Society. In recognition of his outstanding research, Tsing Hua University awarded him the Tsing Hua Chair Professor of Engineering in 2003.

AUTHORS

L. H. Allen
Leslie H. Allen is an Associate Professor in Materials Science at the University of Illinois, Urbana. He received his Ph.D. (1990) in Materials Science at Cornell University with Jim Mayer. Previous to his Ph.D. he worked in industry developing thin film photovoltaics and imaging arrays of IR detectors. His current research is focused on size-dependent phenomena in electronic materials, including spatially confined solid state reactions of silicides, melting point depression in metals, thickness dependence of the glass transition in thin polymer films, and the characterisation of self-assembled monolayer of alkanethiols. His group has developed a new heat capacity technique for thin films – nanocalorimetry that can be used to investigate the unusual thermodynamic characteristics of nanostructures and ultra thin films.

P. Besser
Paul Besser is an AMD Fellow in the Technology Development Group at Advanced Micro Devices in Sunnyvale, California. He earned his B.S. at NCSU (1988) and Ph.D. at Stanford (1993) in Materials Science and Engineering. Since joining AMD, Paul has led numerous metallisation, silicide and reliability teams for AMD, and has researched silicides at IMEC in 1997. He is currently researching stress effects in ultra-fine Cu interconnects and leading the metals group developing silicides, contacts, advanced barriers and novel metals for AMD's 65 nm Technology Platform. He has co-authored over 50 research publications and holds over 100 U.S. patents.

J. E. Burnette
James E. Burnette is currently a researcher in the Department of Physics at North Carolina State University. He completed his Ph.D. in 2004 at North Carolina State University and is a member of the Surface Science Laboratory at NCSU. He is a member of the American Physical Society and the National Society of Black Physicists. His research interests include Silicon-Germanium Heteroepitaxy, thin film growth and characterisation, and nanostructure formation on silicon and silicon-based materials.

S. L. Cheng
Shao-Liang Cheng received the Ph.D. degree in materials science and engineering from the National Tsing Hua University, Taiwan, R.O.C., in 1999. In 2003, he joined the faculty at the National

Central University, Taiwan, as an Assistant Professor in the Department of Chemical and Materials Engineering. His current research interests include the processing and characterisation of nanomaterials, structure of amorphous materials, and interfacial reactions in metal thin films/semiconductor systems. He has authored and co-authored more than 45 journal papers and 70 conference papers. He also holds one patent in Taiwan, R.O.C.

L. J. Chou

Li-Jen Chou was born in Tao-Yuan, Taiwan. He received his B.S. degree from the materials science and engineering department, National Tsing Hua University, Taiwan in 1985. He received his M.S. degree from the ECE department at the University of Missouri at Rolla in 1990. Immediately, he joined the MBE Group in the ECE department at the University of Illinois at Urbana-Champaign in 1991 and accomplished his Ph.D. degree in January, 1998. He is currently an associate professor in the materials science and engineering department at the National Tsing Hua University. His current research interests include growth and characterisation of Si base nano materials, growth and fabrication of GaN nano devices and high-resolution transmission electron microscopy.

C. Detavernier

Christophe Detavernier is a post-doctoral researcher at the Department of Solid-State Physics in Ghent University, where he graduated with a Master's degree in Physics in 1997 and a Ph.D. in Solid-State Physics in 2001. He was a visiting post-doctoral researcher at the IBM T.J. Watson Research Centre from January 2002 until December 2003. His research interests concern microstructural evolution (phase formation, grain growth, agglomeration) and texture in thin films. He has co-authored more than 30 publications and is a co-inventor on five patent applications.

M. Himmerlich

Marcel Himmerlich is a student of Technical Physics at Technical University Ilmenau, Germany. He is currently working on his diploma thesis about Photoelectron Emission Microscopy and laterally resolved Ultraviolet Photoelectron Spectroscopy in the research group of Professor J.A. Schaefer. This includes studies on semiconductor surfaces as well as applications of these techniques to tribological questions. In 2003 he joined the group of Professor Nemanich for a six-month internship at NC State University working on "Growth dynamics of germanium islands on silicon." For his Ph.D. studies he will be focusing on surface properties of wide band gap semiconductor thin films and their interaction with adsorbates in the Centre for Micro- and Nanotechnologies in Ilmenau.

Authors

Y. F. Hsieh

Yong-Fen Hsieh is currently the Chairman of Materials Analysis Technology Inc. She received the Ph.D. degrees in Materials Science and Engineering from National Tsing-Hua University in 1988 and joined AT&T Bell labs, Murray Hill, as a postdoctoral MTS in 1989–1991. She also joined ITRI/MRL, ERSO, United Microelectronics Corp., Unipac Optoelectronics Corp., and AU Optronics Inc. in her career development. She has more than 50 publications and holds 55 patents in IC, TFT-LCD, and III–V compound semiconductors. Dr. Hsieh was awarded the Distinguished Youth Award by Chinese EDMS (Electronic Device and Materials Society) in 1996.

K. Inoue

Ken Inoue received the B.S. and M.S. degrees in electrical engineering from Hosei University, Tokyo, Japan, in 1990, and 1992, respectively. He joined NEC Corporation, Japan, in 1992 and has been engaged in development of ULSI process technology. From 1993 to 1998, he was working on salicide technologies at the Thin Film Process Development Laboratory, ULSI Device Development Laboratories. He is currently working on Embedded DRAM Device Development at the Advanced Technology Development Division.

K. Imai

Kiyotaka Imai received the B.S. and M.S. degrees from Hokkaido University in 1984 and 1986, respectively. He joined NEC Corporation in 1986, where he has been working on research and development of BiCMOS and Si/SiGe HBT. Then he worked at the University of California, Berkeley as visiting researcher during 1995 and 1996. He received Ph.D. degree from Hiroshima University in 2003. Now he has been working on research and development of low power CMOS devices as an engineering manager of Advanced Device Development Division, NEC Electronics Corporation. Dr. Imai is a member of the Institute of Electronics, Information and Communication Engineers of Japan.

T. Kikkawa

Takamaro Kikkawa received the B.S. and M.S. degrees in Electronic Engineering from Shizuoka University, Japan, in 1974 and 1976 and the Ph.D. degree in Electronic System in 1994 from Tokyo Institute of Technology, Japan. He joined NEC Corporation, Japan in 1976. From 1983 to 1984 he was a Visiting Scientist at the Massachusetts Institute of Technology. In 1998, he joined the faculty at Hiroshima University, Japan, where he is Professor at the

Research Centre for Nanodevices and Systems and the Department of Semiconductor Electronics and Integration Science, Graduate School of Advanced Sciences of Matter. His research focuses on material science and reliability physics for low-k dielectrics and metal interconnects, and wireless interconnect technology using silicon integrated antenna for ultra-wide band signal transmission.

S. S. Lau

S.S. Lau is a Professor at the University of California at San Diego. He started at Bell Telephone Laboratory with Thin Film Technology. At the California Institute of Technology he worked with Jim Mayer on ion beam analysis, Rutherford backscattering spectrometry and thin film reactions. At UCSD he has worked on Electronic materials and devices, in particular, wafer/materials integration, group IV and III–V compounds semiconductor contact technology, and Nitride FET processing and devices. He is a fellow of the American Physical Society, a Fulbright Scholar and the co-ordinating editor of Materials Science and Engineering Report.

C. Lavoie

Christian Lavoie is a Research Staff Member at the IBM Watson Research Centre in Yorktown Heights, New York and an Adjunct Professor at the Ecole Polytechnique de Montreal in Canada. He graduated from Polytechnique in Montreal, with a B.Eng.Phys. in 1988 and a M.A.Sc. in 1990 and earned his Ph.D. in Physics in 1995 at UBC in Vancouver, Canada. Since joining IBM in 1995 he has been involved in the development and application of in situ characterisation techniques, conducting research on the optimisation of processes and materials for CMOS devices. He is a co-author of more than 100 publications and a co-inventor on more than 40 patent applications.

Z. Ma

Zhiyong Ma is a senior principle engineer with Logic Technology Development Q&R at Intel Corporation. He received his B.S. degree in Metallurgical Engineering from Shanghai University of Technology, China in 1984, M.S. degree in Materials Engineering from Purdue University in 1990, and Ph.D. degree in Materials Science and Engineering from University of Illinois at Urbana-Champaign in 1994. He worked at Digital Equipment Corporation as a senior process engineer before joining Intel in 1995. His primary research interest includes thin film metallisation, kinetics and analytical methods. He has published more than 20 papers and filed several patents.

Authors

J. W. Mayer
James W. Mayer is a member of the National Academy of Engineering and Regents Professor at Arizona State University. He started at Hughes Research Laboratory with nuclear particle detectors and ion implantation. At the California Institute of Technology he worked on ion beam analysis, with emphasis on Rutherford backscattering spectrometry. He and S.S. Lau were involved with thin film reactions. He was the Bard Professor of Materials Science at Cornell for 12 years and studied silicide formation, solid-phase epitaxy and ion-beam mixing before moving to Arizona State University. He won the Von Hippel Award of the Materials Research Society and is a fellow of the American Physical Society.

R. J. Nemanich
Robert J. Nemanich is Professor, Department of Physics and associate member of the Department of Materials Science and Engineering at North Carolina State University. He completed his Ph.D. at the University of Chicago and joined Xerox PARC in 1976. In 1986 he moved to NC State. He has a long-standing involvement with the Materials Research Society, has served as President in 1998 and is currently the President of the International Union of Materials Research Societies. He is a Fellow of the American Physical Society and has served on the executive committee of the Division of Materials Physics. Nemanich is the Editor-in-Chief of the journal Diamond and Related Materials. His research has centred on Si based materials, carbon-based materials, and wide band gap nitrides and has focused on properties of surfaces, interfaces and thin films.

L. P. Ren
Ms. Liping Ren received two MS degrees in Electrical Engineering from the University of California at Los Angeles, CA, in 1999 and in Materials Science and Engineering from Harbin Institute of Technology, P. R. China. Her graduate work was about the advanced source/drain engineering for sub-0.18 micron CMOS devices and novel silicide technology for sub-0.18 micron CMOS on ultra-thin film SOI. In 1999, she joined the International Rectifier Corporation working in the IC R & D department for advanced device design and technology development. Ms. Ren holds two US patents. Her current research interests include the process integration, device modelling, and circuit design of deep submicron SOI/CMOS for low-power and high-speed application.

Authors

K. N. Tu

K. N. Tu received his Ph. D. degree in Applied Physics from Harvard University in 1968. He spent 25 years at IBM T. J. Watson Research Centre as Research Staff Member in the Physical Science Department. During that period, he also served as Senior Manager of Thin Film Science Department and Materials Science Department for 10 years. In September 1993, he joined the Deptartment of Materials Science and Engineering at UCLA as full professor. He is a Fellow of American Physical Society, The Metallurgical Society (TMS), and an Overseas Fellow of Churchill College, Cambridge University, UK and was president of the Materials Research Society in 1981. He received the Application to Practice Award from TMS in 1988 and Humboldt Award for US Senior Scientists in 1996 and was elected a member of Academia Sinica, Republic of China in 2002. He has over 350 journal publications, edited 13 proceedings, and co-authored a textbook on "Electronic thin Film Science," published by Macmillan in 1992. His research interests are in kinetic theories in thin films and nanoscale structures, metal-Si interfaces, electromigration, and Pb-free solder metallurgy.

Chapter 1

Silicides – an introduction

J.W. Mayer and S.S. Lau

1.1 ORIGINS AND ALUMINIUM METALLISATION

Paul Totta, IBM Fellow, wrote a marvellous report in 1968 on metallisation [1] that sets the stage for the present, over three decades later. The introduction reads [1]:

> The design trend in monolithic integrated circuits has moved rapidly in the direction of using smaller, faster devices and circuits in ever-increasing numbers on a single silicon chip. To further shorten the electrical path between active elements, much of the wiring, which had formerly been done on modules or printed circuit cards, is now in a second or third metallisation level on the chip. One of the present design limitations for the miniaturization trend is the device thin film metallurgy. The extent of reduction of thin film conductors is, in turn, restricted partly by intrinsic metal properties such as electromigration capability or conductivity, and also by processing limitations such as the ability to shape the film by photolithography and subtractive etching.

Origins and aluminium metallisation p.1

Pit formation and Al spikes p.1

Low temperature oxidation: the Au story p.3

Silicides and backscattering spectrometry p.4

Introduction to silicides of nanodimensions p.4

Acknowledgment p.12

References p.12

In this case, aluminium (Al) won over Ag- and Cu-based systems. Even when Al fell out of favour due to electromigration problems, the addition of Cu, "Cu-doped Al", fixed the electromigration problem.

The use of Al as metallisation in contact with Si, ran into problems due to "pit" formation in the Si following heat treatment at temperatures between 400 and 450°C.

1.2 PIT FORMATION AND Al SPIKES

To make electrical connections to the Si devices, a layer (usually 1–2 μm thick) of metal is deposited on a "patterned" Si wafer. The metal makes contact with the Si only in the openings in the oxide layer. The metal layer is then heated to ensure an intimate contact between the Si and metal layer. Al is favoured as the contact metal because it has a high electrical conductivity and it

Silicides – an introduction

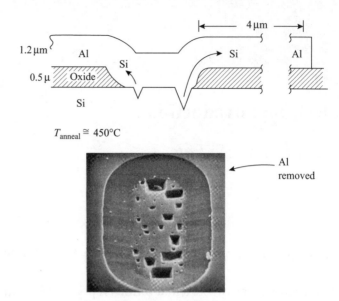

FIGURE 1.1 Pit formation in Si due to the heat treatment of the Al film in contact with the Si (taken from Mayer and Lau [2]).

forms a protective oxide layer on the top surface. There is one major problem associated with the use of Al. Silicon is soluble in Al, and during heat treatment, Si diffuses into Al, forming pits in the Si. The pitting phenomenon is shown schematically in FIGURE 1.1 [2].

The solubility of Si in Al is between 0.5 and 1.0 at.% at temperatures of 450–500°C (below the eutectic temperature of 577°C).

There have been numerous investigations of the diffusion of Si into Al. At a temperature around 450°C, Si can diffuse over 100 μm along the film in 30 min. The high diffusivity of Si into Al is responsible for the Al "spikes". Chemical etching to remove the Al layer also dissolves the regions in Si where the Al has penetrated (the spike region). The spikes are then revealed as pits in the Si; FIGURE 1.1. It is often possible to reduce the spiking problem by increasing the processing temperature rapidly – rapid thermal processing (RTP) – such that intimate contact is formed with only a limited amount of Si dissolving into the Al layer.

Up to the mid-1970s, the formation of pits was not too troublesome because the amount of Si dissolved in the Al could be controlled by designing the Al pattern correctly (referred to as Al design rules in pattern layouts). In recent applications, the Al spikes in the Si have become deep enough to destroy the operation of a transistor. It is common practice to deposit an

Al film containing 3–5 at.% of Si. The excess Si in the Al can precipitate during heat treatment and form crystallites of Si on the Si surface.

1.3 LOW TEMPERATURE OXIDATION: THE Au STORY

Gold is not the right candidate for a metal–silicon, low-resistance, ohmic contact. The eutectic temperature is too low. Further, if an Au layer is deposited on Si and heated in ambient air (or even dipped in boiling water), the surface appears to have a greenish tinge – a silicon oxide layer forms [3].

The formation of the oxide layer is initiated by the release of the silicon atoms from the single-crystal substrate and the subsequent migration of the atoms through the Au layer. This is indicated schematically in FIGURE 1.2. At the surface, the silicon atoms react with oxygen to form an SiO_2 layer. The SiO_2 layer grows at the Au/SiO_2 interface (oxygen diffusing through the oxide layer). The strong ambient effects show that the oxidising species diffuse through the oxide layer.

The Au/Si interface also plays a role in the process. The presence of a thin oxide layer at the interface between Au and Si can prevent the release of silicon. The characteristics of the Au/Si interface may also be responsible for the fact that growth of the oxide layer is about five times faster on $\langle 110 \rangle$ oriented silicon than on $\langle 111 \rangle$ oriented samples.

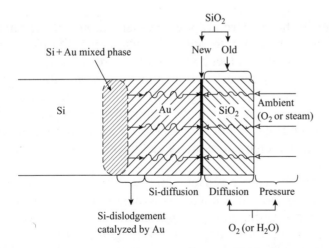

FIGURE 1.2 Model for the mechanism of SiO_2 formation at temperatures below the Si/Au eutectic point (taken from Hiraki et al. [3]).

Silicides – an introduction

The importance of the Au/Si story is that it demonstrated the existence of solid–solid reactions in thin layers of metal on Si at temperatures well below the eutectic [4]. Investigations on silicides soon followed [5].

1.4 SILICIDES AND BACKSCATTERING SPECTROMETRY

Silicides became important in the early 1970s. Researchers at IBM Research Labs, Yorktown Heights, AT&T, Bell Labs, Murray Hill and the California Institute of Technology were at the forefront. These researchers were all fierce competitors and had used ion beam analysis with megaelectronvolt ^{4}He ions to study ion implantation in silicon. They fell upon silicides with a vengeance.

Backscattering spectrometry was the analytical method of choice [6]. The equipment was simple; an accelerator that produced megaelectronvolt He ions (these researchers had all worked with ion accelerators in implantation studies) and a detector/multichannel analyser to record and display the energy from the metal film [7]. The samples were ideal – thin metal (a few thousand Angstroms thick, nanometres were not used) – the spectra were easy to analyse (FIGURE 1.3), and silicide formation obvious.

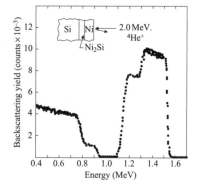

FIGURE 1.3 Backscattering spectrum for 2.0 MeV ^{4}He ions incident on a multilayer sample with Ni$_2$Si formed between Ni and the Si substrate (taken from Chu et al. [7]).

1.5 INTRODUCTION TO SILICIDES OF NANODIMENSIONS

Silicides in the modern era are an essential part of the nanoelectronics and photonics technology, where the silicides serve as interconnects and contact pads. Silicides are also used as contacts for many novel devices such as nanowires/nanotubes of various materials. These structures, grown by various techniques, are often tangled up, and difficult to be manipulated into orderly forms for nanoelectronic applications. We focus in this discussion on the formation, characteristics and the positioning of nanoscale silicides.

In nanoscale science and technology at the molecular level, positional control and self-replication [8] are often required. Positional control signifies the ability to put atoms exactly where they are intended to be; for example, the arrangement of carbon atoms in fabricating diamond. Self-replication is the ability to reproduce itself, thus enabling the products to be made inexpensively. For example: potatoes are complex to make on a molecular level,

Silicides – an introduction

yet they grow by adding water to dirt plus air and sunlight, potatoes are, therefore, self-replicating and inexpensive to grow. These two requirements can sometimes be met in biological nanotechnology on a molecular level, whereas in nanoelectronics both of these requirements are difficult to meet, especially the requirement of self-replication.

In the following, we discuss the issue of positional control of nanoscale silicides, to the extent of silicide-feature location, but not necessarily in atomic positioning of the metal and Si atoms. We begin by introducing the concept of sub-micrometre size silicides used in conventional CMOS technology. Silicides have been widely used as self-aligned silicide (salicide) contacts on the source, drain and poly-silicon (poly-Si) gate regions of CMOS devices to reduce the contact and series resistances [9–12]. It is well known that the resistivity of the commonly used silicides, such as $TiSi_2$ and $CoSi_2$, increases as the gate line-width reduces to sub-micrometre dimensions [11, 12]. Nickel silicide (NiSi) has recently emerged as a more suitable silicide for sub-micrometre CMOS devices [13–17]. We use the results reported in Reference 18 as an example to illustrate the thermal stability of sub-micrometre NiSi as a function of line-width. NiSi was formed on three types of Si substrates by depositing ~300 Å of Ni film on blank poly-Si, amorphous Si (a-Si) films and single crystalline Si substrates [18]. The samples were annealed from 200 to 800°C (in steps of 50°C) for 1 h in N_2 or in forming gas. The electrical properties of the NiSi film were measured using the Hall effect, and the layer structure and composition using Rutherford backscattering spectrometry (RBS).

FIGURE 1.4 shows the sheet resistance of the Ni silicide layer (Ni ~300 Å) on different Si substrates after annealing at 400–800°C for 1 h in N_2. Backscattering measurements on single-crystal Si substrates indicate that a thin NiSi film (~660 Å thick, see FIGURE 1.5) has been formed after annealing at 400–750°C. The silicide layer has a stable sheet resistance of about 1.8 Ω/sq, with a calculated resistivity of about 12 µΩ cm. After annealing at 800°C, the sheet resistance further increases to ~3.3 Ω/sq (resistivity increases to ~18 µΩ cm). This observation suggests that the higher resistivity phase $NiSi_2$ has begun to form.

On poly-Si substrates, a silicide layer with stable sheet resistance of about 2.2 Ω/sq was observed after annealing at 400–700°C. The calculated resistivity is around 15 µΩ cm for a NiSi layer of ~660 Å thickness (see FIGURE 1.5). The slightly higher resistivity compared with that on single-crystalline Si wafers is likely due to a grainier microstructure of the NiSi layer formed on poly-Si. After annealing at 750°C, the resistance increases to about 4.4 Ω/sq, suggesting the beginning of the formation of $NiSi_2$ mixed

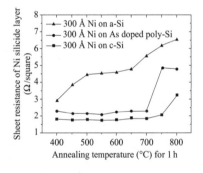

FIGURE 1.4 Sheet resistance of Ni silicide layer (Ni = 300 Å) on amorphous Si (a-Si), poly-crystalline Si (poly-Si) and crystalline Si (c-Si) after the 400–800°C annealing (from Reference 18).

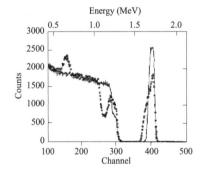

FIGURE 1.5 RBS spectra of 660 Å thick NiSi film on single crystal Si wafers (upper curve) and on poly-Si (lower curve) substrates after annealing at 750°C for 1 h (from Reference 18).

in with the NiSi phase. In addition to the formation of $NiSi_2$, the increase in the resistivity can also be due to the growth of the poly-Si grains into the NiSi layer. The NiSi formed on poly-Si is thermally less stable than that formed on single-crystal Si substrates.

The backscattering results shown in FIGURE 1.5 indicate that the NiSi layer is still in single phase on single-crystal Si wafers after annealing at 750°C for 1 h (within the detection limit of backscattering). On poly-Si substrates, however, the surface Si signal increases (at about channel 300), the peak Ni signal decreases (at about channel 400) and Ni signal spreads to the left, indicating that the silicide layer is no longer a uniform NiSi layer. Computer simulations showed that $NiSi_2$ has been formed, and optical microscopic inspection showed that Si grains also grow

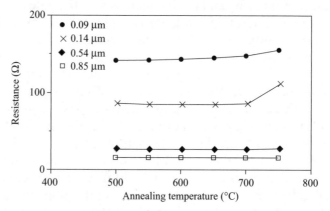

FIGURE 1.6   Resistance of NiSi film formed on arsenic-doped poly-Si lines ($L = 50\,\mu m$, $W = 0.09$–$0.85\,\mu m$) after the 500–750°C/1 h annealing (from Reference 18).

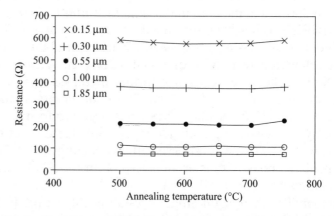

FIGURE 1.7   Resistance of NiSi film formed on boron-doped poly-Si lines ($L = 50\,\mu m$, $W = 0.15$–$1.85\,\mu m$) after the 500–750°C/1 h annealing (from Reference 18).

into the silicide layer, breaking up the silicide layer. These results are consistent with the sheet resistance measurements shown in FIGURE 1.4.

For the case of NiSi formed on a-Si, also shown in FIGURE 1.4, the sheet resistance of the silicide is unstable. After annealing at 400°C for 1 h, the sheet resistance of the silicide film is ~2.9 Ω/sq, indicating a resistivity of ~20 $\mu\Omega$ cm, hence the silicide layer is not a uniform NiSi layer. As the annealing temperature increases to 800°C, the sheet resistance increases to ~7 Ω/sq (resistivity ~50 $\mu\Omega$ cm). Backscattering indicated a non-uniform silicide layer at annealing temperatures as low as 400°C (RBS results are not shown here). Ni has a relatively high diffusivity in a-Si, resulting in Ni reactions with the entire a-Si layer to form various silicides (and mixed in with the a-Si layer) at annealing temperature as low as 400°C. It is known that a metal, such as Ni and Pd, in contact with a-Si enhances recrystallisation of the a-Si and the NiSi reaction rates [19]. These reactions render the Ni films on a-Si thermally very unstable.

FIGURE 1.6 shows the resistance R of the NiSi lines (with length, $L = 50\,\mu$m, thickness, $t = 600$ Å and linewidth, W, ranging from 0.09 to 0.85 μm) formed on arsenic-doped poly-Si lines after annealing at 500–750°C for 1 h. These silicide fine lines were delineated using electron beam lithography. The resistance of the silicide/poly-Si line is roughly equal to the resistance R of the silicide line alone, where $R = \rho L/tW$, ρ being the resistivity. The test pattern of the silicide lines has a four-pad contact structure. R was found by measuring the voltage difference between the two pads located on the left-hand side of the silicide line, while constant current was passing through the two pads located on the right-hand side of the line. By plotting R versus $1/W$, ρ can be deduced from the slope of the plot, provided that the plot is a straight line. It can be seen from FIGURE 1.7 that for each line-width, the resistance (hence the resistivity) of the NiSi line is rather constant with the annealing temperature ranging from 500 to 700°C. After annealing at 750°C, the resistance increases for the NiSi lines with smaller line-widths of 0.09 and 0.14 μm, suggesting thermal instability of fine silicide lines. From a linear plot of resistance versus 1/line-width at fixed annealing temperature, the resistivity for the NiSi lines is found to be around 15 $\mu\Omega$ cm.

The resistance of the NiSi lines formed on boron-doped poly-Si lines (with $L = 50\,\mu$m, $t = 600$ Å, $W = 0.15$–1.85 μm) is shown in FIGURE 1.7. The thermal stability is similar to the NiSi lines on arsenic-doped poly-Si lines. After annealing at 500–750°C for 1 h, the resistance of the NiSi line for each line-width is almost constant with the annealing temperature. The corresponding resistivity

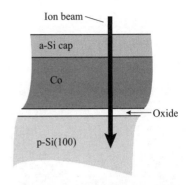

FIGURE 1.8 Layer structure for ion beam to form CoSi$_2$. Nominal layer thicknesses are: a-Si (5 nm), Co (34 nm) and SiO$_x$ (2 nm) (from Reference 21).

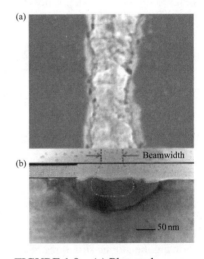

FIGURE 1.9 (a) Plan and (b) cross-section views of a CoSi$_2$ wire formed by ion-beam mixing. The dotted outline in (b) represents the implant profile in the as-deposited layer structure. The inset of (b) represents the as-deposited layers above the Si substrate. The inset layers (from top to bottom) are a-Si, Co and SiO$_x$. The solid line across (b) represents the height of the surface prior to the implant. Arrows depict beam width (from Reference 21).

for the NiSi lines is found to be around $14\,\mu\Omega\,\text{cm}$. NiSi is therefore stable on poly-Si up to $\sim 750°\text{C}$ with a line-width of $\sim 0.1\,\mu\text{m}$.

As we can see from the example above, the formation characteristic and the thermal stability of NiSi depend on the structure of the reacting Si substrate and the line-width. These observations can be generalized to a number of silicides, including $TiSi_2$ [8] and $CoSi_2$ [8, 20], commonly used as fine featured interconnects and contacts in nanoelectronic devices.

In addition to the commonly used lithographic methods for positional control of fine-line silicides, focused ion-beam mixing has been explored for the fabrication of sub-micrometre silicide structure without the use of photoresist-based lithography methods [21, 22]. FIGURE 1.8 shows the schematics of the layer structures of ion-induced $CoSi_2$ formation. For an ion beam focused to the size of a few tens of nanometres, the induced silicides can be of similar size as the ion beam. Ion-beam mixing of the layers forms regions of vacancy and interstitial pairs in the Si substrate. The interfacial barrier oxide is locally disrupted (see FIGURE 1.8). These damaged regions provide fast diffusion paths for Co and Si atoms, reducing the activation barriers for silicide formation. After ion-mixing with As^{2+} ions at 200 keV and a dose of a few times of 10^{15}–$10^{16}/\text{cm}^2$, the samples were processed by rapid thermal annealing (up to $\sim 750°\text{C}$) to form $CoSi_2$ and, perhaps, repair the Si lattice damage. FIGURE 1.9 shows the plan and cross-sectional views of a $CoSi_2$ wire formed by focused ion-beam mixing. The silicide line is broader than the ion-beam due to thermal annealing, which causes lateral silicide reactions. FIGURE 1.10 shows a test structure for measuring the electrical properties of the ion-induced silicide. The $CoSi_2$ silicide lines are formed by ion-beam mixing. The measurement shows that the resistivity of the $CoSi_2$ ranges from 12 to 23 $\mu\Omega\,\text{cm}$, indicating that reasonably low resistivity fine-line silicide can be fabricated using focused ion-beam mixing without photoresist. The value of the resistivity depends on the processing of ion-mixing and subsequent annealing. The positioning of the silicide lines is well controlled by the focused ion beam. Application of this direct-write method to other metal–Si systems, such as Ni, Pd or Cr may also be achieved, and allows for self-aligned formation of silicides directly on gate oxide films.

While the silicide lines discussed above are of sub-micrometre dimensions, one-dimensional effects of electron transport in conductors are generally not observed until sub-10-nm in lateral dimensions are reached [23]. One of the major roadblocks in developing nanoscale electronic circuits is the lack of an

FIGURE 1.10 Micrograph of sheet resistance test structure with deposited contact pad metal. The silicide lines are made using the focused ion-beam mixing process (from Reference 22).

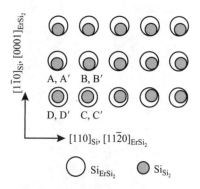

FIGURE 1.11 Schematics showing the atomic structures of $ErSi_2$ and Si crystals, and the orientation relationship between them. The dark large and bright small spheres represent the Si and Er atoms, respectively. The Si atoms proposed to form the interface between the $ErSi_2$ and Si are marked with A, B, C, D, and A', B', C', D', respectively. The unified $ErSi_2$/Si interface is formed by superposing A on A', etc. The plan view illustrates the asymmetry of the lattice mismatches along the principal growth axes (from Reference 25).

efficient technology to fabricate such fine wires. E-beam, X-ray, focused ion-beam and other lithography techniques have been tried, sub-10-nm dimension fabrication proved to be possible, but difficult. We discuss here an alternative method of "self-assembly" technique for the fabrication of nanosilicide wire. This method is based on growing a silicide with an appropriate asymmetric lattice mismatch to Si to achieve sub-10-nm silicide wire formation [24, 25]. In general, epitaxial films as thin as one atomic layer can grow pseudomorphically when the lattice constant of the film matches that of the substrate to within 2%. As the mismatch increases to >2%, the strain energy in the deposited film can be relaxed by the creation of islands of the over-layer. If the epitaxial layer has the same symmetry as the substrate, the over-layer will "self-assemble" into symmetric islands. If the over-layer has small mismatch in one direction, but large mismatch in an orthogonal direction, that is, asymmetric mismatch, this can allow for an unrestricted growth of the over-layer in one direction, but limited growth in the other direction. This situation is illustrated in FIGURE 1.11 where an erbium disilicide, $ErSi_2$, over-layer is grown on a Si(100) substrate. The mismatches along the $[11-20]ErSi_2/[110]Si$ and $[0001]ErSi_2/[1-10]Si$ directions are -1.5 and 6.3%, respectively. As $ErSi_2$ grows on Si(001), the lattice mismatch strain increases much faster along the large mismatch direction of $[0001]ErSi_2$; thus, the growth takes place along the $[11\bar{2}0]ErSi_2$ direction to minimise the strain energy, and the growth along the $[0001]ErSi_2$ direction is limited. The resulting $ErSi_2$ nanowire structure grown on Si(001) is shown in FIGURE 1.12, where two orthogonal sets of $ErSi_2$ fine lines are clearly seen [24]. The width of these elongated nanosilicide lines is 3–5 nm, with a height of about 1 nm and length of 200–400 nm. FIGURE 1.13 shows a high-resolution STM topographic scan of an $ErSi_2$ wire, indicating the well-controlled shape and height. While these nanosilicides can be geometrically arranged on clean Si(001) surfaces, the locations of the formation of wires cannot be precisely controlled. This method of "self-assembled" nanowire can be applied to other silicide systems, possibly the rare-earth silicides and PtSi, with asymmetric lattice match.

In addition to the silicide interconnects and contact pads intended for more conventional electronic devices mentioned above, there is currently intense interest in one-dimensional nanostructures, such as free-standing nanotubes and nanowires, due to their potential use in the testing of fundamental concepts of dimensionality and their possibility of serving as building blocks for nanoscale devices [26]. The electrical and optical properties of

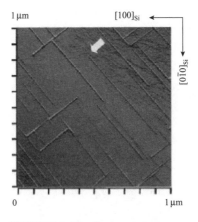

FIGURE 1.12 Low-resolution STM topograph showing $ErSi_2$ nanowires grown on a flat Si(001) substrate (from Reference 24).

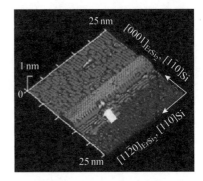

FIGURE 1.13 High-resolution (25 nm × 25 nm) STM topograph showing a section of an $ErSi_2$ nanowire and the Si(001) dimmer rows. The Si terraces increase in height from lower right to upper left (from Reference 24).

Silicides – an introduction

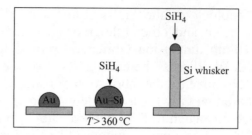

- VLS growth mechanism:
  - Au catalyzes SiH$_4$ decomposition
  - Nucleates growth of single crystal Si wire
  - Diameter of nanowire determined by size of Au particle and growth conditions
- Methods to control size of Au catalyst
  - Nanoscale patterning of Au, laser catalytic growth, thermal evaporation, Au nanoparticles solution phase growth

FIGURE 1.14 Vapour–liquid–solid growth of Si nanowires using the VLS technique (J.M. Redwing, Penn State, private communication).

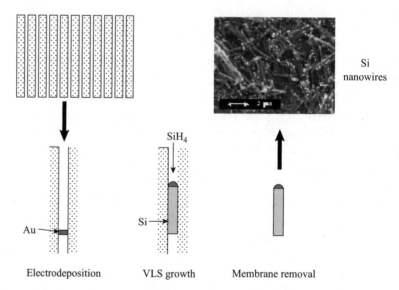

FIGURE 1.15 Template-directed synthesis with VLS growth of Si nanowires (from Reference 30).

semiconductor nanowires depend strongly on the wire diameter [27] as well as the crystallographic orientation [28] and defect structure of the material. We show an example here of the vapour–liquid–solid (VLS) growth method to produce single-crystal Si nanowires. In this technique, a metal, such as gold (Au), is used as a catalytic agent for nucleating whisker growth from a Si-containing vapour [29]. Au and Si form a liquid alloy that

Silicides – an introduction

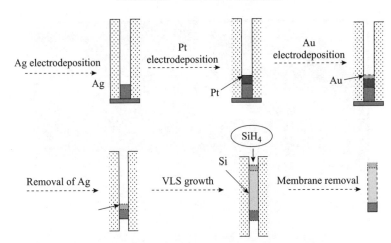

FIGURE 1.16 Formation of metal contacts with Au and AuPtSi on Si nanowires (J.M. Redwing and S.E. Mohney, Penn State, private communication. For more details on the formation of Co Silicides nanowires contacts, see reference 31).

has a eutectic temperature of 363°C, which, upon supersaturation, nucleates the growth of a Si wire. FIGURE 1.14 shows the schematics of the VLS growth technique. By combining template-directed synthesis with VLS growth, this technique provides a simple method for the production of large quantities of crystalline Si nanowires with uniform diameters that are determined by the pore size of the membrane (see FIGURE 1.15). Furthermore, by combining metal electrodeposition with VLS growth in nanoporous membranes, this technique enables the fabrication of metal–Si–metal nanowires and metal contacts on the nanowires (see FIGURE 1.16). FIGURE 1.17 shows an SEM micrograph of a Si nanowire with Au and $Pt_xAu_ySi_z$ contacts, formed by such a method. It can be seen that the Si nanowire is straight and the contacts are well defined. These Si straight wires can be aligned with an electric field, and placed on metal pads for electrical property measurements. FIGURE 1.18 shows an FESEM image of the back-gated test structure used for four-point resistance and gate-dependent conductance measurements of Si nanowires. A 3D schematic of the back-gated test structure is shown as an inset on the top left, along with a high magnification image of the SiNW on the top right. The separation between the D/S electrodes varied from 1.6 to 3.2 μm for different devices on the test structure. The contacts consist of Ti/Au metallisation for ease of processing. To form silicide nanocontacts, annealing is usually needed and may require more processing steps. From the discussion above, it is clear that the field of silicide nanowires research is still in its infancy; many technical issues remain unsolved, and much

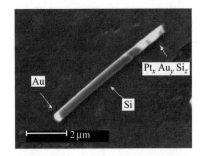

FIGURE 1.17 Si nanowires with Au and $Pt_xAu_ySi_z$ contacts (J.M. Redwing and S.E. Mohney, Penn State, private communication. For more details on the formation of Co Silicides nanowires contacts, see reference 31).

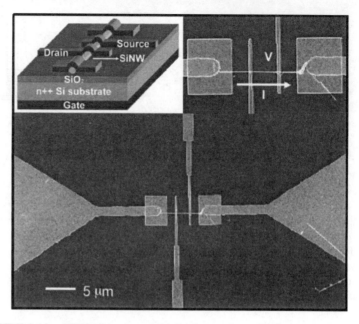

FIGURE 1.18 FESEM image of the back-gated test structure used for four-point resistance and gate-dependent conductance measurements of Si nanowires (from reference 32).

more work is required before silicides can be practically applied to nanoelectronics.

ACKNOWLEDGMENT

J. Mayer and S.S. Lau were supported in part by the NSF (L. Hess).

REFERENCES

[1] P.A. Totta, W.E. Mutter [High Conductivity Metallurgy for Multilevel Integrated Circuits, *IBM Report (USA)* (1968)]
[2] J.W. Mayer, S.S. Lau [*Electronic Materials Science: For Integrated Circuits in Si and GaAs* (Macmillan Publishing Company, New York, 1990)]
[3] A. Hiraki, E. Lugujjo, J.W. Mayer [Formation of silicon oxide over gold layers on silicon substrates, *J. Appl. Phys. (USA)* vol.43 (1972)]
[4] A. Hiraki, M.-A. Nicolet, J.W. Mayer [Low Temperature Migration of Silicon in Thin Layers of Gold and Platinum, *Appl. Phys. Lett. (USA)* vol.18 (1971) p.178–80]
[5] R.W. Bower, J.W. Mayer [Growth Kinetics in the Formation of Metal Silicides on Silicon, *Appl. Phys. Lett. (USA)* vol.20 (1972) p.359–61]
[6] M.-A. Nicolet, J.W. Mayer, I.V. Mitchell [Microanalysis of Materials by Backscattering Spectrometry, *Science (USA)* vol.177 (1972) p.841–9]
[7] W.-K. Chu, J.W. Mayer, M.-A. Nicolet [*Backscattering Spectrometry* (Academic Press, Inc., New York, 1978)]
[8] K. Eric Dexler [*Nanosystems* (Wiley Interscience, New York, 1992)]
[9] S.P. Murarka [*J. Vac. Sci. Technol. B (USA)* vol.4 (1986) p.1325–31]

[10] C.M. Osburn, Q.F. Wang, M. Kellam, et al. [*Appl. Surf. Sci. (Netherlands)* vol.53 (1991) p.291–312]
[11] K. Maex [*Mater. Sci. Eng. (Netherlands)* vol.R11 (1993) p.53–153]
[12] E.G. Colgan, J.P. Gambino, Q.Z. Hong [*Mater. Sci. Eng. (Netherlands)* vol.R16 (1996) p.43–96]
[13] T. Morimoto, T. Ohguro, H.S. Momose, et al. [*IEEE Trans. Electron Devices (USA)* vol.42 (1995) p.915–21]
[14] D.X. Xu, S.R. Das, J.P. McCaffrey, C.J. Peters, L.E. Erickson [*Mater. Res. Soc. Proc. (USA)* vol.402 (1996) p.59–64]
[15] E.G. Colgan, J.P. Gambino, B. Cunningham [*Mater. Chem. Phys. (Netherlands)* vol.46 (1996) p.209–14]
[16] Q.Z. Hong, Stella Q. Hong, F.M. D' Heurle, J.M.E. Harper [*Thin Solid Films (Switzerland)* vol.23 (1994) p.479–84]
[17] J. Chen, J-P. Colinge, D. Flandre, R. Gillon, J.P. Raskin, D. Vanhoenacker [*J. Electrochem. Soc. (USA)* vol.144 (1997) p.2437–2442]
[18] M.C. Poon, F. Deng, M. Chan, W.Y. Chan, S.S. Lau [Resistivity and Thermal Stability of Nickel Mono-silicide, *Appl. Surf. Sci. (Netherlands)* vol.157 (2000) p.29–34]
[19] Z. Jin, G.A. Bhat, M. Yeung, H.S. Kwok, M. Wong [Nickel Induced Crystallisation of Amorphous Silicon Thin Films, *J. Appl. Phys. (USA)* vol.84 (1998) p.194–200]
[20] C.-D. Lien, M. Finetti, M.-A. Nicolet, S.S. Lau [Electrical Properties of Thin Co_2Si, CoSi, and $CoSi_2$ Layers Grown on Evaporated Silicon, *J. Electron. Mater. (USA)* vol.13 (1984) p.95–105]
[21] M.M. Mitan, D.P. Pivin, T. Alford, J.W. Mayer [Direct Patterning of Nanometer-scale Silicide Structures on Silicon by Ion-beam Implantation Through a Thin Barrier Layer, *Appl. Phys. Lett. (USA)* vol.78 (2001) p.2727]
[22] M.M. Mitan, D.P. Pivin, T. Alford, J.W. Mayer [Patterning of Nanometer-scale Silicide Structures on Silicon by Direct Writing Focus Ion-beam Implantation, *Thin Solid Films (Switzerland)* vol.411 (2002) p.219]
[23] M. Sundaram, S.A. Chambers, P.F. Hopkins, A.C. Gossard [New Quantum Structures, *Science (USA)* vol.254 (1991) p.1326]
[24] Y. Chen, D.A.A. Ohleberg, G. Mederiros-Ribeiro, Y.A. Chang, R.S. Williams [Self-assembled Growth of Epitaxial Erbium Disilicide Nanowires on Si (001), *Appl. Phys. Lett. (USA)* vol.76 (2000) p.4004]
[25] C. Dekker [*Phys. Today* vol.52 (1999) p.22]
[26] Y. Chen, D.A.A. Ohleberg, G. Mederiros-Ribeiro, Y.A. Chang, R.S. Williams [Growth and Evolution of Epitaxial Erbium Disilicide Nanowires in Si (001), *Appl. Phys. A (USA)* vol.75 (2002) p.353]
[27] L. Brus [*J. Phys. Chem. (USA)* vol.98 (1994) p.3575]
[28] H. Yorikawa, H. Uchida, S. Muramatsu [*J. Appl. Phys. (USA)* vol.79 (1996) p.3619]
[29] R.S. Wagner, W.C. Ellis [*Appl. Phys. Lett. (USA)* vol.4 (1964) p.89]
[30] K.K. Lew, C. Reuther, A.H. Carim, J.M. Redwing [Template-directed Vapor-liquid-solid Growth of Silicon Nanowires, *J. Vac. Sci. Technol. B (USA)* vol.20 (2002) p.389–92]
[31] A.M. Mohammed, S. Dey, K.K. Lew, J.M. Redwing, S.E. Mohney [Fabrication of Cobalt Silicide nanowire contacts to Si nanowires, *J. Electrochem. Soc.* 150 (9) (2003) G577-G580]
[32] K. Lew, L. Pan, T.E. Bogart, S.M. Dilts, E.C. Dickey, J.M. Redwing, Y. Wang, M. Cabassi, T.S. Mayer, S.W. Novak [Structural and Electrical Properties of Trimethylboron-doped Silicon Nanowires, accepted for publication recently in APL (2004)]

Chapter 2

Silicide formation

L.J. Chen

2.1 INTRODUCTION

Metal thin films are integral parts of all microelectronics devices. They have been used in microelectronics devices such as ohmic contact, Schottky barrier contact, gate electrode, interconnect, diffusion barrier, adhesion layer and anti-reflecting layer.

With the advance in semiconductor device fabrication technology, the shrinkage in linewidth continues at a fast pace. According to the prediction of the roadmap by the Semiconductor Industry Association (SIA) in 2003, in the 0.13 µm generation devices, the gate length and thickness of silicide at the contact window are 50 and 25 nm, respectively. In the year 2007 for the 90 nm generation devices, these numbers will further decrease to 40 and 20 nm, correspondingly (TABLE 2.1) [1]. Further, more transistors will be incorporated in one chip. In addition, owing to the demand for increased integration level, the surface area was not adequate to meet the interconnect demand. Multilevel interconnections provide flexibility in circuit design and a substantial reduction in die size and thus chip cost. FIGURE 2.1 shows a scanning electron microscope (SEM) cross-section of a six-level metal

Introduction p.15

Metal contacts p.16
  Al/Si alloys p.16
  Al/Si/Cu alloys p.17
  Metal silicides p.18

Gate electrodes p.32
  Polycrystalline WSi_x p.32
  SALICIDE p.33

Interconnects p.34
  Al/Cu alloys p.34
  W interconnects p.34
  Cu interconnects p.35

Plug p.35
  Al-plug p.35
  W-plug p.37

Diffusion barrier layer p.39
  PVD-TiN p.40
  CVD TiN and Ti p.42
  Low-pressure chemical vapour deposition p.42

Adhesion and wetting layer p.43

Anti-reflection coating layer p.44

Conclusions p.45

References p.45

TABLE 2.1 Pertinent data selected from the 2003 International Semiconductor Technology Roadmap.

Year of first production	2003	2004	2007	2010	2013	2016
Technology node	100	90	65	45	32	22
DRAM half pitch (nm)	100	90	65	45	32	22
MPU gate length (nm)	45	37	25	18	13	9
Contact X_j (nm)	49.5	40.7	27.5[a]	NA[b]	NA	NA
Contact silicide thickness (nm)	25	20	17	13	19[c]	13

[a] Assuming the use of NiSi
[b] Assuming to be with the drain extension
[c] Assuming the raised source/drain

Silicide formation

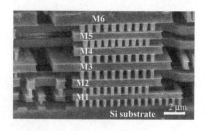

FIGURE 2.1 An SEM cross-section of a six-level metal backend structure (courtesy of UMC).

backend structure. Electrical connection between the various metal layers is provided by vertical interconnects commonly referred to as vias.

With the scaling down of devices, the contact windows and vias have become smaller. However, it is difficult to scale down the thickness of the interlayer dielectric. As a result, the aspect ratios of contact windows and vias increase significantly. Since the filling of plug into vias significantly influences the electrical properties and reliability of devices, to fill the metal into high aspect ratio and small holes is one of the most important challenges in the ultra large-scale integration (ULSI) technology.

Advanced semiconductor device interconnects are on the brink of a dramatic change. With the ever-increasing demand for faster logic devices and higher capacity memories, semiconductor designers will maintain the current trend toward shrinking design ground rules and increased interconnect packing densities. As interconnect lines shrink and move closer together in the sub-quarter-micrometre devices, the resistivity of Al wiring becomes unacceptably high, and other low-resistivity metals must be considered.

As the device size steadily scales down, the packing density increases. That means the number of active devices fabricated on a unit area is increased. In order to connect these devices effectively and to maintain the functions and complexity of the circuits, interconnection lines will occupy a large part of the chip area and the length of the interconnection lines will increase exponentially. The delay of signal propagation through the interconnection lines increases due to the increase in length, the interconnection resistance and parasitic capacitance (thinner layer, smaller lateral size) [2]. These opposing influences on the device access or switching time lead to a so-called interconnection limited circuit. To solve the problems mentioned earlier, an alternative metallisation method must be used. The multilevel interconnection technology is an effective way to overcome the problems.

2.2 METAL CONTACTS

2.2.1 Al/Si alloys

In the early times, Al was used exclusively as the conducting material in IC devices for its low resistivity ($\sim 2.8\,\mu\Omega\,\text{cm}$), good adhesion, ease in etching and good formability of ohmic contact on p^+/n^+ junctions [2]. However, as the size of contact windows shrinks to submicrometre regime, the shrinkage of contact area between metal and Si substrate leads to the increase in contact

Silicide formation

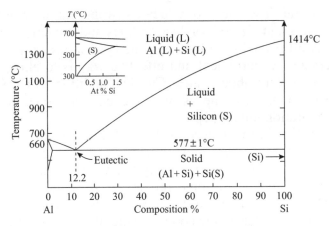

FIGURE 2.2 Al/Si phase diagram.

resistance. The detrimental influences of native oxide and other contaminants on the contact quality become more severe. A pre-cleaning step prior to the deposition of metal thin films becomes necessary.

Since in the backend processing steps, the processing temperatures often reach 450–500°C, the solubilities of Si in Al are in the range of 0.5–1 at.% as shown in the phase diagram of the Al/Si system in FIGURE 2.2 [3]. The Si atoms diffuse rapidly in the Al films and form pits in the silicon. An example is shown in FIGURE 2.3. The pitting in turn leads to the Al penetration or problem manifested by spiking. To overcome this disadvantage, about 1 at.% Si solutes were added to prevent Al spiking during subsequent thermal cycling. It is also possible to reduce the spiking problem by ramping up the processing temperature rapidly by rapid thermal processing (RTP) such that intimate contact is formed with only a limited amount of Si dissolving into the Al layer.

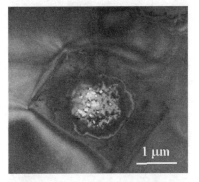

FIGURE 2.3 Pits in the silicon after the Al film is stripped off.

2.2.2 Al/Si/Cu alloys

In addition to forming contact, Al metallisation also serves as interconnects. As the dimensions of devices shrink, the interconnects are also reduced in thickness and width. Already, interconnect RC delay approaches the intrinsic gate delay. In addition, reduced interconnect line cross sections translate into increased current densities. Line failure due to electromigration becomes a serious concern for Al/Cu interconnect lines in the state-of-the-art semiconductor devices.

Electromigration is caused by the interaction of electrical carrier with the metal atoms. Atoms are transported mainly through the grain boundaries, creating voids on one end and hillocks on

Silicide formation

the other end. The ways to reduce electromigration include the increase in grain size of conducting lines and reduction in grain boundary diffusion by stuffing the grain boundaries with an alloying element. A convenient and effective way is to alloy the Al interconnect with about 1 at.% Cu. The addition of Cu solutes was used to increase the ability of conduction lines to resist electromigration failure [2].

Al metal has been replaced by Al/Si/Cu alloy in practical usage for some time. To improve the reliability of such contact structures, deeper junctions immediately below the contact openings were also employed in some processes. Although Al/Si alloy instead of pure Al prevents junction spiking, it also introduces other problems. During the cooling cycle of a thermal annealing, the solid solubility of Si in Al decreases with temperature. Thus Al would become supersaturated with Si, which causes nucleation and growth of Si precipitates from Al/Si alloy. An example is shown in FIGURE 2.4. Such precipitation occurred at both Al/SiO_2 interface and Al/Si interface in the contacts (preferentially around the contact hole edges where solid-phase epitaxial growth occurred easily). These Si precipitates are p-type, since Al is a p-type dopant in silicon. If such precipitates form at Al/n^+-Si contact, it leads to an undesirable increase in contact resistance. In addition, when the precipitates randomly formed within the narrow Al interconnect lines, these precipitates can be largely thought to obstruct a considerable cross-sectional area of the metal lines. At these locations a large divergence of current flux can be produced, thus leading to an early failure of devices by an electromigration-induced open circuit. In addition, the Si nodules may also lead to the etching problem of the anti-reflection coating (ARC)-TiN [4]. Cu often interacts with Al to form high-resistivity Al_2Cu as shown in FIGURE 2.5. As the device feature size scales down, these effects become more pronounced. As a result, more elaborate contact structures that can overcome the limitation of Al/Si contact have to be developed.

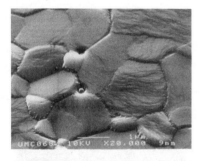

FIGURE 2.4   Silicon nodules in Al/Si alloy (courtesy of UMC).

FIGURE 2.5 Al_2Cu precipitates in Al/Cu alloy.

2.2.3   Metal silicides

There are more than a hundred silicides. However, only near noble, noble and refractory metal silicides are used in IC devices. For metallisation of devices, the general requirements are:

1. low resistivity
2. good adhesion to Si
3. low contact resistance to Si
4. appropriate Schottky barrier height or ohmic with heavily doped Si (n^+ or p^+)

Silicide formation

5. thermal stability
6. appropriate morphology for subsequent lithography and etching
7. high corrosion resistance
8. oxidation resistance
9. good adhesion to and minimal reaction with SiO_2
10. low interface stress
11. compatibility with other processing steps such as lithography and etching
12. minimising metal penetration
13. high electromigration resistance
14. formability at low temperature.

For metal contacts, a blanket metal thin film is usually deposited on a patterned wafer. Silicide is then formed selectively at contact windows after heat treatments. To prevent overlying Al or W from reacting with the silicide film, a diffusion barrier layer, such as TiN or Ti/W layer, is deposited between the silicide layer and overlying metal layer [5].

PtSi and Pd_2Si were used early on for metal contacts to lower the contact resistance of Al alloys as well as to serve as a diffusion barrier layer between Al alloy film and silicon. In the early 1980s, as the linewidth decreased to about 1 μm, many refractory metal silicide films, such as $MoSi_2$, WSi_2, $TiSi_2$ and $TaSi_2$ were used by different manufacturers. For the 0.25 μm technology, $TiSi_2$ was almost used exclusively [6]. For devices with linewidth of 0.18 μm or smaller, $TiSi_2$, $CoSi_2$ and NiSi are possible candidate contact materials [1]. A comparison of these silicides for ULSI applications is listed in TABLE 2.2.

2.2.3.1 Formation of silicides

Many different deposition techniques can be used to deposit metal thin films. Currently, sputtering is used almost exclusively to deposit metal layers for contacts or in the self-aligned silicidation (SALICIDE) process. On the other hand, chemical vapour deposition of WSi_x and W films is the dominant method to form gate electrode or local interconnect and metal plug, respectively. The merits of the different deposition methods will be described in a later section.

The usual steps to form a silicide are:

1. The wafers are first cleaned consecutively by organic solution, dilute HF and deionised water, followed by blown dry with a nitrogen gun or "spin–rinse–dry". An alternative is to dip the wafer in dilute HF then blown dry with a nitrogen gun or "spin dry".

TABLE 2.2 Comparison of C54-TiSi$_2$, CoSi$_2$ and NiSi for ULSI applications.

	C54-TiSi$_2$	CoSi$_2$	NiSi
Formation temperature (°C)	600–700	600–700	400–600
Thin film resistivity ($\mu\Omega$ cm)	13–20	14–20	14–20
Schottky barrier height (n-Si, eV)	0.6	0.64	0.67
Selective etching solution	NH$_4$OH:H$_2$O$_2$	HCl:H$_2$O$_2$	HNO$_3$:HCl
Dominant moving species	Si	Co	Ni
Si consumption ratio	2.27	3.64	1.83
Silicide thickness ratio	2.51	3.52	2.34
Melting point (°C)	1500	1326	992
Eutectic point (°C)	1330	1204	964
Thermal stability temperature (°C)	<950	900	700
Epitaxy on silicon	No	Yes	No
Reduction of SiO$_2$ by metal	Yes	No	No

2. The wafers are immediately placed in the metal deposition chamber, sputter-cleaning of the surface by Ar ions if necessary (Ar-sputtering may cause particle issue).
3. Deposition of metal thin films on silicon at room temperature or at a higher temperature.
4. Heat treatments, either by traditional furnace annealing or by rapid thermal annealing to form silicides.

Prior to the deposition of metal thin films, a 1.5–2 nm thick SiO$_2$ layer was usually present at the Si substrate surface following the etching of the thermal oxide. It is necessary for the metal contact metal layers to penetrate the thin oxide layer to react with the silicon to form silicides. Ti and Ni atoms are capable of penetrating through the thin oxide. On the other hand, Co atoms have difficulty to form silicide with Si if a thin oxide layer is present at the interface. An Ar ion sputter-cleaning step is usually required. Since CoSi$_2$ is widely used in 0.18 μm or smaller in linewidth devices, the formation of CoSi$_2$ is used as an example to illustrate the steps to form silicides on silicon. The deposition of Co thin films by sputtering is kept at room temperature. A mixture of Co$_2$Si and CoSi is formed at 300°C. CoSi$_2$ forms at 550°C [7]. For RTA, the first-step and second-step annealings are conducted at 500–550°C for 30–60 s and at 700–850°C for 30–60 s, respectively.

Interfacial reactions of metal thin films with silicon are rather peculiar in that polycrystalline metal film reacts with single-crystal silicon. The substrate is covalently bonded and the thin film is metallic. As a result, the microstructure of the silicide film and orientation of the substrate may play an important role in influencing the reaction. Some silicides can form at a temperature as low as 100°C. The mechanism for the break-up of silicon bonds at such a low temperature is rather intriguing [7]. Furthermore, the silicide phases formed at relatively low temperature are apparently related more to the growth kinetics rather than dictated by the thermodynamic consideration.

First nucleated phase and simultaneous occurrence of multiphases

In the metal–Si binary phase diagrams, three or more silicide phases can be usually found. However, only selective phases are detected after thermal annealing of metal thin films on silicon. From X-ray diffraction (XRD) and Rutherford backscattering (RBS) data, it was concluded initially that only one phase grows at a time for a clean system. This is consistent with the assertion that the formation of silicides is determined more by the growth kinetics than by energetics. However, more refined analysis by high-resolution transmission electron microscopy (HRTEM) in conjunction with fast Fourier transform (FFT) analysis indicated that formation of multiphases occurred in a number of refractory metal/Si systems [8].

The first phase nucleation is one of the long-standing problems in the interfacial reactions of metal thin films on silicon. Theoretically, difficulties were encountered in predicting the reaction product of a basically non-equilibrium process from equilibrium thermodynamics. Experimentally, conventional techniques, such as XRD and electron diffraction, are more appropriate for the detection of the growth phase rather than for the unambiguous identification of the first nucleated phase owing to their limited sensitivity and resolution in the determination of the phase. For Ti/Si system, as many as four different crystalline phases (Ti_5Si_3, $TiSi$, $C49-TiSi_2$ and $C54-TiSi_2$) were reported to form in thin film reactions [8]. The close proximity and/or overlapping of diffraction rings of these silicides rendered the unambiguous identification of phases during the initial stages of reactions rather difficult. As a result, although the final stable phase is known to be $C54-TiSi_2$, the phase formation sequence is still unclarified. Ti_5Si_3, $TiSi$ and $C49-TiSi_2$ were variously reported to be the first nucleated phase. HRTEM in conjunction with optical diffractometry were used to identify the first nucleated phase in silicide formation. The combined techniques are capable of unambiguous identification of a phase as small as 1 nm in size. The sensitivity compares favourably with conventional selected area electron diffraction and microdiffraction which correlate diffraction patterns with areas about 500 and 100 nm in size in the specimens, respectively. Since Ti is a strong oxygen getter, to minimise the influence of oxygen on the interfacial reactions in Ti/Si system, ultrahigh vacuum (UHV) deposition of the thin films is considered to be most appropriate.

In the Ti/Si system, Ti_5Si_3, located at the Ti/a-interlayer interface, was identified to be the first nucleated phase [8]. Ti_5Si_3, Ti_5Si_5, TiSi and C49-$TiSi_2$ along with amorphous interlayer were observed to be present simultaneously in samples annealed at higher temperatures. Examples are shown in FIGURES 2.6 and 2.7. Fundamental issues in silicide formation need to be addressed in light of the discovery of the formation of amorphous interlayer and as many as four different silicide phases in the initial stages of interfacial reactions of UHV-deposited Ti thin films on silicon.

For metal/silicon systems, the formation of amorphous interlayers during low temperature annealing has been reported to occur in Pt/Si, Ti/Si, Zr/Si, Hf/Si, Nb/Si, Ta/Si, V/Si and a number of rare-earth metal/Si systems [8–15]. Schwartz and Johnson proposed that a large negative heat of mixing and anomalously fast diffusion of one species in the other are necessary for solid-state amorphising reaction [16]. Walser and Bene (WB) proposed a rule for the first phase nucleation. The WB rule states that the first nucleated phase is the highest melting point silicide neighbouring the deepest eutectic point (the most stable congruently melting silicide) in the binary phase diagram [17]. The phenomenological rule is about 80–90% successful in predicting the first nucleated phase, the physical origin of which is not entirely clear at present and will not be elaborated here. The end phase of the reaction is usually the most silicon-rich silicide, often the disilicide, indicating that it is near the thermal equilibrium.

Good correlations were found among difference in atomic size and electronegativity between metal and Si atoms, the calculated free energy of mixing as well as the critical and maximum amorphous interlayer thickness for the Y/Si and a number of refractory metal/Si systems [18].

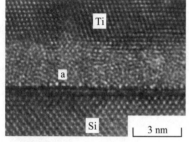

FIGURE 2.6 An amorphous interlayer at the Ti/(001)Si interface in an as-deposited sample.

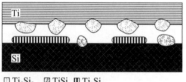

□ Ti_5Si_3 ▨ TiSi ◫ Ti_5Si_4
□ a-interlayer ■ C49-$TiSi_2$

FIGURE 2.7 Schematic diagram showing the formation of multiphases in a Ti/Si sample.

Growth kinetics of silicides

Kinetic data are crucial for a basic understanding of interfacial reactions between metal thin films and silicon. Most silicides are formed at a temperature far lower than the eutectic temperature.

Silicide formation

The growth is often diffusion controlled or interface-reaction controlled. The thickness of the silicide is proportional to the square root of time t and t, respectively. The presence of contaminating or doping impurities was found to influence the growth rate. For Pt films deposited in UHV, the growth rate of PtSi was found to increase significantly. However, the growth law remained the same [19].

Cross-section transmission electron microscopy (XTEM) has been demonstrated to provide direct and accurate kinetic data, such as the sequence of phase formation, the dependence of the phase growth and morphology of phase and interface structure in the growth of silicides on silicon [20].

In cases of $TiSi_2$, $CoSi_2$, $NiSi_2$ and a number of rare-earth silicides, the silicide formation took place within a narrow temperature range and nucleation was suggested as a controlling mechanism [14, 21, 22]. The nucleation effects are eliminated when these phases are formed on an amorphous layer [23]. The importance of nucleation effects in silicide formation has been discussed extensively by d'Heurle [21]. The films produced from nucleation limited reactions are often rather rough.

Dominant diffusing species

In silicide formation, either the metal atoms diffuse across the metal/silicide interface, or Si atoms diffuse across the silicide/Si interface, or both. In order to determine the dominant diffusing species, it is common to introduce an inert marker. In thin film reactions, the markers are usually tens of nanometres in size. It should not influence the growth kinetics of silicide formation. Ideally, the markers should be inert and remain immobile as the diffusing species streams by. An additional constraint is that the marker should be located in the silicide layer to avoid possible influence due to the presence of the interface. There are several ways to introduce the markers:

1. *Implanted marker*: inert gas atoms were implanted into silicon, followed by annealing at about 600°C. On the one hand, the implanted damages were partially removed. On the other hand, inert gas bubbles, 5–10 nm in size, were formed as markers. Metal films are then deposited on the substrate after the removal of SiO_2 and hydrocarbon contaminants on the surface. RBS has often been used to monitor the progress of the interfacial reactions.
2. *Deposited marker*: inert metal islands were first deposited on silicon. Following the cleaning of the surface, a thin silicide-forming metal layer was deposited. The metal markers must adhere well with the Si so that the subsequent cleaning step

will not influence the markers adversely. In addition, the islands should be of small size and do not react with the silicide-forming metal at the growth temperature.
3. *Radioactive markers*: radioactive Si, with a half-life of 2.62 h, was formed following neutron bombardment. By monitoring the movement of the radioactive Si, dominant diffusing species is determined.

Two issues are usually accompanied with a continuous marker film. One is the interface drag and the other is the barrier effect [23]. For the interface drag, the question whether or not the marker can be dragged with the movement of the interface depends upon the relative magnitude of interfacial energies per unit area in the vicinity of the marker. If marker drag occurs, it is not possible to determine which species is moving across the marker. One solution for this problem is to embed a marker into the same phase that will be observed in the later growth. As a result, the interface energy difference can be eliminated. For the barrier effect, the marker between two reacting elements acts as a barrier, resulting in a higher starting reaction temperature and/or a lower rate. Therefore, the appropriate diffusion marker technique prefers to place inert clusters in the interlayer instead of a continuous planar marker.

From the marker experiments it was revealed that for metal-rich silicides such as M_2Si, the dominant diffusing species are mostly metal atoms. On the other hand, in the formation of monosilicide and disilicide, Si atoms are generally the dominant diffusing species. However, there are exceptions. Important silicides in ULSI technology, the dominant diffusing species in the growth of $TiSi_2$, $CoSi_2$, WSi_2 and $NiSi$ are Si, Co, Si and Ni, respectively [6, 7, 21]. For the $TiSi_2$ salicide process, if the temperature, time and ambient for the RTA1 were not optimised, C49-$TiSi_2$ and/or C54-$TiSi_2$, which are not easily removed by ammonia and peroxide solution, are prone to form on the dielectric sidewall between poly-gate and source/drain, resulting in the so-called bridging problem. The bridging problem may lead to device failure. Since Co is the dominant diffusing species in the formation of $CoSi_2$, the bridging problem is less troublesome in the $CoSi_2$ salicide technology.

Formation mechanisms of silicides

From the viewpoint of phase transformation, the formation of silicide results from the reaction of two elemental solids at a temperature far lower than their eutectic point. It is also the product of the reaction between a microcrystalline metal and covalently bonded single-crystal silicon. The formation temperatures of metal-rich silicide, monosilicide and disilicide are about 200,

400 and 600°C, respectively. These temperatures are generally much lower than the eutectic temperature. Since the metal layer is often microcrystalline, fast diffusion of metal and/or atoms along the grain boundaries is expected.

Almost all refractory metals form only disilicides in thin film reactions and at a relatively high temperature of 600°C. There are significant differences among the melting points of these metals. Therefore, it is expected that the supply of metal atoms is not a dominant factor in determining the reaction rate of the silicide formation. On the other hand, the metal-rich silicides are generally formed at about 200°C. At such a low temperature, it is doubtful that Si atoms can be released from the Si lattice. Although the supply of Si atoms is important in the interfacial reactions, the formation mechanism may be very different. At a temperature above 600°C, lattice vibration facilitates the release of Si atoms from the silicon substrate. However, at a temperature as low as 200°C, a different mechanism is expected to be operating [7].

2.2.3.2 Effects of impurities

Oxidising ambient and oxide layer

Oxidising ambient and interfacial oxide layer may affect the formation and growth of silicides. For the processing of devices, these must be taken into consideration.

Oxidising ambient

Oxygen and water vapour are universal contaminants in deposition systems and thermal processing ambient. The presence of oxygen may prevent or retard the growth of silicide. In the annealing ambient, a thin layer of oxide is often formed on top of the silicide layer after annealing. For the formation of $CoSi_2$, a thin layer of SiO_2 is formed on top of the silicide layer [24]. The oxide layer is often used to protect the $CoSi_2$ during the etching process. On the other hand, if excess oxygen were present in forming $TiSi_2$, a thin metal oxide layer would form to degrade the adhesion and conduction properties of $TiSi_2$. The oxygen content must be minimised in the annealing ambient. For the Ti SALICIDE process, it is paramount to form $TiSi_2$ in high purity nitrogen ambient by RTP [25, 26].

Interfacial oxide layer

Many transition metals can react with both Si and SiO_2 [6, 7]. The reactions of metal with SiO_2 can be classified as:

1. Au does not react with SiO_2, its silicide (Au_4Si) is also unstable. However, Au in contact with SiO_2 can cause the dissolution of the SiO_2 when heated above the Au/Si eutectic point.

2. Al, Sn and Pb can form an oxide layer. When Al is deposited on SiO_2, strong adhesion can be obtained.
3. Pt and Pd themselves do not form oxide layers easily, and consequently their adhesion to a thick oxide layer is rather poor. However, when these metals are in contact with a thin native oxide on Si, silicide can still occur at low temperatures.
4. Refractory metals, such as Cr, Ti and V, can react with the SiO_2 to form both silicides and metal oxides. When deposited on thick oxide layers, they develop strong adhesive bonds. As a result, Cr and Ti are often used as a glue layer between the oxide layer and the less adhesive metals. On the other hand, since these metals do form oxide layers, the metal oxide layer becomes a diffusion barrier for further reaction. It has been shown that silicide layers can be formed on thick SiO_2 layers but generally at temperatures of 100–200°C higher than that where the metal reacts with Si. The silicide formed is generally metal-rich, for example, V_3Si is formed rather than VSi_2.

The interaction between Co and Si was found to be rather sensitive to the presence of oxygen contaminant at the silicon surface. As a result, it is important to clean the surface properly. Ti can reduce the native oxide layer. The reaction between Ni and Si was found to be affected not significantly by the presence of the native oxide layer at the Ni/Si interface.

Dopants

Both ion implantation and silicidation have become standard processing steps in the fabrication of ULSI devices. For self-aligned silicidation devices, silicides are formed on source, drain and gate regions simultaneously to ease the lithography requirements and to lower the contact resistance. The presence of dopant is known to influence the silicide formation [6]. Silicide formation, on the other hand, also causes dopant redistribution which would affect the junction properties [27]. In addition, the presence of metal films on doped silicon may alter defect structures commonly observed in implanted samples [28]. As the device dimensions scale down to deep sub-micrometre region, it is essential to understand the interactions of metal contacts with highly doped shallow junctions. In this section, issues pertinent to the metal contacts in Ni, Co and Ti on ion implanted silicon are discussed.

Phase formation and epitaxial growth
In the Ni/(001)Si system, the silicide of particular interest in device applications is the low-resistivity NiSi. In nickel on blank (001)Si, NiSi is formed and is stable at 350–700°C [6]. It has been reported

that dopants do not affect NiSi formation [29]. However, striking effects of B^+ and BF_2^+ implantation on the growth of epitaxial $NiSi_2$ on silicon were observed. As a result of ion implantation into (001)Si, epitaxial $NiSi_2$ was found to grow at 200–280°C instead of the usual formation temperature of about 800°C on blank (001)Si. Both B and F atoms introduced by ion implant into silicon were found to promote the epitaxial growth of $NiSi_2$ on silicon at low temperatures. Little or no effect on the formation of $NiSi_2$ was found on As^+- and P^+-implanted (001)Si. Good correlation was found between the atomic size factor and the resulting stress and $NiSi_2$ epitaxy at low temperatures. The final structure of the silicide layer was found to depend critically on the thickness of the starting Ni overlayer and the annealing temperature. The amorphicity of the substrate apparently played an important role in promoting the formation of polycrystalline $NiSi_2$ at low temperatures [30–32].

For BF_2^+ implanted (001)Si, the implantation amorphous samples were found to favour the formation of $CoSi_2$ at 400°C and laterally uniform growth of the phase at high temperatures. The effects of As^+ implant were found to be less pronounced [33, 34]. In general, the formation of $CoSi_2$ was found to be less sensitive to dopants than the formation of $TiSi_2$.

High concentrations of As, P and BF_2 can retard the formation of $TiSi_2$ [35]. A reduction in silicide thickness is often observed for $TiSi_2$ formed on heavily doped Si or polysilicon, suggesting the formation of $C49$-$TiSi_2$ was retarded. There is evidence that As and P also retard the formation of the $C54$-$TiSi_2$ phase [36].

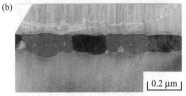

FIGURE 2.8 XTEM images showing the removal of EOR defects by $CoSi_2$ formation in BF_2^+-implanted (001)Si for samples annealed at: (a) 700°C and (b) 800°C.

Removal of end-of-range defects by silicide formation

It has been recognised for some time that residual defects formed in ion implanted silicon can be reduced by silicide formation [37]. The end-of-range (EOR) defects are known to cause large leakage current for shallow junctions [38]. The removal of EOR defects by the formation of NiSi, $CoSi_2$ and $TiSi_2$ on ion implanted silicon was demonstrated. Examples are shown in FIGURES 2.8 and 2.9. It was shown that complete annihilation of EOR defects by silicide formation depends on a number of factors, which include distance between silicide/Si interface and location of original a/c interface and/or EOR defects, annealing temperature and time, defect complexity, grain sizes of silicide and proximity to the sample surface. In addition, continued generation of vacancies during the silicide growth appears to be essential to reduce the EOR defects [39].

A significant reduction of EOR defects was achieved in BF_2^+-implanted (111)Si by the formation and growth of a

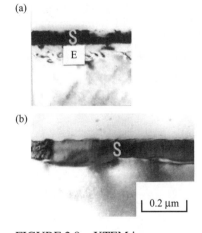

FIGURE 2.9 XTEM images showing the removal of EOR defects by NiSi formation on 110 keV, $5 \times 10^{15}/cm^2$ BF_2^+-implanted (001)Si; (a) as-deposited and (b) samples annealed at 600°C for 1 h.

Silicide formation

polycrystalline NiSi$_2$ overlayer in samples annealed at 850–900°C. In contrast, no apparent reduction in the density of EOR defects was found in BF$_2^+$-implanted silicon with no NiSi$_2$ overlayer or with an epitaxial NiSi$_2$ overlayer, subjected to the same heat treatments. The significant reduction in density of the EOR defects was correlated to the grain growth in the thin polycrystalline NiSi$_2$ overlayer [40].

Thermal stability of silicides on silicon

The surface morphology of TiSi$_2$ was found to be significantly influenced by the implantation in the silicon substrate. The presence of doping impurities was found to improve the surface coverage of TiSi$_2$ on silicon. Simultaneous presence of B and F was found to be most effective in retarding the degradation of surface morphology of the TiSi$_2$ thin films. The presence of a high concentration of doping impurities at the silicide/Si interface is thought to hamper the epitaxial growth of TiSi$_2$ on (111)Si. On BF$_2^+$ implanted silicon, TiSi$_2$ was stabilised by the accumulation of fluorine atoms at the grain boundaries to retard the grain growth. Examples are shown in FIGURES 2.10 and 2.11 [41, 42].

Continuous and pinhole-free CoSi$_2$ films were grown on both blank and B$^+$-, BF$_2^+$-, As$^+$- and P$^+$-implanted (001)Si samples after annealing at 600–900°C for 1 h. However, severe island formation was found in samples annealed at 1000°C for 1 h. Examples are shown in FIGURES 2.12 and 2.13 [32, 33]. Single-crystal CoSi$_2$ films were grown on B$^+$-implanted silicon (111)Si annealed at 1000°C for 1 h. The CoSi$_2$/Si interface was found to be rather flat with undulation less than 10 nm in amplitude. Almost single-crystal CoSi$_2$ films were grown on As$^+$- and P$^+$-implanted samples. On the other hand, the areal fractions of surface coverage of epitaxial CoSi$_2$ were measured to be 70 and 80% in blank and BF$_2^+$-implanted samples.

The surface coverage and grain size of polycrystalline NiSi$_2$ were found to be significantly influenced by the implantation species in silicon substrate. In Si$^+$-, B$^+$-, As$^+$- and P$^+$-implanted samples, agglomeration of NiSi$_2$ became very severe after annealing at 800°C for 1 h. In contrast, almost full surface coverage was found in F$^+$- and BF$_2^+$-implanted samples after annealing at 900°C for 1 h. Sheet resistance data are shown in FIGURE 2.14 [43]. The growth of laterally uniform NiSi$_2$ and resistance to agglomeration at high temperatures in F$^+$- and BF$_2^+$-implanted samples are attributed to the retardation of the growth of NiSi$_2$ grains by the presence of large fluorine bubbles at the grain boundaries. It is worthwhile noting that fluorine bubbles were observed to

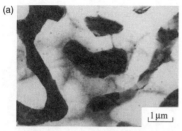

FIGURE 2.10 Planview micrographs of TiSi$_2$ samples annealed at 1000°C for 1 h with 50 nm thick Ti on: (a) blank, (b) BF$_2^+$-implanted and (c) BF$_2^+$-implanted (001)Si and then annealed at 1000°C for 0.5 h.

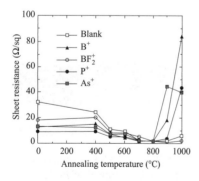

FIGURE 2.11 Sheet resistance versus annealing temperature for TiSi$_2$ films by reacting 50 nm thick Ti film with various ion-implanted (111)Si.

Silicide formation

distribute throughout the NiSi$_2$ film. However, large bubbles were concentrated at the grain boundaries.

Effects of nitrogen ambient and N$^+$ implantation

Nitrogen is commonly used in the Ti salicide process as the annealing ambient. Ti silicidation in N$_2$ ambient often results in a thin TiN layer on top of the TiSi$_2$ layer. The top TiN layer can be readily removed with a selective etch. For Ti films on silicon, the film thickness and the competing reaction with Si limit the amount of TiN formed. The formation of TiSi$_2$ is enhanced at a higher anneal temperature whereas the formation of TiN is slightly reduced. Nitrogen has little effect on NiSi and CoSi$_2$ formation since Co and Ni do not react with nitrogen at temperatures for silicide formation [36].

FIGURE 2.12 XTEM image showing the island formation in a CoSi$_2$ sample on 110 keV, 5×10^{15}/cm^2 BF$_2^+$-implanted (001)Si annealed at 1000°C.

Nitrogen implantation was found to stabilise TiSi$_2$ in device applications [44]. In addition, it has been used in deep MOSFET sub-micrometre devices to suppress the B and As diffusion as well as hot-carrier degradation [45]. For Ti on 30 keV BF$_2^+$–20 keV N$_2^+$ and 30 keV As$^+$–20 keV N$_2^+$ implanted samples, a continuous low-resistivity TiSi$_2$ layer was found to form in all samples annealed at 700–900°C. The formation of TiSi$_2$ was retarded by the presence of nitrogen in the silicon substrate. For Ti on 1×10^{15}/cm^2 N$_2^+$- and As$^+$-implanted samples, EOR defects were completely eliminated in all samples annealed at 700–900°C. The results indicated that with appropriate control, N$^+$-implantation can be successfully implemented in forming low-resistivity TiSi$_2$ contacts on shallow junctions in deep sub-micrometre devices [46, 47].

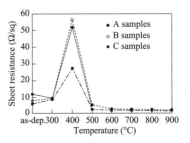

FIGURE 2.13 Sheet resistance versus annealing temperature for CoSi$_2$ film samples annealed at 1000°C for 1 h with 50 nm thick Co on: (a) blank, (b) BF$_2^+$-implanted and (c) BF$_2^+$-implanted (001)Si and then annealed at 1000°C for 0.5 h.

2.2.3.3 Oxidation of silicides on silicon

NiSi$_2$, CoSi$_2$ and TiSi$_2$ on silicon

To successfully utilise metal silicides in device applications, a good understanding of their oxidation behaviours is required. Self-passivating silicon dioxide layer can be thermally grown over silicides, without any degradation of chemical or electrical properties of the silicide film when there is sufficient silicon underlying the silicide to provide a source for SiO$_2$ growth [6]. The ability to grow a high-quality SiO$_2$ film on top of the silicide film in the oxidising ambient is also advantageous for the circuit design of multilevel metallisation.

For dry oxidation of C54-TiSi$_2$, activation energies for parabolic and linear growth were found to be 1.97 and 2.5 eV, respectively. On the other hand, activation energies for parabolic and linear growth were found to be 1.88 and 2.10 eV, respectively, for wet oxidation. The closeness of linear activation energy with

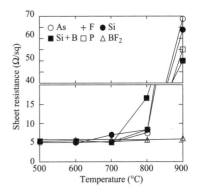

FIGURE 2.14 Sheet resistance versus annealing temperature for NiSi$_2$ films by reacting 30 nm thick Ni film with various ion-implanted (001)Si.

that of pure silicon is thought to be due to the fact that both are related to the breaking Si–Si bonds at the silicon surface [48].

For dry oxidation of $NiSi_2$, activation energies for parabolic and linear growth were found to be 1.87 and 1.94 eV, respectively. On the other hand, activation energies for parabolic and linear growth were found to be 1.72 and 1.59 eV, respectively, for wet oxidation. Compared to the oxidation of $TiSi_2$ on silicon and silicon, a model based on the dominant diffusing species through silicide, that is, metal and Si for $NiSi_2$ and $TiSi_2$, respectively, has been proposed to explain the substantial difference and closeness in linear activation energies of wet oxidation between $NiSi_2$, $TiSi_2$ and pure silicon, respectively [49]. Crystallinity and thickness of silicide layer as well as substrate orientation were found to exert strong influences on the oxidation of $NiSi_2$ on silicon [50].

For dry oxidation of $CoSi_2$ on (111)Si, activation energies for parabolic and linear growth were found to be 1.91 and 2.01 eV, respectively. For wet oxidation, activation energies for parabolic and linear growth were found to be 1.75 and 1.68 eV, respectively [51]. The oxidation kinetics was found to be practically the same for $CoSi_2$ and $TiSi_2$ in the parabolic growth regime, but substantially different from those of $TiSi_2$ on silicon and pure silicon. The similarity in growth kinetics of cubic CaF_2 structure $CoSi_2$ and $TiSi_2$ on silicon with small mismatches to silicon is correlated to essentially the same stress level in these two silicides during the oxidation.

Room temperature oxidation of Cu_3Si on silicon

Oxidation of pure silicon is also known to occur at room temperature; however, its growth rate is less than 3 nm in 2 weeks. Low temperature oxidation of silicon can be greatly enhanced, if metals are used. Silicon atoms in a metallic state are expected to be highly mobile and readily oxidised at low temperatures [52]. The room temperature oxidation of silicon in the presence of Cu_3Si was reported by Harper et al. [53]. It was proposed that Cu_3Si oxidises according to the following reaction:

$$Cu_3Si + 2O \Rightarrow SiO_2 + 3Cu$$

which leads to the growth of silicon dioxide films up to 1 μm thick by room temperature aging after several weeks in air. The Si in the intermixed region or in compound form is present in a different metallic environment, inducing different oxidation behavior. It is also commonly recognised that catalytic action appears to be responsible for the low temperature oxidation process of Si.

The extent of oxidation was found to depend critically on the starting thickness of Cu_3Si. The oxidation was found to be more restricted on (111)Si than that in (001)Si. The SiO_2 layer thickness was found to decrease with the average grain size of the starting Cu_3Si layer. HRTEM revealed that the oxidation is initiated at the grain boundaries. An oxide film as thick as 4.5 μm was grown at room temperature over a period of 2 weeks in (001)Si samples. Examples are shown in FIGURES 2.15 and 2.16. The growth of thick oxide film was achieved by minimising the grain size of Cu_3Si through a reaction of Cu and an intermediate amorphous silicon layer at 200°C [54 56].

2.2.3.4 Epitaxial growth of silicides

Epitaxial silicides possess several attractive characteristics for consideration. Epitaxial silicides belong to a special class of silicides which exhibit a definite orientation relationship with respect to the silicon substrate. A silicide is expected to grow epitaxially on silicon if the crystal structures are similar and the lattice mismatch between them is small. The impetus for the study of epitaxial silicides mainly stemmed from several favourable characteristics of epitaxial silicides in comparison with their polycrystalline counterparts, including greater stability and a lower stress at the interface, alleviation of grain boundary effects as well as conductivity enhancement [57].

$NiSi_2$ and $CoSi_2$ can be grown in single-crystal form on silicon [58]. Both C49- and C54-$TiSi_2$ were found to grow on silicon [59]. In particular, C54-$TiSi_2$ can be grown to tens of micrometres in grain size [60–62]. Initial studies on the epitaxial growth of silicides on silicon were mostly on the growth of silicides on a large area. However, in device applications, silicides were grown on laterally confined silicon. Lateral confinement was found to exert significant influence on the epitaxial growth of $NiSi_2$ and $CoSi_2$ on silicon.

Epitaxial growth of $NiSi_2$ on (111)Si inside 0.1–10 μm oxide openings prepared by electron beam lithography has been studied. Striking effects of size and shape of deep sub-micrometre oxide openings on the growth of $NiSi_2$ epitaxy were observed. Epitaxial growth of $NiSi_2$ of single orientation on (111)Si was found to occur at a temperature as low as 400°C inside contact holes of 0.2 μm or smaller in size. Contact holes were found to be more effective in inducing the epitaxial growth of $NiSi_2$ of single orientation than that of linear openings of the same size. The effects of size and shape of lateral confinement on the epitaxial growth of $NiSi_2$ on (111)Si are correlated with the stress level inside oxide openings [63, 64].

The self-aligned formation of $CoSi_2$ was achieved on the selective epitaxial growth (SEG) silicon layer on (001)Si inside

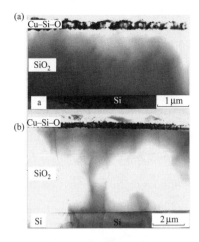

FIGURE 2.15 XTEM images of Cu(100 nm)/(001)Si samples annealed at 200°C for 1 h and exposed in air for: (a) 1 week and (b) 2 weeks.

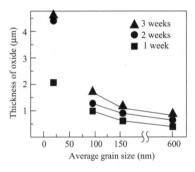

FIGURE 2.16 Oxide thickness versus grain size of starting Cu_3Si layer of Cu(200 nm)/(001)Si samples annealed at 300–700°C for 1 h and exposed in air for 1–3 weeks. Also shown is a Cu(100 nm)/(001)Si sample annealed at 200°C and exposed in air for 1–3 weeks.

Silicide formation

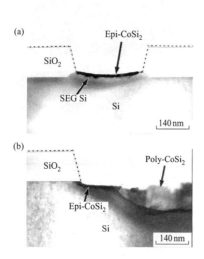

FIGURE 2.17 XTEM images of epitaxial and a mixture of epitaxial and polycrystalline CoSi$_2$ inside 0.25 and 1.8 µm in size linear openings, respectively, in samples heated by a two-step RTA process.

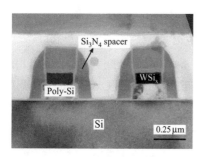

FIGURE 2.18 WSi$_x$/poly-Si polycide structure in a 16M DRAM. Also seen are W-plug and TiN/Ti barrier/adhesion layers.

0.1–0.6 µm oxide openings. Self-aligned CoSi$_2$ film without lateral growth of silicide was grown on the SEG Si layer by rapid thermal annealing at 700°C in N$_2$ ambient [65]. Striking effects of oxide opening size on the growth of CoSi and CoSi$_2$ on silicon inside 0.2–2 µm linear oxide openings and contact holes were found. The formation of CoSi at low temperature appeared to be retarded by the local compressive stress near the edge of linear oxide openings. A thin, uniform epitaxial CoSi$_2$ was grown inside 0.5 µm or smaller linear openings and 0.7 µm or smaller contact holes by both one- and two-step rapid thermal annealing processes. On the other hand, epitaxial and polycrystalline CoSi$_2$ were found to form near the edges and central regions, respectively, of 0.6 µm or larger linear openings. The size effect of oxide openings is correlated to the distribution of local stress induced at the oxide edge. The relative ease in the epitaxial growth of CoSi$_2$ near the oxide edge of linear openings and of 0.7 µm and smaller contact holes is attributed to the thinness of the CoSi layer. Examples are shown in FIGURE 2.17 [66, 67].

2.3 GATE ELECTRODES

2.3.1 Polycrystalline WSi$_x$

Highly doped polycrystalline silicon possesses the advantages of high temperature stability and compatibility with the conventional processing. As a result, it has been commonly used as the gate electrode of the metal–oxide–semiconductor field effect transistor (MOSFET) devices. However, since the resistivity of the poly-Si is higher than 400 µΩ cm, it has gradually been replaced by the silicides with a resistivity of one order of magnitude lower.

Polycide refers to a layered structure with a thin silicide layer on top of poly-Si. An example is shown in FIGURE 2.18. To maintain the stability and reliability of the thin gate oxide layer, the thickness of the silicide is kept to be less than one-half times the underlying poly-Si thickness. Among all silicides, only TaSi$_x$, MoSi$_x$, WSi$_x$ and TiSi$_x$ possess high temperature stability, low resistivity and low stress, which are required for the gate electrode silicide layer [68–71]. The x denotes the ratio of Si/M atoms in the chemical formula. In addition to the high temperature stability, the ability to grow a high-quality SiO$_2$ film on top of the silicide film in the oxidising ambient is also advantageous for the circuit design of multilevel metallisation.

WSi$_x$ is of particular interest since it can be deposited with low-pressure chemical vapour deposition (LPCVD). The general conditions are with a pressure of 100–500 mTorr, WF$_6$ reacts with

Silicide formation

SiH$_4$ or SiH$_2$Cl$_2$ to form WSi$_x$ at 300–600°C. It can also be deposited in a high vacuum sputtering chamber to enhance the adhesion. However, the step coverage was relatively poor compared with the LPCVD films. For LPCVD films, a 900°C anneal is usually required to lower the resistivity of the film to about 70–100 μΩ cm. In the meantime, the Si/W ratio is also lowered from 2.5–2.8 to 2.2–2.3 [2, 72].

Both CoSi$_2$ and TiSi$_2$ are vulnerable to processing chemicals, particularly to HF solutions. The most difficult obstacles, however, are in polycide patterning. TiSi$_2$ is attacked by Cl-based chemistry during polysilicon etch, and this makes the line-width control difficult at the gate level. CoSi$_2$, on the other hand, lacks a volatile by-product and is difficult to pattern by dry etching [2].

2.3.2 SALICIDE

Self-aligned silicidation (SALICIDE) process is used to form silicides simultaneously at source/drain and gate regions without the need of an additional lithographic step [73]. An example is shown in FIGURE 2.19. The formation of low-resistivity silicides at the source/drain regions reduces the MOSFET series resistance and contact resistance between the contacting silicide and the junction. An additional advantage is the reduction of the number of contact windows to facilitate the design of interconnection in the layout of the device so that the total area of the chip may be reduced. The metals viable for SALICIDE process include Ti, Co, Ni, Pd and Pt (TABLE 2.2). These metals can form self-aligned silicides with two-step RTA and an intermittent selective etching step to remove the unreacted metals and other compounds.

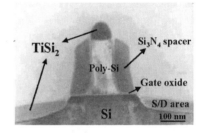

FIGURE 2.19 A 0.1 μm TiSi$_2$ SALICIDE structure.

Low-resistivity TiSi$_2$ is the most commonly used silicide in SALICIDE process owing to its low resistivity, high temperature stability and ease in deposition and chemical properties. Traditional steps in SALICIDE process are to form a dielectric sidewall spacer between gate and source/drain regions first. Subsequently, a 20–60 nm thick Ti is deposited followed by the first 600–700°C RTA (RTA1) in N$_2$ or NH$_3$ ambient. Titanium silicide is then formed at the gate and source/drain regions. An etching step is carried out to remove the unreacted Ti and/or TiN on the dielectric layer. The final step is to anneal by RTA at about 800°C (RTA2) to transform high-resistivity C49-TiSi$_2$ to low-resistivity C54-TiSi$_2$ at the gate and source/drain regions. The key steps involve removal of native oxide at the Si regions; deposition of metal; temperature, time and ambient control in two-step annealing; temperature, concentration and time control of selective etching [2, 68].

The first step RTA for self-aligned $CoSi_2$ is usually done at 450–480°C to form CoSi. A second step RTA at 700–750°C is used to transform CoSi to $CoSi_2$.

2.4 INTERCONNECTS

The packing density in deep sub-micrometre integrated circuits is dependent on metal interconnect density. Interconnect design rules are, therefore, scaled very aggressively from one generation of technology to the next. In addition, more levels of metal are added to provide a larger area for interconnects, along with greater flexibility in routing.

2.4.1 Al/Cu alloys

Aluminium has been used as interconnects since the beginning of the IC age. Currently, Al and its alloys are the major interconnect materials because the physical and chemical properties of aluminium are compatible with current VLSI processing: aluminium forms a thin protective oxide film that withstands various thermal processes; it has relatively low electrical resistivity and halide compounds with a relatively high vapour pressure which are suitable for reactive ion etching (RIE), and it is an inexpensive material. However, Al-based interconnects in the deep sub-micrometre regime are susceptible to electromigration failure, a current density of 2×10^5 A/cm^2 is considered an upper limit for Al interconnect lines. It has gone through the era of pure Al, Al/Si alloy, Al/Si/Cu alloy and Al/Cu alloy [74]. Cu is known to distribute near the grain boundaries of Al and suppress the electromigration under high current. However, it tends to form high-resistivity Al_2Cu precipitates after thermal treatments.

2.4.2 W interconnects

W is also used as the local interconnect for its superior electromigration resistance as well as higher reliability than that of Al/Cu alloy. In addition, its resistivity is lower than that of polycrystalline Si and $TiSi_2$, which are used as local interconnects [75]. However, for long distance interconnects, Al/Cu is still more suitable since low resistivity is demanded.

The resistivity and stress of W interconnects should be lower than $10\,\mu\Omega\,cm$ and 8×10^9 $dyne/cm^2$, respectively. To lower the resistance further, a layer of Al/Cu is often deposited on top of the W layer. Low stress is required to prevent the peeling of the interconnecting layer from the dielectric layer.

Silicide formation

TABLE 2.3 Comparison of various metal interconnects.

	Cu	Al	Au	Ag	W
Electrical resistivity ($\mu\Omega$ cm)	1.67	2.66	2.35	1.59	5.65
Young's modulus (10^{11} dyne/cm^2)	12.98	7.06	7.85	8.27	41.1
Thermal conductivity (W/cm)	3.98	2.38	3.15	4.25	1.74
Specific heat capacity (J/kg/K)	386	917	132	234	138
Melting point (°C)	1085	660	1064	962	3387
Corrosion resistance	Poor	Good	Excellent	Poor	Good
Adhesion to SiO$_2$	Poor	Good	Poor	Poor	Poor
Electromigration resistance	High	Low	Very high	Low	Very high

2.4.3 Cu interconnects

For 0.18 μm or even more advanced devices, there is a strong impetus to introduce interconnects with resistivity lower than that of Al and its alloy. While copper has long been recognised as a superior electrical conductor, it has been difficult to adapt to semiconductor manufacturing, leaving aluminium as the material of choice for over 30 years. The resistivity of Cu is 35% lower than that of Al. In addition to its good electromigration and corrosion resistance, a number of other factors contribute to the favourable status of Cu as a replacement. TABLE 2.3 compares various metal interconnects. Major manufacturers have already implemented the Cu interconnection scheme in the fabrication of devices. The process enables the manufacturers to shrink electronic circuitry to smaller dimensions and fit more computer logics onto a single chip. The use of copper enhances the electrical properties of semiconductors, and should yield significant performance gains.

2.5 PLUG

Plug is the stud in the contacts to the underlying layer or vias of dielectric layer between various metal layers. Although there are many different schemes to form plugs in the multilevel interconnection, Al- and W-plugs are the two main choices.

2.5.1 Al-plug

The step-coverage of PVD-Al is rather poor owing to the directionality of sputtered metal atoms. As a result, thicker Al layers are deposited along the horizontal direction at vias and contact windows than those at the sidewalls. The uneven deposition of films results in the formation of voids which leads to reliability problems. To alleviate the void formation in PVD, both Al reflow and high temperature deposition of Al are effective [76, 77]. On the other hand, CVD can produce low-resistivity Al films with good step-coverage [78]. It is a very promising technique to form satisfactory Al interconnects.

Based on the RC delay consideration, it is desirable to replace the W-plug with the Al-plug. However, the Al-plug is now only considered at the contact level, and mostly for DRAMs. Al planarisation has become a critical issue since the exacerbation of the lithography depth-of-focus and etching problems for multilevel devices with relatively large topographical difference. In addition, the chemical mechanical polishing (CMP) process, which can achieve the formation of Al-plug and Al planarisation at the same time, has gained importance. Three production-worthy processes are discussed in the following sections.

2.5.1.1 High temperature flow

Aluminium film is first deposited in a medium/low temperature sputtering chamber at high power. The wafer is then transferred to a high temperature flow chamber. As the wafer is heated at high temperature, Al flows into the vias by solid-state diffusion to form an Al-plug. The main driving force for the Al reflow is the surface tension (surface energy). The advantage of the process is that the process parameters involved are small in number. Therefore, the process control is relatively simple. However, the process needs to be carried out at a high temperature ($\sim 500°C$) for a long time. As a result, considerable thermal budget is required. In addition, the barrier metal has to be robust; usually, TiN is used [38].

2.5.1.2 High-pressure force-fill

The process is similar to the high temperature reflow process. In this process, all contact windows are covered by Al film leaving the vias. The residual pressure inside the vias is about the same as the operating pressure (about 3 mtorr). The wafers are then transferred to a high-pressure chamber maintained at about 400°C. A very high pressure (>60 MPa) exerted by Ar gas is then applied to the wafers. The high pressure difference at the contact windows, accompanying the ductile property of Al at high temperature, force-fills Al into the contact windows to form the Al-plug [79]. The process involves high-pressure gas. Therefore, special cautions should be exercised in the design and safety of the high-pressure chamber. For devices with various sizes of contact holes, it is a challenge to control the pressure so that the filling of the holes can be achieved without damaging the devices.

2.5.1.3 Cold/hot Al sputtering and planarisation

The process consists in two-step cold Al and hot Al deposition (in a single chamber or separate chambers) [38]. Prior to the deposition of Al, it is common to deposit a Ti layer of several tens of nanometres thickness as the wetting layer to ensure the uniform adhesion of Al on Ti surface and avoidance of the "beading" phenomenon. The aluminium layer of several hundreds of nanometre thickness is then sputtered at low temperature with high power. The Al layer is fine-grained and uniform and serves as the nucleation layer for the subsequently deposited Al layer at high temperature and low power. The new layer continues the nucleation and growth process at the nuclei formed during the cold deposition. With the driving force of surface tension, the Al flows into the holes and forms plugs and achieves the planarisation at the same time. The process can be carried out at a temperature lower than that of high temperature reflow process without the requirement of high pressure. It is therefore a rather promising technique. However, the uniformity of the heater stage for the wafers, adequate vapour degas and prevention of micro-contamination need to be taken care of to ensure the reproducibility of the process. The process has been successfully implemented in a number of fabrication factories around the world.

2.5.2 W-plug

CVD-W possesses excellent step coverage in contact windows and vias. It is currently the best filling technique for high aspect ratio, small size holes.

2.5.2.1 Blanket tungsten deposition

Blanket W is widely used for the fabrication of the W-plug [80]. The first step is to deposit an adhesion layer on contact windows or vias. A blanket W layer is then deposited on the adhesion layer and fills the contacts or vias. Finally, the surface layer of W on top of the dielectric layer is removed and leaves only W-plugs. Since the adhesion between W and the dielectric layer is rather poor, the deposition of an adhesion layer prior to that of W film is necessary. After the removal of the surface W film, the remaining adhesion layer can serve as the barrier/adhesion layer for subsequently deposited Al interconnects.

For blanket W deposition, it is important to control reaction conditions to ensure high step coverage. Using H_2 to reduce WF_6 was

Silicide formation

found to produce better step coverage than that of SiH_4. In a 380°C, 90 torr reaction chamber, the step coverage of about 95% for W film can be achieved with a mixture of 90 sccm WF_6, 700 sccm H_2 and other diluting gas. The step coverage can be further increased with the reduction in reaction temperature and partial pressure of H_2 as well as the increase in partial pressure of WF_6. The low temperature process should take care of the stress issue.

Tungsten etchback (WEB) can be used to remove the surface W film. WEB adopts the plasma etch with SF_6. The reaction products are volatile, which are then removed. The surface W film is removed and leaves only the W-plug. WEB can be selected to stop at the adhesion layer. The process requires subsequent removal of W and W compounds on the adhesion layer to avoid the short circuiting of the interconnects. WEB can also be stopped at the dielectric layer after the complete removal of the adhesion layer. Chlorine compound plasma etch is often required for the removal of the TiN adhesion layer.

CMP is an alternative method to remove the W film. The method can also be applied to achieve planarisation. In the CMP process, W is gradually removed by the SiO_2 or Al_2O_3 powders in an etching and buffered solution environment. In a mixed slurry environment, W reacts with the buffered solution to form a protective oxide layer, which is then mechanically removed by the SiO_2 or Al_2O_3 powders. Fresh W is then exposed to the etching solution. The removal of surface W film is achieved with such a repeated dynamical equilibrium. Owing to the planarisation by the CMP, stacked vias can be formed, which favours the increase in packing density; see FIGURE 2.20.

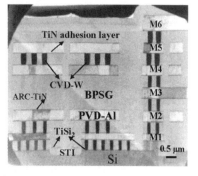

FIGURE 2.20 W-plug in six-level metallisation with CMP process.

2.5.2.2 Selective tungsten deposition

Selective tungsten deposition allows the direct filling of the contact windows and vias [81]. The process steps are then reduced. The process does not require the deposition of an adhesion layer on the dielectric film and WEB. Instead, W grows directly from the bottom of contact windows and vias until the filling of the holes. Since W can be deposited on Si, metal silicides and metals, but not easily on the oxide layer, it is possible to control the reaction mechanism to achieve the selective growth on specific areas.

Although selective tungsten deposition can be used to fill very small contact windows and vias, there are difficulties in its implementation. The wafers need appropriate cleaning to ensure the deposition selectivity. Otherwise, tiny W patches are deposited on the dielectric layer. Furthermore, too much Si is consumed during the reaction of WF_6 with the Si at the source or drain. Excessive consumption of Si at source or drain may result in unacceptable junction leakage.

The traditional method for surface cleaning is not effective for selective tungsten deposition. However, a mixture of dilute HF cleaning, plus NF_3 and H_2 plasma cleaning as well as pyrolysis of ClF_3 gas cleaning has led to better performance.

2.6 DIFFUSION BARRIER LAYER

In device applications, in order to suppress or reduce the interactions between various contacting layers such as Al/metal silicide, W/Al and Al/WSi_x, a diffusion barrier is often deposited between the layers.

The characteristics of an ideal diffusion barrier are:

1. It does not react with the metal overlayer and Si/silicide underlayer at the processing temperature.
2. It is more stable than the possible compounds in all layers.
3. It prevents the penetration of metal and Si atoms.
4. It makes low contact resistances with Si, metal or silicide. The bulk resistivity of the barrier material is usually not that important compared with the contact resistivity at the interfaces. In general, the contribution of the resistivity of a thin barrier layer, typically about 100 nm thick or less, to the total contact resistance is insignificant. In contrast, the contact resistivities at both the aluminium and the silicon (or silicide) interfaces are usually the main contributors to the total contact resistance. Therefore, they have to be kept as low as possible.
5. It has good adhesion with Al, metal or silicide.
6. It has low stress so that peeling or delamination can be avoided [3, 6].

It is obvious that an ideal barrier does not exist in practice. Therefore, compromise has to be made. For example, limited reaction of neighbouring layers is allowed, but cannot be consumed completely. A further example is a small amount of metal or Si may diffuse across the barrier layer as long as it still meets the specification of the various layers.

The use of TiN thin films as a diffusion barrier in silicon technology was proposed by Nelson in 1969 [82]. An example is shown in FIGURE 2.21. Since then, a great deal has been accomplished in this area. The contact resistivity of TiN on low-resistivity silicon was found to be of the order of $10^{-5}\,\Omega\,cm^2$, depending on the resistivity of the substrate material. The barrier height of TiN on high-resistivity silicon has been reported to be 0.49 eV.

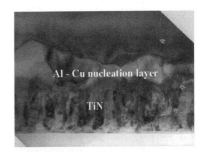

FIGURE 2.21 TiN barrier layer.

Silicide formation

Measurements showed that the Al/TiN/TiSi$_2$/Si contact system has an acceptable contact resistivity. In contrast, the contact resistivity of the Al/TiN/Si contact system was higher than $10^{-4}\,\Omega\,cm^2$ and thus difficult to measure accurately [5].

2.6.1 PVD-TiN

The deposition of TiN is commonly carried out in a UHV chamber. Ti is sputtered by Ar ion bombardment and reacts with the flowing N$_2$ to form TiN on silicon wafers. The process is called reactive ion sputtering (RIS) [83]. With the increase in packing density and shrinkage in size of contact windows/vias and increase of its aspect ratio, the step coverage is declining steeply. To keep Al (or F from WF$_6$) and F from diffusing to contact regions in silicon and the metal underlayer, respectively, a minimum of TiN thickness at the bottom of the windows must be maintained. Therefore, the thickness of TiN on the dielectric layer surrounding the contact windows/vias is often two or three times that required at the contact bottom, which creates an overhang problem at the contact windows as illustrated in FIGURE 2.22. The overhang, in turn, will influence the filling of Al or W in the subsequent process adversely.

2.6.1.1 *Collimated sputtering*

In order to increase bottom coverage, a collimator is often placed between the metal target and wafers. The process is called collimated sputtering [84]. It consists of honeycomb cells with fixed height/diameter ratio. With the collimator in place, only metal atoms sputtered from the target within a small angular range from the normal direction can pass through the collimator and deposit on the wafers. The other sputtered atoms along the more oblique directions are thus filtered out. In other words, the collimator plays the role of filter. The efficiency of filtering depends on the height/diameter ratio of the honeycomb cell in a collimator. The higher the ratio, the larger fraction of incident atoms will be filtered out. On the other hand, it leads to more straight atomic flow on the wafers and improved bottom coverage. However, since a large number of metal atoms are filtered out, the deposition rate will suffer accordingly. In addition, the diameter of the honeycomb cell inevitably shrinks with the deposition time, the deposition rate will further degrade as a result. Furthermore, owing to the difference in materials properties between the metal and collimator, the resulting mechanical stress and/or thermal stress often leads to peeling and becomes a potential particle source. The thin film properties of the deposited metal, particularly Ti, such as stress and uniformity, are very sensitive to the states of the collimator.

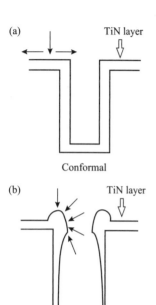

FIGURE 2.22 Schematic diagram illustrating the overhang problem surrounding the contact windows/vias.

Silicide formation

In practice, for collimated sputtering, it is necessary to undertake elaborate bake-out and burn-in steps to maintain the process stability. The steps escalate the cost of the operation following preventive maintenance (PM). It is pretty much a standard process used worldwide.

2.6.1.2 Long throw sputtering

The principle of the long throw sputtering is to increase the mean-free-path of the sputtered atom so that the possibility of collision with other atoms to deflect the sputtered atoms from the incident direction is decreased [85]. To achieve the purpose, the sputtering gas pressure is reduced. In order to maintain the thickness uniformity, the distance between the target and wafers needs to be increased (to about twice that of common sputtering apparatus). The bottom coverage is significantly improved with the highly directional and normal to the wafer atomic flux. The deposition rate of long throw sputtering is rather low. The thickness uniformity of deposition at the centre and edge of the same wafer is also marginal. With the expected increase in wafer size from 200 to 300 mm, the distance between the target and wafers needs to be further increased. The development will increase the dimension of the deposition chamber, thus aggravating the difficulties in hardware design, set-up and maintenance. The further decrease in sputtering rate will also be seriously scrutinised in the productivity analysis.

For both collimated sputtering and long throw sputtering, it is feasible that adequate bottom coverage can be achieved for sub-0.25 μm contact holes. In order to prolong the practical applications of metal sputtering technologies, other alternatives need to be developed.

2.6.1.3 Ionised metal plasma (IMP) sputtering

Ionised metal plasma sputtering utilises plasma density 10–100 times that of the common metal sputtering. Metal atoms are first sputtered with magnetron DC power. The sputtered atoms are thermally activated by high-pressure gas in the sputtering chamber. At the same time, the collisions between atoms and electrons are aided by the application of an independent RF power. A large fraction of metal atoms is thus ionised, which differs from the neutral atoms in the common sputtering process [86]. The plasma density in the IMP process is of the order of 10^{11}–10^{12} cm^{-3}, far higher than those of the common sputtering process. The ionised metal atoms are then accelerated by the plasma-induced self-bias or separate wafer bias towards the wafers. Excellent bottom coverage can be achieved with the atomic flux with high directionality generated by the IMP. The deposition rate is also better than marginal. The application of an additional RF bias at the wafers was found to improve the bottom coverage and alter the crystal structure of the deposited film favourably.

The probability of ionisation of metal atoms is determined by its duration in the plasma. The longer the duration, the higher the ionisation probability. If all the sputtered metal atoms were of high energy ($\sim$1–10 eV), the duration of high-velocity atoms in the plasma would be relatively short. Therefore, IMP must slow down the sputtered atoms with high pressure ($\sim$10 mtorr) gaseous atoms.

IMP was found to produce films with lower resistivity and higher uniformity than that of traditional sputtering. Improved bottom coverage leads to the possibility of depositing thinner film and still meeting the device requirement. As a result, the cost of metal deposition is decreased and the throughput is increased. The cost of ownership (COO) is lower than that of traditional sputtering.

2.6.2 CVD TiN and Ti

Chemical vapour deposition can be classified into three categories; low-pressure chemical vapour deposition (LPCVD), metal–organic CVD (MOCVD) and plasma-enhanced CVD (PECVD). Much attention has been placed on the LPCVD and MOCVD TiN.

2.6.3 Low-pressure chemical vapour deposition

In LPCVD [87–89], the wafers are first subjected to a pre-clean with Ar ion to remove the oxide and/or other contaminants on the surface. The depositions of Ti and TiN are carried out in CVD Ti and CVD TiN chambers, respectively. The reactant gases for CVD TiN are $TiCl_4$ and NH_3. For CVD Ti, $TiCl_4$ is reduced by H_2 to form Ti. The reactant gases pass through gas dispersion rings and shower head at the top of the reactor so that uniform distribution of the gases can be achieved. In the deposition of TiN, excellent step coverage can be obtained without plasma enhancement. On the other hand, deposition temperature can be lowered with plasma enhancement. From the growth rate versus temperature data, an activation energy of 0.41 eV is obtained for TiN growth. In general, if the activation energy is higher than 0.25 eV, the reaction is surface reaction limited, which can lead to almost 100% surface coverage. For the $TiCl_4 + NH_3$ reaction to form TiN, reaction occurs only after multiple bombardment of reactant gases with the contact holes, the rate is limited by surface reaction. The disadvantages of CVD-TiN include difficulties in controlling deposition rate and particle density. On the other hand, the source material cost is low and the electrical resistivity of the deposited film is not excessively high. CVD-Ti is deposited almost exclusively with PECVD Ti.

2.6.3.1 Metal–organic chemical vapour deposition

For MOCVD-TiN [90, 91], TDMAT/TDEAT and TDMAT+NH$_3$/TDEAT+NH$_3$ are two main precursors. Without NH$_3$ as the catalyst, the TiN films are grown at relatively high temperature, with high electrical resistivity and high concentrations of C and O. The addition of NH$_3$ can lower the reaction temperature and the electrical resistivity of the film. The activation energy of MOCVD-TiN was found to be lower than 0.1 eV and is diffusion limited. The deposition rate is controlled by the diffusion rate of reactant gases to the wafer surface. Therefore, the surface coverage is about that of metal sputtering. As the reactants diffuse and bond to the inner wall of the holes, low surface coverage results.

In addition to covering the dielectric surface, PVD-TiN should also cover the bottom of the contact windows. The demand for the bottom coverage of TiN arises from the need to protect the silicon surface from chemical attack by SiF_6 to maintain the reliability of the devices. In order to lower the contact resistance, an additional layer of Ti is deposited to form $TiSi_2$. The TiN overlayer thus also protects the underlying Ti from the chemical attack by SiF_6. "Volcanoes" can form when the via profiles, Ti thickness, and TiN thickness are incompatible. A CVD-TiN film alleviates this concern. Current CVD TiN processes are adequate for contact fill. However, for the $TiCl_4$-based CVD TiN, the temperature (400°C) for via applications is too high.

2.7 ADHESION AND WETTING LAYER

An underlayer is required prior to via-fill, a glue layer for CVD W, and a wetting layer in the case of Al-plug-fill. There are two options for the glue layer in the TiN/AlCu/TiN metallisation scheme, a single TiN layer or a Ti/TiN stack. A Ti layer is preferred as a wetting layer for Al-plug, as it helps to reduce any remaining oxide at the via bottom; however, a single TiN process is simpler. Ti, because of its high reactivity, also results in good adhesion between the oxide and the metal layers.

In order to improve the adhesion properties between W and SiO_2, a metal glue layer is required [92, 93]. The stresses of the W film are of the order of $1-2 \times 10^{10}$ dyne/cm^2, enough to

Silicide formation

break the integrity of the multilayered film. TiN has been known to be a very useful diffusion barrier layer, and anti-reflecting layer for aluminium interconnect metallisation in VLSI manufacturing process [94]. TiN also meets the requirement for improving the adhesion properties between CVD-W and SiO_2 insulating layer because it shows good adhesion to SiO_2. The thermal stability temperature of TiN/SiO_2 system is relatively high (>1000°C), and the W/TiN structure has been shown to be a stable low-resistance interconnection. It is known that the crystal orientation of the TiN films is determined by the deposition method and film thickness. The crystal structure and electrical properties of the subsequently deposited CVD-W could be influenced by the initial growth of TiN adhesion layer. The preferred orientation of TiN is also known to change with the TiN thickness. An example is shown in FIGURE 2.23 [95].

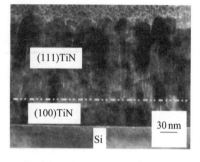

FIGURE 2.23 XTEM image showing the change of TiN orientation with thickness.

2.8 ANTI-REFLECTION COATING LAYER

Owing to the high reflectivity of aluminium alloys layer, the photoresist imagings are often degraded through reflective light scattering. TiN can also be used as an efficient ARC layer on top of the aluminium alloys to reduce the effect of reflective notching and widen the exposure window in photolithographic patterning [96]. An example is shown in FIGURE 2.20.

The photoresist developer solution (containing $(CH_3)_4NOH$) was found to penetrate into the Al/Cu films through the cracks of ARC TiN resulted by the large interface stress between Al/Cu and TiN thin films and etch the Al/Cu alloy during the later development process. On the other hand, for the TiN-ARC/Al/Cu/Si system, the Si nodules were generally found to precipitate at the Al alloy grain boundaries during cooling. As a result, trench holes with high aspect ratio were created on top of Si nodules. The poor bottom and sidewall coverage of TiN-ARC at these small holes has been accepted as inevitable. During the subsequent development process, these poor coverage sites were found to provide the photoresist developer solution a pathway to penetrate through and to etch the Al/Cu/Si alloys. The etched Al surface was oxidised during the later rinse process. The Al_2O_3 layer thus formed, it acted as an etch mask during metal reactive ion etch (RIE) process and resulted in the ring defects. In addition, the density of ring defects was found to increase with the deposition temperature of Al alloy. The effects can be alleviated by using a thicker Al/Cu layer. For the TiN-ARC/Al/Cu/Si system, the deposition of a thin Ti layer prior to the deposition of the TiN layer was found to reduce the ring defects significantly [97].

2.9 CONCLUSIONS

In multilevel interconnection, TiSi$_2$ had been the main SALICIDE material until the mass production of 0.18 μm generation devices. For more advanced devices, CoSi$_2$ and NiSi are gaining in importance. Consecutively deposited poly-Si and WSi$_x$ POLYCIDE shall continue to be used in many deep sub-micrometre devices.

CVD Al is expected to replace the cold Al process in cold/hot Al process since it offers better step coverage and continuity. In addition, the CVD Al produces a smoother surface, which facilitates the nucleation and flow of the overlying hot Al layer. As a result, the hot Al temperature can be lower than 400°C.

With the introduction of low-k dielectric, the processing temperature for PVD is expected to be lowered. It becomes a challenge for Al-plug and planarisation processes. It is therefore imperative to develop a low temperature PVD process.

Copper is an attractive material for wiring in sub-0.13 μm devices. Also, stray capacitance of wiring needs to lower the permittivity of interlayer insulators such as inorganic films and other organic materials. Multilevel wiring of six to eight layers is necessary for large-scale logic ULSIs. Global planarisation of insulators and metals will be essential in the fabrication of multilevel wiring. Cu RIE, Cu CMP and Cu CVD will require new equipments that can be used to form copper films, and for Cu patterning.

Copper wiring systems can be built at minimum expense by matching processes and equipment used for Al wiring. Copper wiring processing having a wide operating range without copper contamination into the bulk silicon during CVD, CMP and RIE will be necessary for many applications. The introduction of copper wiring to ULSIs will begin with the logic ULSIs used in microprocessor and similar devices that require super high speed, and then will be extended to memory ULSIs.

REFERENCES

[1] Semiconductor Industry Association [*The International Technology Roadmap for Semiconductors*, 2003 edition (Semiconductor Industry Association, San Jose, CA, 2003)]
[2] J.W. Mayer, S.S. Lau [*Electronic Materials Science: For Integrated Circuits in Si and GaAs* (MacMillan, New York, 1990)]
[3] M. Hansen [*Constitution of Binary Alloys* (McGraw-Hill, New York, 1958)]
[4] Y.C. Peng, L.J. Chen, Y.R. Yang, W.Y. Hsieh, Y.F. Hsieh [*J. Korean Phys. Soc. (Korea)* vol.35 (1999) p.80-3]
[5] C.Y. Ting, M. Wittmer [*Thin Solid Films (Netherlands)* vol.96 (1982) p.327-45]

[6] M.A. Nicolet, S.S. Lau [*Materials Process and Characterization* Ed. N.G. Einspruch, G.R. Larrabee (Academic, New York, 1983) p.329–464]
[7] K.N. Tu, J.W. Mayer [*Thin Films – Interdiffusion and Reactions* Ed. J.M. Poate, K.N. Tu, J.W. Mayer (Wiley, New York, 1978) p.359–405]
[8] M.H. Wang, L.J. Chen [*J. Appl. Phys. (USA)* vol.71 (1992) p.5918–25]
[9] L.J. Chen, I.W. Wu, J.J. Chu, C.W. Nieh [*J. Appl. Phys. (USA)* vol.63 (1988) p.2778–82]
[10] A.E. Morgan, E.K. Broadbent, K.N. Ritz, D.K. Sandana, B.J. Burrow [*J. Appl. Phys. (USA)* vol.64 (1988) p.344–53]
[11] J.Y. Cheng, L.J. Chen [*J. Appl. Phys. (USA)* vol.68 (1990) p.4002–7]
[12] J.H. Lin, L.J. Chen [*J. Appl. Phys. (USA)* vol.77 (1995) p.4425–30]
[13] T.L. Lee, L.J. Chen [*J. Appl. Phys. (USA)* vol.73 (1993) p.5280–2]
[14] L.J. Chen, J.M. Liang, C.S. Liu, et al. [*Ultramicroscopy (Netherlands)* vol.54 (1994) p.156–65]
[15] L.J. Chen [*Mater. Sci. Eng. R (Netherlands)* vol.29 (2000) p.115–52]
[16] W.L. Johnson [*Prog. Mater. Sci. (USA)* vol.30 (1986) p.81–134]
[17] R.M. Walser, R.W. Bene [*Appl. Phys. Lett. (USA)* vol.28 (1976) p.624–6]
[18] J.C. Chen, G.H. Shen, L.J. Chen [*J. Appl. Phys. (USA)* vol.84 (1998) p.6083–7]
[19] C.A. Crider, J.M. Poate [*Appl. Phys. Lett. (USA)* vol.36 (1980) p.417–19]
[20] J.Y. Cheng, H.C. Cheng, L.J. Chen [*J. Appl. Phys. (USA)* vol.61 (1987) p.2218–23]
[21] F.M. d'Heurle [*J. Mater. Res. (USA)* vol.3 (1988) p.167–95]
[22] L.S. Hung, J. Gyulai, J.W. Mayer, S.S. Lau, M.A. Nicolet [*J. Appl. Phys. (USA)* vol.54 (1983) p.5076–80]
[23] S.Q. Wang, J.W. Mayer [*Thin Solid Films (Netherlands)* vol.207 (1992) p.37–41]
[24] G.J. Huang, L.J. Chen [*J. Appl. Phys. (USA)* vol.76 (1994) p.865–70]
[25] C.M. Osburn, H. Berger, R.P. Donovan, G.W. Jones [*J. Environ. Sci. (USA)* vol.31 (1988) p.45–57]
[26] G.J. Huang, L.J. Chen [*J. Appl. Phys. (USA)* vol.72 (1992) p.3143–9]
[27] M. Wittmer, C.Y. Ting, K.N. Tu [*J. Appl. Phys. (USA)* vol.54 (1983) p.699–705]
[28] D.S. Wen, P.L. Smith, C.M. Osburn, G.A. Rozgonyi [*Appl. Phys. Lett. (USA)* vol.51 (1987) p.1182–4]
[29] T. Morimoto, T. Ohguro, H. Momose, et al. [*IEEE Trans. Electron Devices (USA)* vol.42 (1995) p.915–22]
[30] S.W. Lu, C.W. Nieh, L.J. Chen [*Appl. Phys. Lett. (USA)* vol.49 (1986) p.1770–2]
[31] L.J. Chen, C.M. Doland, I.W. Wu, J.J. Chu, S.W. Lu [*J. Appl. Phys. (USA)* vol.62 (1987) p.2789–92]
[32] W.J. Chen, L.J. Chen [*J. Appl. Phys. (USA)* vol.70 (1991) p.2628–33]
[33] W. Lur [Ph.D. Thesis, National Tsing Hua University, Hsinchu, Taiwan (1989)]
[34] W. Lur, L.J. Chen [*J. Appl. Phys. (USA)* vol.64 (1988) p.3505–11]
[35] T. Hara, S.C. Chen [*Microelectron. Engng. (Netherlands)* vol.37–38 (1997) p.67–74]
[36] J.P. Gambino, E.G. Colgan [*Mater. Chem. Phys. (Netherlands)* vol.52 (1998) p.99–146]
[37] L. van den Hove, R. Wolters, K. Maex, R.F. de Keersmaecker, G.J. Declerck [*IEEE Trans. Electron Devices (USA)* vol.34 (1987) p.554–61]

[38] R. Liu [*ULSI Technology* Ed. C.Y. Chang, S.M. Sze (McGraw-Hill, New York, 1996) p.371–471]
[39] W. Lur, J.Y. Cheng, C.H. Chu, et al. [*Nucl. Instrum. Methods B (Netherlands)* vol.39 (1989) p.297–301]
[40] W.J. Chen, L.J. Chen [*J. Appl. Phys. (USA)* vol.69 (1991) p.7322–4]
[41] W. Lur, L.J. Chen [*J. Appl. Phys. (USA)* vol.66 (1989) p.3604–11]
[42] J.F. Chen, L.J. Chen, W. Lur [*Nucl. Instrum. Methods B (Netherlands)* vol.96 (1995) p.361–5]
[43] W.J. Chen, L.J. Chen [*J. Appl. Phys. (USA)* vol.71 (1992) p.653–8]
[44] S. Shimizu, T. Kuroi, K. Kusunoki, Y. Okumura, M. Inuishi, H. Miyoshi [*Jpn. J. Appl. Phys. (Japan)* vol.35 (1996) p.802–6]
[45] T. Kuroi, M. Kobayashi, M. Shirahata, et al. [*Jpn. J. Appl. Phys. (Japan)* vol.34 (1995) p.771–5]
[46] T. Murakami, T. Kuroi, Y. Kawasaki, M. Inuishi, Y. Matsui, A. Yasuoka [*Nucl. Instrum. Methods B (Netherlands)* vol.121 (1997) p.257–61]
[47] S.L. Cheng, L.J. Chen, B.Y. Tsui [*J. Mater. Res. (USA)* vol.14 (1999) p.213–21]
[48] G.J. Huang, L.J. Chen [*J. Appl. Phys. (USA)* vol.74 (1993) p.1001–7]
[49] G.J. Huang, L.J. Chen [*J. Appl. Phys. (USA)* vol.78 (1995) p.929–36]
[50] A. Hiraki [in *Progress in the Study of Point Defects* Ed. M. Doyama, S. Yoshida (University of Tokyo, Tokyo, Japan, 1977) p.393]
[51] C.A. Hewett, S.S. Lau [*Appl. Phys. Lett. (USA)* vol.50 (1987) p.827–9]
[52] C.R. Chen, L.J. Chen [*J. Appl. Phys. (USA)* vol.78 (1995) p.919–25]
[53] J.M.E. Harper, A. Charai, L. Stolt, F.M. d'Heurle, P.M. Fryer [*Appl. Phys. Lett. (USA)* vol.56 (1990) p.2519–21]
[54] C.S. Liu, L.J. Chen [*J. Appl. Phys. (USA)* vol.74 (1993) p.3611–13]
[55] C.S. Liu, L.J. Chen [*J. Appl. Phys. (USA)* vol.75 (1994) p.2730–2]
[56] C.S. Liu, L.J. Chen [*Thin Solid Films (Netherlands)* vol.262 (1995) p.187–98]
[57] L.J. Chen, K.N. Tu [*Mater. Sci. Rep. (Netherlands)* vol.6 (1991) p.53–140]
[58] R.T. Tung [*Mater. Chem. Phys. (Netherlands)* vol.32 (1992) p.107–33]
[59] M.S. Fung, H.C. Cheng, L.J. Chen [*Appl. Phys. Lett. (USA)* vol.47 (1985) p.1312–14]
[60] I.C. Wu, J.J. Chu, L.J. Chen [*J. Appl. Phys. (USA)* vol.60 (1986) p.3172–5]
[61] J.J. Chu, I.C. Wu, L.J. Chen [*J. Appl. Phys. (USA)* vol.61 (1987) p.549–51]
[62] L.J. Chen, I.W. Wu, J.J. Chu, C.W. Nieh [*J. Appl. Phys. (USA)* vol.63 (1988) p.2778–82]
[63] C.S. Chang, C.W. Nieh, L.J. Chen [*Appl. Phys. Lett. (USA)* vol.50 (1987) p.259]
[64] J.Y. Yew, L.J. Chen, K. Nakamura [*Appl. Phys. Lett. (USA)* vol.69 (1996) p.999–1001]
[65] J.Y. Yew, H.C. Tseng, L.J. Chen, Y. Nakamura, C.Y. Chang [*Appl. Phys. Lett. (USA)* vol.69 (1996) p.3692–4]
[66] H.F. Hsu, L.J. Chen, J.J. Chu [*J. Appl. Phys. (USA)* vol.69 (1991) p.4282–5]
[67] J.Y. Yew, L.J. Chen, W.F. Wu [*J. Vac. Sci. Technol. B (USA)* vol.17 (1999) p.939–44]
[68] B.L. Crowder [*Thin Solid Films (Netherlands)* vol.93 (1982) p.358–68]
[69] E. Nagasawa, H. Okabayashi, M. Morimoto [*Jpn. J. Appl. Phys. Lett. (Japan)* vol.22 (1983) p.L57–L59]
[70] S.P. Murarka [*Solid State Technol. (USA)* vol.28–29 (1985) p.181–5]
[71] S.L. Zhang, M. Ostling [*Crit. Rev. Solid State Mater. Sci. (USA)* vol.28 (2003) p.1–129]

[72] K.Y. Lee, Y.K. Fang, C.W. Chen, M.S. Liang, J.C. Hsieh [*IEEE Electron Device Lett. (USA)* vol.18 (1997) p.181–3]
[73] Y.N. Chen, M.W. Lippitt, H.Z. Chew, W.M. Moller [*IEEE Trans. Electron Devices (USA)* vol.50 (2003) p.2120–5]
[74] I. Ames, F.M. d'Heurle, R.E. Horstmann [*IBM J. Res. Dev. (USA)* vol.44-1 (2000) p.89–91]
[75] M. Bakli, L.M. Baud, H. Saad, D. Pique, P. Rabinzohn [*Microelectron. Engng. (Netherlands)* vol.33 (1997) p.175–88]
[76] K. Hoshino [*Jpn. J. Appl. Phys. (Japan)* vol.39 (2000) p.994–8]
[77] K. Sakurai, A. Onoyama, T. Fujii, K. Yamanishi, S. Fujii, H. Morita [*IEEE Trans. Semicond. Manuf. (USA)* (2002) p.118–26]
[78] W.J. Lee, S.K. Rha [*Jpn. J. Appl. Phys. (Japan)* vol.42 (2003) p.3372–6]
[79] J.F. Jongste, X. Li, J.P. Lokker, G.C.A.M. Janssen, S. Radelaar [*Microelectron. Engng. (Netherlands)* vol.37–38 (1997) p.319–27]
[80] C.K. Wang, H.S. Wu, N.T. Ou, H.C. Cheng [*Jpn. J. Appl. Phys. (Japan)* vol.41 (2002) p.5120–4]
[81] M. Takahashi, T. Ohno, Y. Sakakibara, K. Takayama [*IEEE Trans. Electron Devices (USA)* vol.48 (2001) p.1380–5]
[82] M. Wittmer [*J. Vac. Sci. Technol. A (USA)* vol.3 (1985) p.1797–1803]
[83] S.M. Rossnagel, D. Mikalsen [*J. Vac. Sci. Technol. A (USA)* vol.9 (1991) p.261–5]
[84] S.M. Rossnegal [*J. Vac. Sci. Technol. A (USA)* vol.21 (2003) p.S74–S87]
[85] H. Wolf, R. Streite, W. Tirschler, H. Giegengack, N. Urbansky, T. Gessner [*Microelectron. Engng. (Netherlands)* vol.63 (2002) p.329–45]
[86] S. Li, Y.K. Lee, W. Gao, T. White, Z.L. Dong, K.M. Latt [*J. Vac. Sci. Technol. B (USA)* vol.19 (2001) p.388–96]
[87] H. Gris, B. Caussat, D. Cot, J. Durand, J.P.Couderc [*Chem. Vapor Depos. (Netherlands)* vol.8 (2002) p.213–19]
[88] T. Kim, S.M. Suh, S.L. Girshick, et al. [*J. Vac. Sci. Technol. A (USA)* vol.20 (2002) p.413–23]
[89] M.J. Buiting, A.F. Otterloo [*J. Electrochem. Soc. (USA)* vol.139 (1992) p.2580–4]
[90] A. Sabbadini, F. Cazzaniga, S. Alberici, et al. [*Microelectron. Engng. (Netherlands)* vol.55 (2001) p.205–11]
[91] S. Riedel, S.E. Schulz, J. Baumann, M. Rennau, T. Gessner [*Microelectron. Engng. (Netherlands)* vol.55 (2001) p.213–18]
[92] E. Sabouret, C. Schaffnit, J.F. Jongste, G.C.A.M. Janssen, S. Radelaar [*Microelectron. Engng. (Netherlands)* vol.37–38 (1997) p.353–63]
[93] K.M. Chang, T.H. Yeh, S.W. Wang, C.H. Li [*J. Electrochem. Soc. (USA)* vol.144 (1997) p.996–1001]
[94] C.Y. Ting [*Thin Solid Films (Netherlands)* vol.119 (1984) p.11–21]
[95] Y.C. Peng, L.J. Chen, W.Y. Hsieh, Y.R. Yang, Y.F. Hsieh [*J. Vac. Sci. Technol. B (USA)* vol.16 (1998) p.2013–18]
[96] C.N. Ho, Y.K. Lim, H. Gerald, W.L. Goh, M.S. Tse, S. See [*J. Electron. Mater. (USA)* vol.30 (2001) p.1595–1601]
[97] Y.C. Peng, L.J. Chen, Y.R. Yang, W.Y. Hsieh, Y.F. Hsieh [*J. Korean Phys. Soc. (Korea)* vol.35 (1999) p.S80–S83]

Chapter 3
Titanium silicide technology

Z. Ma and L.H. Allen

3.1 INTRODUCTION

Titanium silicide is one of the first few silicides considered for application in ultra-large-scale integrated circuits (ULSI) owing to its low resistivity, good thermal stability and compatibility with Si processes. There are two ways of using titanium silicide in Si-based devices: polycide gate electrode/interconnect and contact/interconnect in self-aligned silicide (SALICIDE) process. This chapter will first describe the methods and mechanisms of titanium silicide formation and then address the relevant materials issues, integration concerns and scaling limits for technological implementation in deep-sub-micrometre device applications. Various schemes for extending the practical limit of titanium silicide will also be briefly mentioned.

3.2 FORMATION OF TITANIUM SILICIDES

There are several forms of titanium silicides with different crystal structures and materials properties. Among them titanium disilicide is of great interest to Si microelectronics as it offers the lowest resistivity and exhibits good thermal integrity during post-silicidation processing. Titanium disilicide can be formed by annealing of (a) co-evaporated or co-sputtered $TiSi_x$ film ($x \sim 2$) deposited over Si; or (b) reacting a Ti thin film with either single crystalline Si or polysilicon at elevated temperatures. The first method is used in the so-called polycide technology in which an amorphous titanium silicide alloy ($TiSi_x$, $x \sim 2$) is co-deposited on top of the doped polysilicon. The composite structure is then annealed at 700–800°C to form the low resistivity titanium disilicide phase. This approach is presently used for gate metallisation in DRAM process.

The Ti/Si thin film reaction approach is now the standard technological implementation in MOS device fabrication because of

Introduction p.49

Formation of titanium silicides p.49

Fundamental aspects of Ti/Si thin film reaction p.50
    Phase formation sequence p.50
    Early stage of the Ti/Si thin film reaction p.52
    Formation of metastable C49-$TiSi_2$ p.53
    C49-to-C54 $TiSi_2$ polymorphic transformation p.54
    Agglomeration p.59

Integration concerns and scaling limits of titanium disilicide p.61
    Effect of doping and dopant redistribution p.61
    Reactions of Ti with SiO_2 and Si_3N_4 p.63
    Lateral silicide encroachment p.63
    Thermal stability p.64
    Chemical stability p.65
    Contact resistivity p.65
    Limitations of SALICIDE scaling p.66

Methods of enhancing the C54-$TiSi_2$ phase formation p.68
    High temperature deposition p.68
    Pre-amorphisation implant p.69
    Metal impurity or interlayer p.70
    Recessed spacer approach p.72

Conclusions p.73

Acknowledgments p.73

References p.73

Titanium silicide technology

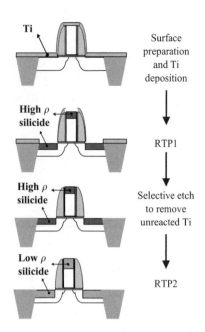

FIGURE 3.1 Schematic representation of the self-aligned silicide (SALICIDE) process using a two-step rapid thermal processing (courtesy of J.A. Kittl, Texas Instruments, Inc.).

the widespread use of the self-aligned silicidation (SALICIDE) technology to simultaneously reduce transistor contact and series resistance. In the SALICIDE process, a Ti thin film is deposited over a MOSFET structure having nitride or oxide sidewall spacers along the polysilicon gate and reacted with the exposed Si regions on the source, drain and gate, in order to form desired silicides in these three areas. The unreacted Ti is then selectively etched away, and a high temperature anneal follows to complete the silicide formation and reduce the sheet resistance of the silicide. A schematic representation of the SALICIDE process is shown in FIGURE 3.1.

3.3 FUNDAMENTAL ASPECTS OF Ti/Si THIN FILM REACTION

3.3.1 Phase formation sequence

There are five equilibrium phases in the Ti–Si binary phase diagram: Ti-rich Ti_3Si, Ti_5Si_3, Ti_5Si_4, $TiSi$ and Si-rich $TiSi_2$, as shown in FIGURE 3.2 [1]. In the case of Ti/Si thin film reaction couple not all the phases appear during heat treatment and only the selected phases are formed in a sequential manner. The phase formation sequence depends critically upon processing conditions

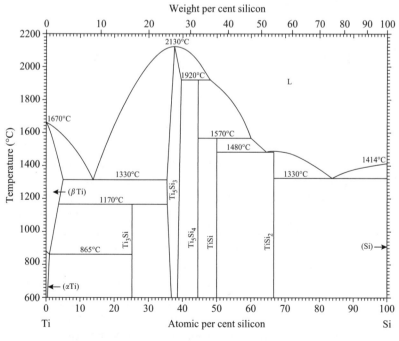

FIGURE 3.2 Equilibrium Ti–Si binary phase diagram.

such as interface property (interface treatment, contaminant, etc.) and annealing conditions and typically follows a reaction path like this: Ti/Si → α-TiSi$_x$ → C49-TiSi$_2$ → C54-TiSi$_2$. Both amorphous silicide (α-TiSi$_x$) and crystalline C49 titanium disilicide phases do not exist in the equilibrium phase diagram and are formed inevitably prior to the equilibrium disilicide phase (C54-TiSi$_2$). The metastable C49 phase usually has a high resistivity of 80–100 $\mu\Omega$ cm while the equilibrium C54 phase offers a very low resistivity of 13–20 $\mu\Omega$ cm, depending on the method of fabrication and level of impurity incorporation.

FIGURE 3.3 shows a typical *in situ* resistance versus temperature curve for the Ti/poly-Si bilayer deposited over oxidised Si during furnace annealing at a heating rate of 10°C/min (~0.17°C/s). There are four distinct changes in resistance during annealing: (1) a linear increase in resistance with increasing temperature from 25 to 350°C, (2) the region with a steep increase in resistance subsequent to region 1, (3) a sharp decrease in resistance at ~550°C, followed by (4) another sharp drop in resistance at ~700°C. Earlier work has shown that the four regions correlate to the characteristic phase changes occurring during annealing [2]. FIGURE 3.4 shows cross-sectional transmission electron microscopy (TEM) micrographs of the samples heated to specific temperatures in the four regions of the curve, cooled

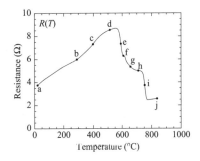

FIGURE 3.3 A typical resistance curve of the Ti/poly-Si bilayer as a function of annealing temperature recorded at 10°C/min using *in situ* four-point probe measurement.

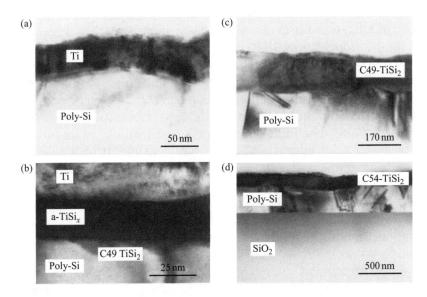

FIGURE 3.4 Cross-sectional TEM micrographs showing interfacial evolution of the Ti/poly-Si bilayer reaction: (a) as-deposited sample, (b) annealed up to ~510°C, (c) annealed up to ~710°C and (d) annealed up to ~850°C. Images (b), (c) and (d) correspond to points d, h and j, respectively, on the resistance curve shown in FIGURE 3.3.

rapidly to ambient temperature. Region 1 corresponds to the temperature coefficient of resistance of the Ti film, while region 2 is a result of formation of the amorphous $TiSi_x$ phase and the incorporation of oxygen into the Ti film. The first steep drop in resistance (region 3) is due to the formation of the metastable $C49$-$TiSi_2$ and the second drop in resistance (region 4) owing to the C49-to-C54 phase transformation. Although the example shown here is obtained from furnace annealing conditions it should be noted that the general shape of curve and their corresponding microstructural characteristics can be extended to the cases of rapid thermal annealing [3, 4].

As amorphous silicide and $C49$-$TiSi_2$ always precede the formation of the desired low resistance C54 phase, it is critical to understand the driving forces and fundamental mechanisms for their formation. This would also be beneficial for practising engineers to diagnose silicide-related process yield issues and design improved processes by manipulating the phase formation sequence and, in particular, promote the formation of desired phases. In Sections 3.3.2 and 3.3.3, we will briefly discuss the basic aspects of early stage of the Ti/Si bilayer reaction and of the metastable C49 phase formation.

3.3.2 Early stage of the Ti/Si thin film reaction

Initial stage of the Ti/Si thin film reaction was studied by several research groups [5–8]. Low temperature interdiffusion or intermixing was observed at the Ti/Si interface when annealed at temperatures below 400°C. An in-depth investigation was done by Holloway and Sinclair [6, 7] using an alternating Ti–Si multilayer structure with the bilayer periodicity of $\sim$10 nm. In the multilayer thin films, reaction takes place at many interfaces simultaneously so it provides much enhanced information on either a phase forming to a limited extent, or, a small volume of interdiffusion. After a 30 s anneal at 455°C, significant interdiffusion occurs over >5 nm distance at the α-Ti/α-Si interfaces. The formation of an amorphous $TiSi_x$ alloy was observed through interfacial reaction. The amorphous alloy phase has a graded composition across the layer and can remain stable up to 550°C, depending upon the composition of the multilayer, that is, the relative thickness of Ti and Si layers [7–9].

The formation of metastable amorphous phase is also observed in several other thin film systems including Mo–Si, Ni–Si and Ni–Zr, etc. [10–12]. The common nature of these reactions is that the metastable amorphous phase is obtained by a slow and gentle kinetic process at low temperatures. Several models were proposed to elucidate the thermodynamic and kinetic aspects of the so-called solid-state amorphisation [11, 12]. Its occurrence is kinetically driven due to the inability of forming crystalline phases and the tendency for energy reduction at low temperatures. Two necessary conditions must be met for this to occur. First, the heat of mixing of the two elements must be negative, which provides the thermodynamic driving force. Second, one element must be a sufficiently fast moving species in the other element and must diffuse substantially into the other at temperatures below the crystallisation temperature of the amorphous phase alloy. This is a kinetic requirement for intermixing. In the case of Ti–Si solid-state reaction, both conditions are satisfied: (1) the heat of formation of the amorphous phase was measured to be -56.7–66.5 kJ/g atom using differential scanning calorimetry [13], (2) marker experiment indicates that Si is the dominant moving species and diffuses faster than Ti [14], which is evidenced by the presence of Kirkendall voids on the α-Si side of the interface [7, 8]. A detailed review of solid-state amorphisation can be found in the literature [12].

3.3.3 Formation of metastable C49-TiSi$_2$

Although several crystalline phases were reported to appear as the first crystalline phase after the amorphous phase [15], it is generally accepted that the dominant crystalline phase to form is the high resistivity, metastable C49 titanium disilicide phase, which has a base-centred orthorhombic structure ($a = 0.362$ nm, $b = 1.376$ nm, $c = 0.360$ nm). Interface impurity or contaminants such as carbon and oxygen can have significant effect on the phase formation sequence and the first crystalline phase. Nucleation and growth of the C49 phase formation was carefully investigated by Ma et al. [9]. FIGURE 3.5 shows the morphological evolution of the C49-TiSi$_2$ phase at the reaction interface. It is clearly seen that the metastable C49 phase individually nucleated along the interphase boundary between amorphous silicide (α-TiSi$_x$) and crystalline Si (c-Si). Upon annealing to a higher temperature or at 530°C for a longer time, the C49 grains grew laterally (in a direction parallel to the α-TiSi$_x$/Si interface) as well as vertically (in a direction perpendicular to the interface) at the expense of the amorphous phase and Si. The lateral growth (lengthening) appears faster than the vertical growth (thickening). Further simultaneous lateral and vertical growth resulted in the coalescence of the individual grains originally separately formed at the interface. This initial growth stage proceeds very fast, as illustrated in FIGURE 3.5(b) for a sample annealed up to $\sim$560°C. The individual growing C49 grains connected each other to result in a continuous layer of polycrystalline disilicide before Ti and α-TiSi$_x$ phase are exhausted (FIGURE 3.5(c)). This initial

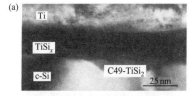
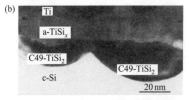
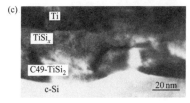

FIGURE 3.5 Bright-field TEM images showing the nucleation and growth of the metastable C49-TiSi$_2$ at the interface of the Ti/poly-Si bilayer upon annealing at 10°C/min up to: (a) $\sim$510°C, (b) $\sim$560°C and (c) $\sim$585°C.

fast reaction was then taken over by a diffusion-controlled growth stage [8]. The metastable C49-TiSi$_2$ has a distinct microstructure and is often heavily faulted along the [0k0] direction [16], which partly contributes to the high resistivity of the phase.

3.3.4 C49-to-C54 TiSi$_2$ polymorphic transformation

The C49-to-C54 titanium disilicide phase transformation has drawn a great deal of attention and research effort because of its practical and scientific interests. The phase change is polymorphic in nature, that is, no compositional change during crystal structure transition, and is accomplished by local atomic transport and rearrangement. The C49 phase has a base-centred orthorhombic structure and a stacking sequence of ABABAB while the C54 phase has a face-centred orthorhombic structure with a stacking sequence of ABCDABCD. As shown in FIGURE 3.6 [17], both structures exhibit similar atomic arrangements and possess a common rectangular base layer consisting of Ti atoms surrounded hexagonally by Si atoms. Two crystal structures can be derived by displacing this base layer in one direction to form an intermediate layer (direction b in the case of the C49 structure) or in both directions (directions a and b in the C54 structure). Different stacking sequence in the c direction gives rise to different crystal structures in this case but the nearest-neighbour configurations are preserved.

Similar to other polymorphic transformation, the C49-to-C54 transformation proceeds via a nucleation and growth process.

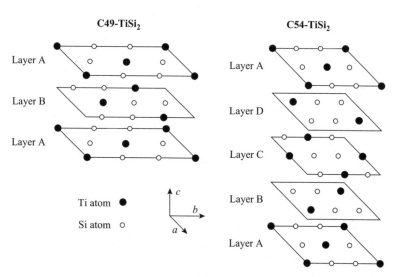

FIGURE 3.6 Crystal structure and atomic arrangement of: (a) C49-TiSi$_2$ ($a = 0.362$ nm, $b = 1.376$ nm, $c = 0.360$ nm) and (b) C54-TiSi$_2$ ($a = 0.825$ nm, $b = 0.478$ nm, $c = 0.854$ nm) within the unit cell.

The thermodynamic driving force for the phase change is usually very small and mainly determined by the change in configurational entropy associated with the structural change:

$$\Delta G_{C49 \to C54} = \Delta G_{C54} - \Delta G_{C49} = (\Delta H_{C54} - \Delta H_{C49})$$
$$- T(\Delta S_{C54} - \Delta S_{C49}) \sim -T(\Delta S_{C54} - \Delta S_{C49})$$

where ΔH_{C49} and ΔH_{C54}, ΔS_{C49} and ΔS_{C54}, represent the heats of formation and configurational entropy for the C49 and C54 phases, respectively. The rate of the transformation highly depends on the type of nucleation, number of such nucleation sites available, and growth rate of the stable C54 phase. In most cases, polymorphic transformation of intermetallic compounds is accompanied by a very small heat effect, typically ranging from 0.01 to 0.03 eV/atom [18].

Kinetics of the phase transformation was extensively studied and measured using various *in situ* and *ex situ* techniques [19–21]. *In situ* resistance measurement appears to be the most popular one due to the excellent resistance response associated with the phase change and real-time monitoring of the transformation process. In this technique, samples are typically prepared with a bilayer of titanium/polycrystalline silicon deposited onto thermally oxidised (001) silicon wafers and pre-annealed to form the C49 phase. When the C49-TiSi$_2$ film is thermally annealed, it gradually transforms to the stable C54 disilicide and the film resistance continuously decreases with annealing time or temperature, that is, $R = f(t, T)$, due to the increase in volume fraction of the low resistivity C54 phase formed by consuming high resistivity C49 phase. By relating the change in resistance to the progress of the transformation, Thompson et al. [19] and Ma and Allen [20] derived an activation energy for the C49-TiSi$_2$ to C54-TiSi$_2$ transformation of 4.5 ± 0.1 eV under furnace annealing conditions for both amorphous TiSi$_2$ and 55 nm Ti/poly-Si bilayer samples. Using the Kissinger approach, Clevenger et al. [21] reported an activation of 3.8 ± 0.5 eV by measuring resistance versus temperature at heating rates up to 3000°C/min. Compared to the C49 phase formation, the large activation energy for the formation of C54-TiSi$_2$ suggests that both nucleation and growth kinetics control its formation.

Study of microstructural evolution of the phase transition revealed heterogeneous nature of the C54 phase nucleation. In addition to grain boundary nucleation, the C54-TiSi$_2$ was also found to nucleate at triple grain junctions of the C49 disilicide, and a limited number of C54 nuclei found at the available grain boundaries or triple grain junctions of the C49 phase, as

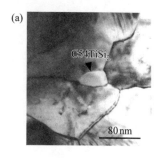

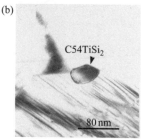

FIGURE 3.7 Planar TEM micrographs showing the location of the C54-TiSi$_2$ nuclei inside the C49-TiSi$_2$ thin films formed from samples with: (a) 25 nm Ti and (b) 100 nm Ti. The C54 nuclei were identified using electron microdiffraction as well as dark-field imaging indicated by arrows.

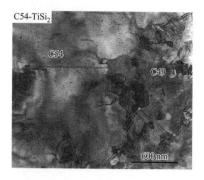

FIGURE 3.8 Bright-field TEM image showing that the growth of the C54 phase proceeds by moving the characteristically ragged incoherent interface across the C49 grains. Note that the C54 grain grew across and at the expense of many small C49 grains.

shown in FIGURE 3.7 [20, 22]. Statistical examinations show that nucleation at triple grain junctions is more prevalent. It has also been reported that among available heterogeneous sites including grain boundaries and triple grain junctions, only about 15 per cent of the sites were occupied. This may be indicative of the difficulty of nucleating the C54 phase and imply the role of geometrical factors associated with various sites in the nucleation event. According to Clemm and Fisher [23], the nucleation energy barrier associated with different geometrical sites can be described by:

$$\Delta G^* = \frac{4}{27} \frac{(b\sigma_{C49/C54} - a\sigma_{C49/C49})^3}{c^2(\Delta G_V - \Delta G_S)^2}$$

where $\sigma_{C49/C54}$ is the interfacial energy between C49 and C54 disilicides, $\sigma_{C49/C49}$ the grain boundary energy, ΔG_V the difference in volume free energy between C49 and C54 phases and the driving force for the phase transition, ΔG_S the volume strain energy associated with the nucleation of C54 phase, and the coefficients a, b and c have a set of specific values depending upon the geometry of the nucleus. Triple grain junctions are energetically more favourable sites compared to grain boundaries. Microstructural development of the transformation exhibits two distinct aspects: interphase boundary between C49 and C54 phases are mostly incoherent and the growth of the C54 phase often proceeds by moving the characteristically ragged incoherent interface across the C49 grains, as illustrated in FIGURE 3.8.

3.3.4.1 Effect of film thickness

With the device dimensions being further scaled down, the thickness and width of local interconnection lines are also reduced to compatible sizes. Several factors were discovered to fundamentally affect the C49-to-C54 phase transformation in constrained dimensions. Thickness and linewidth of the silicide show the most significant impact on shallow junction scaling and integration [17, 20, 24–28]. The effect of Ti film thickness or C49-TiSi$_2$ thickness on the C49-to-C54 transformation was systematically investigated using *in situ* resistance measurement and transmission electron microscopy by Ma et al. [20, 22, 28].

Three different thicknesses of titanium films, namely, 25, 55 and 100 nm, were sputter deposited over phosphorus-doped polycrystalline Si grown using low pressure chemical vapour deposition. The Ti/poly-Si bilayers were pre-annealed to form varying thicknesses of polycrystalline C49-TiSi$_2$. As seen in both plan-view and cross-sectional images (FIGURE 3.9), the metastable C49 disilicides are heavily faulted. The average grain size of the C49 phase

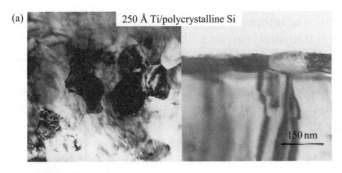

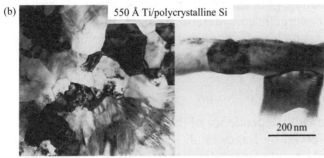

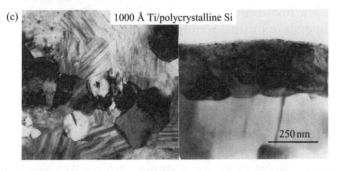

FIGURE 3.9 Plan-view (left) and cross-sectional (right) TEM micrographs of the C49-TiSi$_2$ films formed by reacting poly-Si with Ti films of different thicknesses, showing the microstructure and morphology of the polycrystalline C49 phase prior to transforming to the C54 phase: (a) 25 nm Ti, (b) 55 nm Ti and (c) 100 nm Ti.

is about 150 nm for 25 nm Ti, 180 nm for 55 nm Ti and 230 nm for 100 nm Ti, respectively. It is noted from cross-sectional views of the C49 microstructure that the C49 phases formed with 25 and 55 nm Ti/poly-Si bilayer reactions are essentially a layer of two-dimensional polycrystalline structure with the grain boundaries running through the entire film thickness. On the contrary, the C49 microstructure has more than one layer of polycrystalline grains formed over the remaining poly-Si. Thickness effect on the kinetics of the phase transition was investigated using isothermal annealing at 640–740°C. It was found that the transformation temperature increases with decreasing initial C49 film

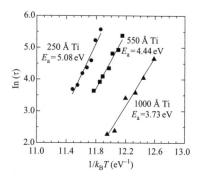

FIGURE 3.10 Arrhenius plots of $\ln(\tau)$ versus $1/k_B T$ for determining the activation energies for the transformation for three different Ti thicknesses. The parameter τ is the time required to complete 50% of the transformation, k_B the Boltzmann constant and T the annealing temperature.

thickness. The temperature difference in initiating the C49-to-C54 structural transition can be as large as 50°C due to the change in C49 film thickness in this case. Using Johnson–Mehl–Avrami analysis [29], the activation energies were determined to be 3.73, 4.44 and 5.08 eV for 100, 55 and 25 nm Ti, respectively, as shown in FIGURE 3.10. It is worth noting that the microstructure and morphology of the C49 phase significantly influenced the nucleation behavior of the stable C54 phase. Nucleation at C49 triple grain junctions became more prevalent in thinner C49 films compared to grain boundary nucleation in thick films, which can be explained in terms of energetic consideration of nucleation at different geometrical sites and largely increased surface-to-volume ratio associated with thickness scaling [28].

3.3.4.2 Effect of linewidth

The linewidth dependence of the C49-to-C54 phase transition was first observed by Lasky et al. [24] using 1, 2, 4 and 40 μm wide lines and later studied in detail by Kittl et al. [26] and Saenger et al. [27] with a linewidth ranging from 0.1 to 1.0 μm. It was found that temperature or time required to complete the transformation increased monotonically as linewidth decreased. But interestingly, the activation energies derived from kinetics analysis are roughly the same, 4.0 ± 0.2 eV, for various linewidths studied in this work (0.26, 0.34, 0.50 and 0.80 μm), implying the same mechanism for the phase transformation (FIGURE 3.11). A model based on nucleation site density argument was proposed to explain the linewidth dependence. It was assumed that nucleation of the C54 phase on narrow lines takes place by a similar mechanism as on blanket

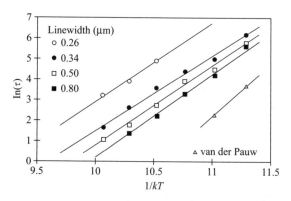

FIGURE 3.11 Arrhenius plots used to determine the activation energies for the C49-to-C54 phase transformation with various linewidths (0.26, 0.34, 0.50 and 0.80 μm). The activation energies for various linewidths are very close to each other, ranging from 3.7 to 4.0 ± 0.2 eV, compared to the one, 5.0 ± 0.5 eV, derived from van der Pauw structure, 40 μm side square.

films, that is, at triple grain junctions or grain boundaries. In contrast to blanket film case, the microstructure of the narrow lines is being confined to the linewidth and density of triple grain junctions is a function of linewidth. With linewidth decreasing to sizes comparable to C49 grain size (typically about 0.1–0.2 μm), the number of nucleation sites is greatly reduced as a result of lateral confinement. As the C49-to-C54 transformation kinetics is dependent upon nucleation site density (i.e. triple grain junctions), temperature and time, the observed linewidth dependence is attributed to reduced nucleation events caused by limited nucleation sites. It is worth pointing out that the C54 grains, once nucleated, span very quickly across the narrow linewidth and grow mainly in one dimension along the patterned line. A detailed treatment of the linewidth-dependent C49-to-C54 phase transformation can be found in References 26 and 30.

3.3.5 Agglomeration

Although titanium disilicide has a very high melting point (1540°C) and often exhibits good thermal stability, morphological instability has been observed at temperatures above 900°C [31–34]. It is noticed that sheet resistance of the $TiSi_2$ films on single crystal Si or polysilicon will increase after annealing at 900°C or higher and the increase is caused by agglomeration of the titanium disilicide films, a morphological degradation where polycrystalline disilicide films become disintegrated, forming an "island" structure [32, 33]. As agglomeration leads to an increase in film sheet resistance and junction leakage, it imposes an upper temperature limit for $TiSi_2$ processing. FIGURE 3.12 shows the sheet resistance of the polysilicon films and the silicided bilayers as a function of the annealing temperature. It is seen from FIGURE 3.12(a) that the resistance of unsilicided polysilicon films remains almost unaffected by additional heat treatments up to 900°C, irrespective of the type of dopants. Annealing at 950°C for 30 min results in a slightly lower resistance for heavily boron- and arsenic-doped wafers. In the case of the $TiSi_2$/polysilicon bilayer, the sheet resistance remains roughly unchanged upon annealing up to 900°C. Above this temperature significant increase in sheet resistance is observed for the undoped sample and the sample implanted with $2 \times 10^{15}/cm^2$ of arsenic, as shown in FIGURE 3.12(b). The resistance of the remaining samples is only slightly changed by the 950°C anneal, indicating that a high doping concentration in the polysilicon layer tends to improve the thermal stability of the silicide.

The driving force for agglomeration is the reduction of surface and interface energy. Kinetically it is accomplished via

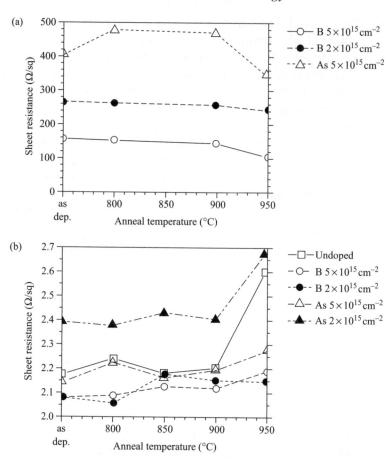

FIGURE 3.12 The sheet resistance of: (a) implanted poly-Si lines as a function of annealing temperature. All wafers received an RTP treatment at 1100°C for 10 s to activate the implanted dopants, and (b) the TiSi$_2$/poly-Si bilayer as a function of annealing temperature. All temperature treatments were performed in nitrogen for 30 min.

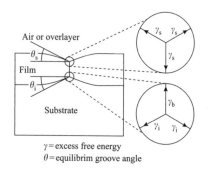

FIGURE 3.13 Energy balances at upper and lower grain boundary grooves.

interface and grain boundary diffusion. The degradation starts at the intersections of the grain boundaries with the top surface and bottom interface, where the surface (γ_s), interface (γ_i), and grain boundary energies (γ_b) tend to equilibrate, that is, $\gamma_b = 2\gamma_s \sin\theta_s$, $\gamma_b = 2\gamma_i \sin\theta_i$, as is schematically represented in FIGURE 3.13. This phenomenon is known as "grain boundary grooving" and was modelled by various researchers [35, 36]. The grooves will eventually grow through the entire silicide layer and the grains will disconnect and form individual islands. Depending on whether the surface has a capping layer or not, these silicide islands will further agglomerate and develop into morphologies as shown in FIGURE 3.14. It is believed that the agglomeration is accompanied by Si regrowth via solid phase epitaxy [37]. It has been theoretically demonstrated that agglomeration can be prevented or retarded by reducing grain size and grain boundary energy or increasing film thickness and surface/interface energies [36].

3.4 INTEGRATION CONCERNS AND SCALING LIMITS OF TITANIUM DISILICIDE

Implementation of titanium SALICIDE process in CMOS technology needs to consider many boundary conditions and its interaction with other transistor modules such as source/drain with different dopant types, oxide or nitride spacer, rapid thermal processing, etc. Its successful integration will require a careful selection of process windows based on a good understanding of (a) effect of dopant type on the phase formation temperature and dopant redistribution during silicidation; (b) effect of impurities on the phase formation; (c) effect of formation ambient; and (d) reactivity of titanium with oxide or nitride, etc. In this section, we will discuss in detail some of these key aspects and how they will impact silicide integration and device performance. With continuous shrinkage of device dimension, the SALICIDE technology also needs to be scaled correspondingly to meet the requirement for shallow junctions. The fundamental scaling limit for titanium silicide systems will also be discussed.

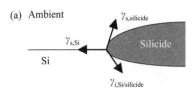
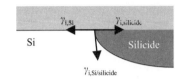

FIGURE 3.14 Schematic cross-section view of equilibrium morphology of silicide/ single crystal Si structure after grooving: (a) in ambient and (b) with a capping layer. The energy balances are shown in terms of surface energies for silicide and Si ($\gamma_{s,silicide}$, $\gamma_{s,Si}$), interface energy between Si and silicide ($\gamma_{i,Si/silicide}$), and interface energies between capping layer and Si or silicide ($\gamma_{i,Si}$, $\gamma_{i,silicide}$).

3.4.1 Effect of doping and dopant redistribution

Fabrication of CMOS devices requires simultaneous formation of low resistance C54-TiSi$_2$ phase on both n-type doped and p-type doped Si or poly-Si. This is typically achieved by reacting a pure titanium film with heavily doped silicon. The n-type dopants generally used are arsenic, phosphorus, antimony, etc. while boron is used for p-type dopant. The dopants are either implanted into Si or introduced into Si during crystal growth (in the case of single crystalline Si) or film deposition (in the case of polycrystalline Si). It is found that titanium silicide formation and growth kinetics are strongly influenced by the type of dopant present and doping level in the silicon [38–40].

FIGURE 3.15 shows a temperature derivative of the *in situ* resistance versus temperature for 57.5 nm of titanium deposited on top of 300 nm of polycrystalline Si, either undoped or doped with boron, arsenic or phosphorus. The phosphorus doping level is $3.0 \times 10^{20}/cm^3$ while the boron and arsenic level are both $2.7 \times 10^{20}/cm^3$. Two sharp minima in the plot are indicative of C49-TiSi$_2$ and C54-TiSi$_2$ phase formation, as verified by X-ray

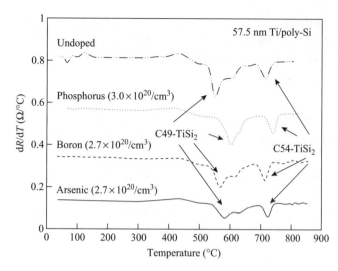

FIGURE 3.15 Temperature derivative of the *in situ* resistance versus temperature for 57.5 nm Ti on top of 300 nm poly-Si at a heating rate of 12°C/min. The poly-Si was either undoped, or doped with $2.7 \times 10^{20}/cm^3$ boron or arsenic, $3.0 \times 10^{20}/cm^3$ phosphorus.

diffraction analysis. From the minima in the curves, the formation temperature for both C49 and C54 phases increases in the order of undoped, boron-, arsenic- and phosphorus-doped polycrystalline silicon. In other words, the presence of either n-type or p-type dopant at this level retards the phase formation and reaction kinetics. The uniformity of the Ti/Si reaction is not seriously affected by heavy silicon doping. But dopant precipitation in a form of Ti compound is often found at this doping level. In the case of arsenic-doped Si substrates ($3.0 \times 10^{21}/cm^3$), TiAs precipitates were observed at C49-TiSi$_2$ grain boundaries, and the C49-to-C54 transformation temperature increased to 850°C due to the influence of precipitates on the C54 phase nucleation.

Dopant diffusion and redistribution during silicidation is another important concern for integration as it directly affects source/drain and contact resistance. Grain boundary diffusion is reported to be the predominant mechanism for dopant (As, P and B) diffusion and redistribution in titanium sicilide and induces a very rapid homogenisation of dopant throughout the silicide [41, 42]. Accumulation of dopants (As, P) at the C54-TiSi$_2$/Si interface has been observed. The accumulation is attributed to the diffusion of dopant and to interface adsorption effects. Electrical nature of the accumulation and its contribution to free carriers at the interface were investigated by Topuria et al. [43] using combined atomic resolution Z-contrast imaging and electron energy loss spectroscopy (EELS). They showed that arsenic segregation occupied Si substitutional sites and was electrically active. Its presence supplies additional charge carriers at the interface and therefore reduces silicide/Si interfacial resistance.

3.4.2 Reactions of Ti with SiO$_2$ and Si$_3$N$_4$

A standard titanium salicide process involves: (1) deposition of a thin titanium film onto an MOS transistor structure having either silicon oxide or nitride sidewall spacers around the polysilicon gate; (2) a low temperature (<700°C) anneal to induce silicidation over the exposed Si areas of the source, drain and gate; (3) removal of the unreacted Ti with a selective wet chemical etch; and finally (4) an 800–900°C anneal to complete the C49-to-C54 phase transformation and minimise the sheet resistance. It is very important to understand and minimise the reaction between Ti and oxide or nitride during silicidation. Otherwise reaction residues could remain on the spacers, resulting in either gate-to-source/drain leakage or short.

Ti/SiO$_2$ reaction path and products are slightly dependent upon the annealing ambient, Ar or N$_2$, as shown in FIGURE 3.16 [44, 45]. During 700–900°C anneal, Ti will reduce some SiO$_2$ to form a Ti-rich silicide (TiSi$_x$ or Ti$_5$Si$_3$) interfacial layer. Depending on the oxygen contamination of Ti film and annealing ambient, either a Ti oxide mixture or TiN$_x$O$_{1-x}$ is formed on top of the interfacial silicide. This is in general agreement with the Ti–Si–O ternary phase diagram [46] where the Ti$_5$Si$_3$–TiO–SiO$_2$ three phase region are at equilibrium (FIGURE 3.17). As shown in FIGURE 3.18, interface evolution is more complex in the case of Ti/Si$_3$N$_4$ reaction [44, 47]. Annealing of a 90 nm Ti film deposited over a 332 nm Si$_3$N$_4$ layer at 800°C in Ar for 30 s resulted in formation of TiN and Ti$_5$Si$_3$ interfacial layers with the remaining Ti film being uniformly contaminated with oxygen from the annealing ambient and nitrogen from decomposed Si$_3$N$_4$. A three-layer microstructure developed at 1000°C anneal, 30 nm TiN$_x$O$_{1-x}$/40 nm C54-TiSi$_2$/50 nm TiN/Si$_3$N$_4$.

3.4.3 Lateral silicide encroachment

An important observation was made in the study of TiSi$_2$ formation using rapid thermal processing in Ar ambient: formation of silicide took place rapidly along the Ti grain boundaries at temperatures around 700°C. This phenomenon is called lateral overgrowth or "lateral encroachment". The preferential grain boundary silicidation is accompanied by a fast vertical diffusion of Si (Si is the dominant moving species). In combination with a lateral surface migration it leads to the formation of a silicide surface layer before the silicidation front has reached the surface. This phenomenon is of particular interest to MOS transistor application.

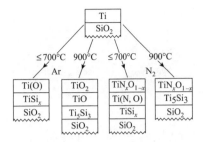

FIGURE 3.16 Schematic of Ti/SiO$_2$ reaction products at various rapid thermal annealing temperatures in Ar or N$_2$ ambients.

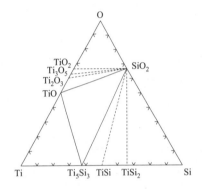

FIGURE 3.17 Ternary representation of the Ti/SiO$_2$ reaction. Tie lines determined (solid) and inferred (dashed) from the observed Ti–SiO$_2$ reaction products. $T = 950°C$.

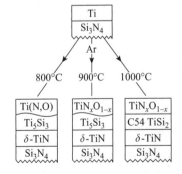

FIGURE 3.18 Schematic of Ti/Si$_3$N$_4$ reaction products at various rapid thermal annealing temperatures in Ar.

Titanium silicide technology

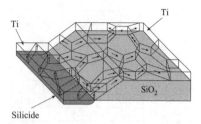

FIGURE 3.19 Model explaining the rapid lateral silicidation along the Ti grain boundaries on top of SiO$_2$ upon annealing in Ar ambient.

In the region near the shoulder or foot of an oxide or nitride spacer this rapid transport of Si may continue laterally along grain boundaries of Ti layer over the sidewall spacer, resulting in silicide formation on the spacer near contact or gate edge (over field oxide). This is illustrated by the schematic in FIGURE 3.19 [48]. If the lateral silicidation develops quickly during rapid thermal annealing, a potential short circuiting will occur between adjacent gate and source/drain regions.

This lateral silicide growth or encroachment can be suppressed by performing the silicidation in a N$_2$ ambient [49–52]. A bilayer reaction product results from a competition between interfacial silicidation and surface nitridation. The uptake of nitrogen apparently inhibits the preferential silicidation along the Ti grain boundaries by blocking the fast grain boundary diffusion paths for Si. In this case a TiN/C54-TiSi$_2$ bilayer typically formed on top of Si and a trilayer structure of TiN/Ti(N,O)/TiSi$_x$ ($x < 1$) over oxide. This trilayer will be stripped after the first silicidation anneal step. Rapid thermal annealing (RTA) is preferred over furnace annealing to minimise or avoid oxidation of Ti during the initial anneal, which would hinder nitrogen incorporation and prevent Ti wet etching.

3.4.4 Thermal stability

In Section 3.3.5 we have discussed the morphological instability or agglomeration under high temperature processing and the fundamental reasons for the degradation. It directly impacts the salicide processing windows by putting an upper limit on the temperature which the system can tolerate without agglomeration. In addition, the onset of agglomeration will shift to lower temperatures with decreasing silicide film thickness, further reducing processing temperature window [17]. Morphological degradation can also contribute to the shallow junction leakage due to the roughened silicided/silicon interface.

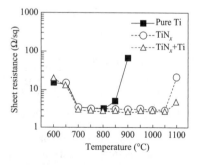

FIGURE 3.20 Sheet resistance curves for pure Ti, TiN$_x$ and TiN$_x$ + Ti samples annealed at different temperatures for 30 s.

Various improvement schemes have been proposed to extend its thermal stability to higher temperature regime. It is found that addition of N$_2$ to Ar during Ti sputtering deposition improves the thermal stability of TiSi$_2$ while incorporating O in Ar degrades it [33, 53]. Ogawa et al. [32] and Chang et al. [54] demonstrated that the TiSi$_2$ film was stable up to 1050°C using a gas mixture of N$_2$: Ar (1 : 45) during Ti film deposition or initial Ti sputtering for 2 s. FIGURE 3.20 shows the dependence of sheet resistance on annealing temperature for pure Ti, TiN$_x$ and TiN$_x$ + Ti samples. The unreacted Ti and TiN layers were removed prior to resistance measurement. Compared to the pure Ti case, the silicide formed with Ti film deposited using nitrogen-containing Ar gas shows little resistance change after annealing up to 1050°C. As discussed previously, the agglomeration process starts with grain boundary thermal grooving and involves the transport of Si and Ti in the silicide via interface and grain boundaries. Improved thermal stability with nitrogen addition is thought to be due to the stuffing of silicide grain boundaries and TiN/TiSi$_2$ interfaces, therefore retarding the Ti and Si diffusion.

3.4.5 Chemical stability

In the fabrication of MOS devices several chemicals are used for cleaning and etching purposes. After the first RTA in N$_2$ ambient the unreacted Ti and TiN are etched away using a mixture of H$_2$O$_2$, NH$_4$OH, H$_2$O without attacking TiSi$_2$, which is fairly resistant to the standard chemicals used for cleaning, for example the solutions used in the RCA cleaning. However, TiSi$_2$ is soluble in HF-containing solutions. This presents a process challenge for using HF-based solutions to clean the contact hole prior to contact metallisation such as Ti/TiN followed by CVD tungsten deposition. In a typical contact etch process, the contact hole is opened using a dry etching process with very high etch selectivity between TiSi$_2$ and SiO$_2$. A buffered HF solution is often used to remove any etch residues from the bottom of the contact. The etch rate of the silicide must be carefully characterised and controlled to avoid contact punch-through, where the W contact lands directly onto Si instead of the silicide, giving rise to high contact resistance [48].

3.4.6 Contact resistivity

The electrical function of silicides is to provide a low resistance path for the device. The initial path of current through the device is vertical but spreads laterally in the junction and channel. The current generates small voltage drops over each segment of the path between the source and drain. These segments include: (1) the contact resistance between silicide and junction; (2) the lateral resistance of the junction beneath the silicides; (3) the resistance beneath the spacer; and (4) the effective resistance of the channel itself. Typical values of the total series resistance, R_s, of the MOS device is of the order of $\sim 100\,\Omega$.

The specific contact resistivity ρ_c ($\Omega\,cm^2$) specifies the effective resistance (normalised to junction area) of the silicide/junction (cm^2) region. The region includes both the actual metallurgical silicide/Si interface as well as the depletion region below the interface. The entire width of the region is extremely narrow in highly doped source/drain regions in the order of 3 nm for dopant concentration of $N_d \sim 10^{20}$ atoms/cm^3. Since this region is so narrow, consideration must be given to issues such as the local uniformity of dopants and the geometrical roughness of the metallurgical interface. The electronic properties of the Si are also degenerate at these high doping levels as the concentration is larger than the density of states of the conduction band, $N_c \sim 3 \times 10^{19}$ atoms/cm^3 of the Si. Because the region is so narrow, conduction of the carriers across the region occurs mainly through tunnelling mechanisms. The specific contact resistivity is strongly dependent on the doping level given by:

$$\rho_c \propto \exp\left(\frac{4\pi \phi_B}{h}\sqrt{\frac{m^* \varepsilon_{Si}}{N_d}}\right)$$

where m^* is the effective mass of the carrier, N_d is the dopant concentration, ε_{Si} is the permittivity of Si and ϕ_B is the Schottky barrier height [55].

Experimental values for the specific contact resistivity of TiSi$_2$ on both n-type and p-type Si are given in FIGURE 3.21. Values range down to 10^{-8} and 10^{-7} (Ω cm^2) for n-type and p-type materials, respectively [56–62]. The importance of the contact resistance to device performance lies in that it comprises a significant portion (20–40 Ω) of the R_s of the device and may not be easily scalable as compared to other components of the series resistance.

3.4.7 Limitations of SALICIDE scaling

The SALICIDE process module begins after transistor source and drain regions are formed. Conventional scaling of the salicide module is typically achieved by reducing Ti film thickness. However, this results in silicide with great susceptibility to many problems, the most notable being incomplete C49-to-C54 phase transformation and enhanced agglomeration. As silicide dimensions decrease laterally as well as vertically, the minimum temperature needed to fully transform titanium silicide to the low resistivity C54 phase shifts upward, while the maximum temperature before the onset of agglomeration moves downward. This translates into a shrinking temperature process window for the formation of a robust low resistivity silicide, as shown schematically in FIGURE 3.22. FIGURE 3.23 illustrates salicided poly lines with unacceptable high resistances due to either incomplete transformation at lower temperature or agglomeration at higher temperatures [63].

At deep sub-micrometre dimensions, a decreased density of active nucleation sites and reduced number of C54 phase nuclei results in inadequate phase transformation. This is especially true

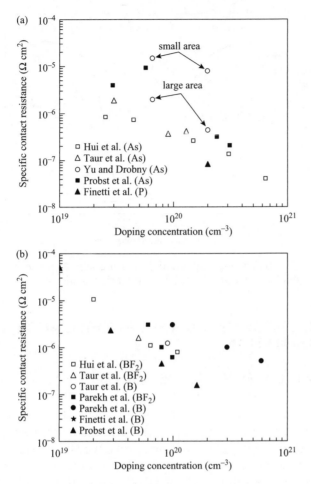

FIGURE 3.21 Experimental values of specific contact resistance of TiSi$_2$/Si versus doping concentration: (a) n-type Si, 300 K, $q\phi_B = 0.6$ eV; (b) p-type Si, 300 K, $q\phi_B = 0.51$ eV.

when grain sizes become comparable to the narrower silicide linewidths. Narrow line resistance goes up significantly when polysilicon linewidth drops below 0.18 μm. Enhanced nucleation and phase transition can be promoted using increased thermal budget but with greater risks of lateral encroachment and agglomeration. Other marginality problems with thinner Ti include film coverage uniformity in stacked-gate regions, contact etch selectivity to silicide, and the ability to achieve adequate transformation of n$^+$ polycide without causing lateral encroachment of p$^+$ polycide, etc. Increased silicide thickness, on the other hand, offers lower sheet resistivities and better defect margin, but at the cost of increased Si consumption. Large Si consumption will likely increase the risks of device junction leakage and higher device external resistances and is incompatible with ultra-shallow junction formation.

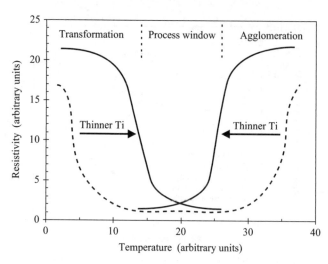

FIGURE 3.22 Schematic diagram of decreasing salicide process window for reduced Ti thickness (courtesy of J. Tsai, Intel Corporation).

3.5 METHODS OF ENHANCING THE C54-TiSi$_2$ PHASE FORMATION

In order to extend the life of titanium silicide beyond 0.18 μm, tremendous effort has been made to enhance the low resistivity C54 silicide formation by promoting nucleation, manipulating the reaction path and improving the process integration [64–76]. Some of these approaches are briefly reviewed here to highlight the novelty, concept and capability developed to improve the process window and meet the need for ULSI applications and continued Moore's law scaling. These various techniques can be integrated or selectively combined to realise the maximum benefits for specific applications.

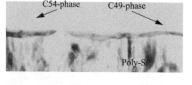

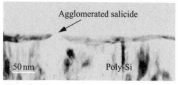

FIGURE 3.23 Cross-sectional TEM images of (upper) incomplete phase transformation at 735°C with 30 nm Ti and (lower) agglomeration at 875°C with <20 nm Ti (courtesy of J. Tsai, Intel Corporation).

3.5.1 High temperature deposition

In a conventional SALICIDE process, Ti film is usually deposited by sputtering onto a Si wafer without intentional heating. The deposition is then followed by annealing in a nitrogen ambient, which leads first to a precursor interfacial amorphous silicide layer before any crystalline silicide phases are formed [6, 8, 9, 15]. It has been shown that deposition of Ti and co-deposition of TiSi$_x$ at elevated temperatures led to the direct growth of silicides and that the temperature and composition of the deposition and the crystallinity of the substrate have a strong influence on the phases of the silicide layer [64–67]. A thin interfacial amorphous TiSi$_x$ layer, with $x \sim 0.5–1$, deposited between Ti film and Si led to a significant reduction in the observed C49-to-C54 transformation temperature.

Titanium silicide technology

TABLE 3.1 Phase identification and sheet resistance of silicide layers grown by co-deposition of 15 nm TiSi$_x$ on amorphous silicon.

Deposition composition	$T_{dep} = 470°C$	$T_{dep} = 520°C$	$T_{dep} = 550°C$
Ti	TiSi 52 Ω/sq	C54 + C49 8.9 Ω/sq	C54 > C49 6.6 Ω/sq
TiSi$_{0.5}$	TiSi 78 Ω/sq	C54 > C49 6.4 Ω/sq	C54 > C49 7.4 Ω/sq
TiSi$_{1.8}$	C54 + C49 14 Ω/sq	C54 + C49 13 Ω/sq	C54 > C49 16 Ω/sq[a]

[a] Discontinuous layer

Nucleation of the low resistance C54 disilicide was observed on amorphous silicon at temperatures as low as 500°C. TABLE 3.1 summarises the phase identification and sheet resistance of silicide layers grown by deposition of 15 nm TiSi$_x$ on amorphous Si. The lowered C54 phase formation temperature was speculated to arise from the avoidance of amorphous silicide layer and nucleation of silicides at Si surface.

Recently, Chang et al. [67] investigated the enhanced formation of C54-TiSi$_2$ by high temperature sputtering of Ti onto pre-amorphised Si and found that the dominant effect of high temperature sputtering is to increase the density of crystallites in the amorphous TiSi$_x$ layer, which serve as nucleation sites for the C49-TiSi$_2$. As a result, the average grain size of the C49 phase is smaller. This, in turn, leads to the lowered transformation temperature.

3.5.2 Pre-amorphisation implant

Based on the fact that the temperature of C49-to-C54 phase transformation decreases with reducing C49 grain sizes and C49 grain size limits the scaling of Ti silicide to ~0.25 μm features for processes using crystalline Si [77], an improved SALICIDE process was developed to achieve smaller C49 grains by incorporating pre-amorphisation implant (PAI) [68–73]. The PAI is typically done with Si, Ge, As and B implant. By reacting Ti with pre-amorphised Si C49 grains as small as 0.07 μm can be obtained. For a given finite linewidth smaller C49 grain sizes translate into increased heterogeneous sites for C54 phase nucleation and lowered transformation temperature. FIGURE 3.24 shows bright-field planar TEM images of the polycrystalline C49-TiSi$_2$ with and without PAI treatment. The C49 grains formed with PAI treatment are small and heavily faulted and quite large

(a)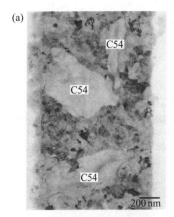

(b)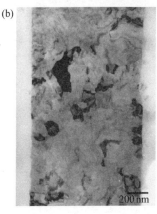

FIGURE 3.24 Bright-field planar TEM images of:
(a) the polycrystalline C49 structure including C54 grains with PAI treatment and (b) the C49 structure without PAI treatment. Linewidth: 0.94 μm.

C54 grains are also found sporadically in the matrix of C49 grains even after 650°C, 30 s anneal. Without PAI, the C49 grains are typically quite large and show relatively less faulted structure. No C54 phase was observed under this condition. With PAI process, feasibility of using titanium silicide has been demonstrated down to 0.1 μm linewidths [69]. But transient enhanced diffusion of dopants induced by PAI are reported and can result in degradation of contact resistance, source/drain series resistance and drive current [73].

3.5.3 Metal impurity or interlayer

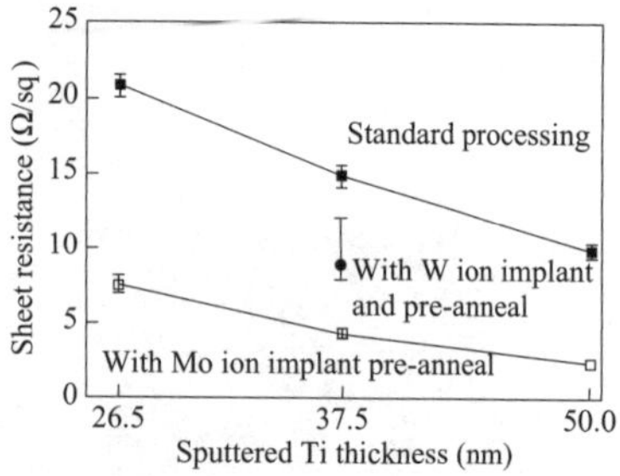

In addition to high temperature deposition and pre-amorphisation implant, other alternative approaches have also been explored. The most notable ones are the addition of metal impurities to the Ti/Si interface [74–76] or a thin metal interlayer between Ti and Si [78]. Mann et al. [74] discovered that ion implantation of a small dose of refractory metal such as Mo or W (1×10^{13}–1×10^{14} ions/cm^2 at 45 keV) into Si substrate before the deposition of Ti film lowered the C49-to-C54 transformation temperature by as much as 150°C. The effect of the implanted refractory metal on the sheet resistance after 600°C, 30 min anneal is shown in FIGURE 3.25 for initial Ti thicknesses of 25, 35 and 55 nm. The samples with Mo implant show a resistivity of 16 μΩ cm, indicating the completion of the transformation, versus the samples without Mo implant, which have a resistivity of >60 μΩ cm and are still dominated by the high resistance C49 phase. The W-implanted samples show a resistivity of about 40 μΩ cm and contain a mixture of both C49 and C54 disilicides.

To further examine the phase formation characteristics, samples consisting of 55 nm Ti/300 nm poly-Si deposited on thermally grown oxide were prepared for *in situ* resistance measurement. One of the samples received a Mo implant and a 900°C, N$_2$ anneal prior to the Ti deposition, while the control sample was similarly treated but did not have a Mo implant. From the *in situ* resistance data shown in FIGURE 3.26(a), significant differences are seen between the two samples in the phase formation characteristics. The sample without the Mo implant forms C49 at ~580°C and remains in this phase until about 780°C, where it undergoes an abrupt transition to C54 disilicide. In contrast, the Mo-implanted sample shows no distinct resistance plateau, which is characteristic of the C49 phase formation, between 700 and 800°C. Instead, this sample transformed directly into the C54 phase at around 700°C. This was confirmed by thin-film X-ray diffraction analysis shown in FIGURE 3.26(b). Similar enhancement effect is also seen with a thin Mo interposed layer of 0.5–1.0 nm between Ti and Si.

FIGURE 3.25 Sheet resistance after 600°C formation anneal and selective etch for samples with and without a metal impurity implant.

Titanium silicide technology

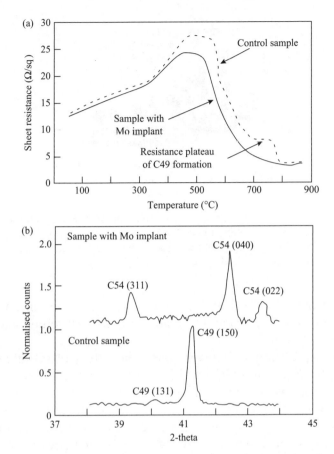

FIGURE 3.26 *In situ* resistance as a function of ramp temperature (a) and the corresponding thin-film X-ray analysis taken on samples cooled from 700°C (b) for a sample with and without the implanted molybdenum.

Two possible mechanisms, template phenomenon and grain size effect, have been proposed to explain the enhanced transformation by Mo doping or Mo interlayer [79–82]. In the template model, the Ti/Si reaction path is believed to be altered. X-ray diffraction studies following the evolution from the early stages of the silicide phase formation [75] show that the reaction of Ti with Mo-doped poly-Si bypasses the high resistivity C49-TiSi$_2$ phase. Ti$_5$Si$_4$ and two Mo silicides, MoSi$_2$ and Mo$_5$Si$_3$, are the first to form at the interfaces. MoSi$_2$ and Mo$_5$Si$_3$ formed at the reaction interface act as template phases for the C54 phase nucleation and local epitaxial growth (see FIGURES 3.27(a) and (b)). On the contrary, the reaction between Ti and Mo-doped single crystal Si proceeds by the conventional phase formation sequence, that is, the metastable C49 phase is not bypassed. The reason for this deviation is still unclear at this point. Mouroux et al. [81, 83] suggest that the template mechanism invokes the formation of an interfacial (RM, Ti)Si$_2$ of the C40 or C11b structure with RM interlayer as Mo, W, Nb or

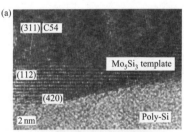

FIGURE 3.27 High resolution TEM images of: (a) a Mo$_5$Si$_3$-based template and (b) a MoSi$_2$ template. Note the orientation relationships between C54 and template phases.

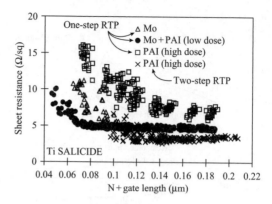

FIGURE 3.28 Gate sheet resistance as a function of gate length for Ti salicide processes. A combination of Mo and pre-amorphisation implant and one-step RTP achieves low sheet resistance down to ultra-narrow 0.06 μm linewidth. All other Ti salicide processes fail below 0.1 μm.

Ta, since the C40, C11b and C54 structures share the same atomic arrangement in their stacking planes but with different stacking sequences [84, 85].

Ohmi and Tung [79] investigated the effect of ultrathin Mo and $MoSi_x$ layer on titanium silicide reaction using transmission electron microscopy and diffraction analysis. They argued that Mo did not alter the sequence of Ti/Si reaction. Rather, the most obvious effect of molybdenum was a reduction of the grain size of the $C49\text{-}TiSi_2$, which could lead to an increase in the nucleation density of the C54 phase and account for the enhanced transformation.

It is worth noting that by combining Mo implant with pre-amorphisation implant, Kittl et al. [86] have demonstrated the scalability of the $TiSi_2$ down to ultra-narrow 0.06 μm gate widths. FIGURE 3.28 shows that the one-step RTP Mo + PAI process achieves low sheet resistance at 0.06 μm gate length with no linewidth dependence down to ~0.07 μm and can still maintain <10 Ω/sq sheet resistance at 0.05 μm. For comparison, optimised Mo only and PAI only one-step RTP processes as well as an optimised PAI only two-step RTP process are also included.

3.5.4 Recessed spacer approach

An integration approach using recessed spacer was developed by Tsai et al. [87] to extend the capability of titanium silicide down to 0.08 μm regime. In this process, as shown schematically in FIGURE 3.29, the entire oxide and nitride spacer stack is recessed to increase the effective reaction area between Ti and poly-Si. The nitride layer is etched first to endpoint using an anisotropic etch, followed by a timed overetch that also uses an

Titanium silicide technology

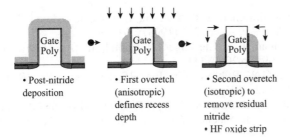

FIGURE 3.29 Schematic representation of the recessed spacer process (courtesy of J. Tsai, Intel Corporation).

anisotropic etch chemistry. A second timed overetch is used to laterally remove the residual spacer nitride with an isotropic etch chemistry. An HF-based clean is then applied to clear the polysilicon surface and upper sidewalls of the remaining oxide to fully expose the top of the polysilicon line. FIGURE 3.30(a) shows the recessed spacer structure following the isotropic etch step and oxide strip. The recessed spacer etch process is an easily integrated solution for improving silicide performance and has been demonstrated on polysilicon gate linewidths as narrow as 0.08 µm (FIGURE 3.30(b)).

3.6 CONCLUSIONS

Fundamental materials aspects of the titanium silicide technology have been reviewed with specific emphasis on its application to SALICIDE process. Various process integration concerns are discussed in light of its applicability and scalability to sub-quarter-micrometre CMOS technology. Methods of extending the life of titanium salicide technology to below ~0.10 µm are also briefly reviewed.

ACKNOWLEDGMENTS

The authors would like to thank Eric Olson for redrawing some of the figures and plots referenced in the chapter. Also, one of the authors (Dr. L.H. Allen) would like to acknowledge support from Dr. LaVerne Hess and support from the National Science Foundation-NSF-DMR 0108694.

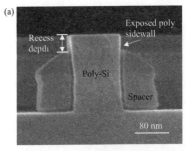

FIGURE 3.30 (a) Cross-sectional SEM image of recessed spacer process after the second overetch and HF-based clean, (b) cross-sectional TEM image of 0.08 µm poly-Si with recessed spacer process and low resistance $TiSi_2$.

REFERENCES

[1] T.B. Massalski [*Binary Alloy Phase Diagrams* (ASM, Ohio, 1986)]
[2] Z. Ma, L.H. Allen, S. Lee [*Mater. Res. Soc. Symp. Proc. (USA)* vol.237 (1992) p.661-5]

[3] R. Ganapathiraman, S. Koh, Z. Ma, L.H. Allen, S. Lee [*Mater. Res. Soc. Symp. Proc. (USA)* vol.303 (1993) p.63–8]
[4] E.G. Colgan, L.A. Clevenger, C. Cabral, Jr. [*Appl. Phys. Lett. (USA)* vol.65 (1994) p.2009–11]
[5] R. Butz, G.W. Rubloff, T.Y. Tan, P.S. Ho [*Phys. Rev. B (USA)* vol.30 (1984) p.5421–9]
[6] K. Holloway, R. Sinclair [*J. Appl. Phys. (USA)* vol.61 (1987) p.1359–64]
[7] K. Holloway, R. Sinclair [*J. Less-Common Metals (Netherlands)* vol.140 (1988) p.139–48]
[8] I.J.M.M. Raaijmakers, K.-B. Kim [*J. Appl. Phys. (USA)* vol.67 (1990) p.6255–64]
[9] Z. Ma, Y. Xu, L.H. Allen, S. Lee [*J. Appl. Phys. (USA)* vol.74 (1993) p.2954–6]
[10] S.B. Newcomb, K.N. Tu [*Appl. Phys. Lett. (USA)* vol.48 (1986) p.1436–8]
[11] R.J. Highmore, A.L. Greer, J.A. Leake, J.E. Evetts [*Mater. Lett. (UK)* vol.6 (1988) p.401–5]
[12] L.J. Chen [*Mater. Sci. Engng. (UK)* vol.R29 (2000) p.115–52]
[13] E. Ma, L.A. Clevenger, K.N. Tu [*Mater. Res. Soc. Symp. Proc. (USA)* vol.187 (1990) p.83–8]
[14] W.K. Chu, S.S. Lau, J.W. Mayer, H. Müller, K.N. Tu [*Thin Solid Films (Switherland)* vol.25 (1975) p.393–402]
[15] M.H. Wang, L.J. Chen [*Appl. Phys. Lett. (USA)* vol.59 (1991) p.2460–2]
[16] R. Beyers, R. Sinclair [*J. Appl. Phys. (USA)* vol.57 (1985) p.5240–5]
[17] H. Jeon, C.A. Sukow, J.W. Honeycutt, G.A. Rozgoni, R.J. Nemanich [*J. Appl. Phys. (USA)* vol.71 (1992) p.4269–76]
[18] P.M. Robinson, M.B. Bever [*Intermetallic Compounds* Ed. J. H. Westbrook (Wiley, New York, 1967) p.38]
[19] R.D. Thompson, H. Takai, P.A. Psaras, K.N. Tu [*J. Appl. Phys. (USA)* vol.61 (1987) p.540–4]
[20] Z. Ma, L.H. Allen [*Phys. Rev. B (USA)* vol.49 (1994) p.13501–11]
[21] L.A. Clevenger, J.M.E. Harper, C. Cabral, Jr., C. Nobili, G. Ottaviani, R. Mann [*J. Appl. Phys. (USA)* vol.72 (1992) p.4978–80]
[22] Z. Ma, L.H. Allen, D.D.J. Allman [*J. Appl. Phys. (USA)* vol.77 (1995) p.4384–8]
[23] P.J. Clemm, J.C. Fisher [*Acta Metall. (UK)* vol.3 (1955) p.70–8]
[24] J.B. Lasky, J.S. Nakos, O.J. Cain, P.J. Geiss [*IEEE Trans. Electron Devices (USA)* vol.38 (1991) p.262–9]
[25] R.A. Roy, L.A. Clevenger, C. Cabral, Jr., et al. [*Appl. Phys. Lett. (USA)* vol.66 (1995) p.1732–4]
[26] J.A. Kittl, D.A. Prinslow, P.P. Apte, M.F. Pas [*Appl. Phys. Lett. (USA)* vol.67 (1995) p.2308–10]
[27] K.L. Saenger, C. Cabral, Jr., L.A. Clevenger, R.A. Roy, S. Wind [*Appl. Phys. Lett. (USA)* vol.78 (1995) p.7040–4]
[28] Z. Ma, L.H. Allen, D.D.J. Allman [*Thin Solid Films (Switzerland)* vol.253 (1994) p.451–5]
[29] J.W. Christian [*The Theory of Transformations in Metals and Alloys, part I*, 2nd Ed. (Pergamon, Oxford, 1975)]
[30] S. Privitera, C. Spinella, F. La Via, M.G. Grimaldi, E. Rimini [*Appl. Phys. Lett. (USA)* vol.78 (2001) p.1514–16]
[31] H. Norström, K. Maex, P. Vandenabeele [*J. Vac. Sci. Technol. B (USA)* vol.8 (1990) p.1223–31]
[32] S.-I. Ogawa, T. Yoshida, T. Kouzaki [*Appl. Phys. Lett. (USA)* vol.56 (1990) p.725–7]

[33] K. Shenai [*J. Mater. Res. (USA)* vol.6 (1991) p.1502–11]
[34] C.Y. Ting, F.M. d'Heurle, S.S. Iyer, P.M. Fryer [*J. Electrochem. Soc. (USA)* vol.133 (1986) p.2621–5]
[35] W.W. Mullins [*J. Appl. Phys. (USA)* vol.28 (1957) p.333–9]
[36] T.P. Nolan, R. Sinclair, R. Beyers [*J. Appl. Phys. (USA)* vol.71 (1992) p.720–4]
[37] S.S. Lau, Z.L. Liau, M.-A. Nicolet [*Thin Solid Films (Switzerland)* vol.47 (1977) p.313–22]
[38] R. Beyers, D. Coulman, P. Merchant [*J. Appl. Phys. (USA)* vol.61 (1987) p.5110–17]
[39] L.A. Clevenger, R.W. Mann, R.A. Roy, K.L. Saenger, C. Cabral, Jr., J. Piccirillo [*J. Appl. Phys. (USA)* vol.76 (1994) p.7874–81]
[40] X.-H. Li, J.R.A. Carlsson, S.F. Gong, H.T.G. Hentzell [*J. Appl. Phys. (USA)* vol.72 (1992) p.514–19]
[41] P. Gas, G. Scilla, A. Michel, F.K. LeGoues, O. Thomas, F.M. d'Heurle [*J. Appl. Phys. (USA)* vol.63 (1988) p.5335–45]
[42] P. Gas, V. Deline, F.M. d'Heurle, A. Michel, G. Scilla [*J. Appl. Phys. (USA)* vol.60 (1986) p.1634–9]
[43] T. Topuria, N.D. Browning, Z. Ma [*Appl. Phys. Lett. (USA)* vol.83 (2003) p.4432–4]
[44] A.E. Morgan, E.K. Broadbent, K.N. Ritz, D.K. Sadana, B.J. Burrow [*J. Appl. Phys. (USA)* vol.64 (1988) p.344–53]
[45] C.-Y. Ting, M. Wittmer, S.S. Iyer, S.B. Brodsky [*J. Electrochem. Soc. (USA)* vol.131 (1984) p.2934–8]
[46] R. Beyers, R. Sinclair, M.E. Thomas [*J. Vac. Sci. Technol. B (USA)* vol.2 (1984) p.781–4]
[47] M. Paulasto, J.K. Kivilahti, F.J.J. van Loo [*J. Appl. Phys. (USA)* vol.77 (1995) p.4412–16]
[48] L. Van den hove, R.F. De Keersmaecker [*Reduced Thermal Processing for ULSI*, Ed. R.A. Levy (Plenum, New York, 1988)]
[49] A. Kikuchi, T. Ishiba [*J. Appl. Phys. (USA)* vol.61 (1987) p.1891–4]
[50] T. Brat, C.M. Osburn, T. Finstad, J. Liu, B. Ellington [*J. Electrochem. Soc. (USA)* vol.133 (1986) p.1451–7]
[51] A.E. Morgan, E.K. Broadbent, A.H. Reader [*Mater. Res. Soc. Symp. Proc. (USA)* vol.52 (1985) p.279]
[52] A.H. Perera, J.P. Krusius [*Appl. Phys. Lett. (USA)* vol.57 (1990) p.1410–12]
[53] C.Y. Ting, F.M. d'Heurle, S.S. Iyer, P.M. Fryer [*J. Electrochem. Soc. (USA)* vol.133 (1986) p.2621–5]
[54] S.M. Chang, S.L. Cheng, L.J. Chen, C.H. Luo [*J. Appl. Phys. (USA)* vol.90 (2001) p.1779–83]
[55] S.M. Sze [*Physics of Semiconductor Devices*, 2nd Ed. (John Wiley & Sons, New York, 1981) p.304]
[56] K. Varahramyan, E.J. Verret [*Solid-State Electron. (UK)* vol.39 (1996) p.1601–7]
[57] M. Finetti, A. Scorzoni, I. Suni [*Le Vide, Les Couches Minces (France)* vol.42 (1987) p.99]
[58] J. Hui, S. Wong, J. Moll [*IEEE Electron Device Lett. (USA)* vol.6 (1985) p.479]
[59] Y. Taur, J.Y.C. Sun, D. Moy, et al. [*IEEE Trans. Electron Devices (USA)* vol.34 (1987) p.575]
[60] Y.C.S. Yu, V.F. Drobny [*J. Electrochem. Soc. (USA)* vol.136 (1989) p.2076–81]

[61] N.S. Parekh, H. Roede, A.A. Bos, A.G.M. Jonkers, R.D.J. Verhaar [*IEEE Trans. Electron Devices (USA)* vol.38 (1991) p.88–94]

[62] V. Probst, H. Schaber, A. Mitwalsky, et al. [*J. Appl. Phys. (USA)* vol.70 (1991) p.693–707]

[63] J. Tsai, N. Stenton, D. Vook, Z. Ma, C.-H. Jan, K. Weldon [*Intel Process Technology Conference*, 1996]

[64] M.D. Naeem, W.A. Orr-Arienzo, J.G. Rapp [*Appl. Phys. Lett. (USA)* vol.66 (1995) p.877–8]

[65] R.T. Tung [*Appl. Phys. Lett. (USA)* vol.68 (1996) p.1933–5]

[66] R.T. Tung, K. Fujii, K. Kikuta, S. Chikaki, T. Kikkawa [*J. Appl. Phys. (USA)* vol.70 (1997) p.2386–8]

[67] S.M. Chang, H.Y. Huang, H.Y. Yang, L.J. Chen [*Appl. Phys. Lett. (USA)* vol.74 (1999) p.224–6]

[68] M. Okihara, N. Hirashita, K. Tai, M. Kageyama, Y. Harada, H. Onoda [*J. Appl. Phys. (USA)* vol.85 (1999) p.2988–90]

[69] J.A. Kittl, Q.Z. Hong, M. Rodder, D.A. Prinslow, G.R. Misium [*VLSI Tech. Digest (USA)* (1996) p.14–15]

[70] K. Tai, M. Okihara, M. Kageyama, Y. Harada, H. Onoda [*J. Appl. Phys. (USA)* vol.85 (1999) p.3132–8]

[71] I. Sakai, H. Abiko, H. Kawaguchi, T. Hirayama, L.E.G. Johansson, K. Okabe [*VLSI Tech. Digest (USA)* (1992) p.66–7]

[72] Z.G. Xiao, H. Jiang, J. Honeycutt, C.M. Osburn, G. McGuire, G.A. Rozgonyi [*Mater. Res. Soc. Symp. Proc. (USA)* vol.181 (1990) p.167]

[73] S. Peterström, B.G. Svensson [*J. Appl. Phys. (USA)* vol.71 (1992) p.1215–18]

[74] R.W. Mann, G.L. Miles, T.A. Knotts, et al. [*Appl. Phys. Lett. (USA)* vol.67 (1995) p.3729–31]

[75] J.A. Kittl, M.A. Gribelyuk, S.B. Samavedam [*Appl. Phys. Lett. (USA)* vol.73 (1998) p.900–2]

[76] S.L. Cheng, J.J. Jou, L.J. Chen, B.Y. Tsui [*J. Mater. Res. (USA)* vol.14 (1999) p.2061–9]

[77] H.J.W. van Houtum, I.J.M.M. Raaijmakers, T.J.M. Menting [*J. Appl. Phys. (USA)* vol.61 (1987) p.3116–18]

[78] A. Mouroux, S.-L. Zhang, W. Kaplan, S. Nygren, M. Östling, C.S. Peterson [*Appl. Phys. Lett. (USA)* vol.69 (1996) p.975–7]

[79] S. Ohmi, R.T. Tung [*J. Appl. Phys. (USA)* vol.86 (1999) p.3655–60]

[80] S.-L. Zhang, F.M. d'Heurle [*Appl. Phys. Lett. (USA)* vol.76 (2000) p.1831–3]

[81] A. Mouroux, S.-L. Zhang, C.S. Peterson [*Phys. Rev. B (USA)* vol.56 (1997) p.10614–20]

[82] M.A. Gribelyuk, J.A. Kittl, S.B. Samavedam [*J. Appl. Phys. (USA)* vol.86 (1999) p.2571–5]

[83] A. Mouroux, T. Epicier, S.-L. Zhang, P. Pinard [*Phys. Rev. B (USA)* vol.60 (1999) p.9165–8]

[84] H.J. Goldschmidt [*Interstitial Alloys* (Butterworths, London, 1967), p.322–30]

[85] I. Engström, B. Lönnberg [*J. Appl. Phys. (USA)* vol.63 (1988) p.4476–84]

[86] J.A. Kittl, Q.Z. Hong, C.P. Chao, H. Yang [*Future Fab Int. (London)* vol.5 (1998) p.247–58]

[87] J. Tsai, A. Myers, C.-H. Jan, K. Whitehill, S. Keating, S. Yang [*Intel Process Technology Conference*, 1998]

Chapter 4

Cobalt silicide technology

T. Kikkawa, K. Inoue and K. Imai

4.1 SCOPE OF THE CHAPTER

According to the scaling rule [1], the device dimensions of complementary metal–oxide–semiconductor (CMOS) transistors have been reduced to improve the performances for ultra-large-scale integrated circuits (ULSI). The shrinkage of the feature size has caused the increase of resistances in both source/drain regions and poly-silicon gate electrodes of the CMOS transistors, resulting in degradation of electrical characteristics. In order to reduce the parasitic resistances in scaled CMOS transistors, self-aligned silicide (salicide) [2] technologies have been developed. Although titanium (Ti) silicide was used for 0.35–0.18 μm technology nodes [3, 4], cobalt (Co) silicide has been introduced as a replacement of Ti silicide due to its scaling limitation [5–10].

This chapter describes the basic properties of Co silicide as well as the advantages and issues of Co salicide processes. First, crystallographic and material properties are described, and then fabrication technologies and electrical characteristics are covered.

Scope of the chapter p.77

Material properties p.77
  Introduction p.77
  Crystal structure p.77
  Silicide phase diagram p.78

Fabrication technology p.78
  Introduction p.78
  Co salicide process p.79
  Co single layer deposition p.80
  TiN/Co bilayer deposition (TiN cap) p.85
  Ti/Co bilayer deposition (Ti cap) p.88
  Co/Ti bilayer deposition (Ti at interface) p.89

Electrical characteristics p.90
  Introduction p.90
  Effect of scaling p.91
  Leakage current p.92
  I–V characteristics p.92
  Applicability p.92

Conclusions p.93

References p.93

4.2 MATERIAL PROPERTIES

4.2.1 Introduction

Since Co silicide is formed by solid phase reaction between Co and silicon (Si), it is important to understand the material properties of Co and its silicide phases. From this point of view, the crystal structures of Co, Si and $CoSi_2$ and phase diagram of various Co silicides are shown in this section.

4.2.2 Crystal structure

Cobalt has two crystal structures, cubic and hexagonal. α-Co possesses a cubic close-packing (ccp) structure and its lattice constant

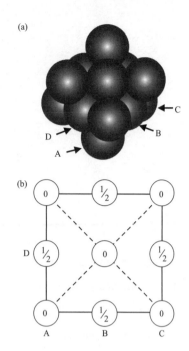

FIGURE 4.1 Schematic diagram of a cubic close-packed (ccp) Co crystal structure. (a) A perspective packing drawing of ccp Co. (b) The position of the atoms in the unit cell of Co projected on a cube face (after [12]).

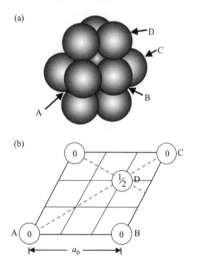

FIGURE 4.2 Schematic diagram of a hexagonal close-packed (hcp) Co crystal structure. (a) A perspective packing drawing of hcp Co. (b) The position of the atoms in the unit cell of Co projected on a plane (after [12]).

is $a_0 = 0.35446$ nm [11], or 0.3548 nm [12]. Co atoms are in a face-centred array and each Co atom has 12 equidistant neighbours. FIGURES 4.1(a) and (b) show a perspective packing drawing and the positions of the atoms in the unit cell of ccp Co projected on a cube face. On the other hand, ε-Co has a hexagonal close-packing (hcp) arrangement and its lattice constants are $a_0 = 0.25071$ nm and $c_0 = 0.40695$ nm [11]. FIGURE 4.2 shows corresponding schematic diagrams of hcp Co structures [12].

The unit cell of Si with the diamond structure has eight atoms in the coordinates as shown in FIGURE 4.3 [12]. Its lattice constant is $a_0 = 0.543070$ nm. Each atom is surrounded by four equidistant neighbours at the centre of a regular tetrahedron.

The reaction of Co with Si forms Co silicide which has $CoSi_2$ arrangement as a low-resistivity phase. FIGURES 4.4(a) and (b) are a perspective packing view showing the distribution of the atoms of $CoSi_2$ within the unit cube and the position of the atoms in the unit cell of $CoSi_2$ projected on a cube face, respectively [12]. Each Co atom is at the centre of eight Si atoms situated at the corners of a surrounding cube, and each Si atom has a tetrahedron of Co atoms. Atomic diameters of Co and Si are 0.2497 and 0.2351 nm, respectively, and the lattice constant of $CoSi_2$ is $a_0 = 0.5356$ nm [12].

4.2.3 Silicide phase diagram

The phase diagram is characterised by the existence of five intermediate phases: Co_2Si (19.24 wt% Si), Co_2Si_2 (24.11 wt% Si), $CoSi$ (32.28 wt% Si), $CoSi_2$ (48.80 wt% Si) and $CoSi_3$ (58.84 wt% Si) as shown in FIGURE 4.5 [13]. The melting points of Co_2Si, $CoSi$ and $CoSi_3$ are 1327, 1395 and 1306°C, respectively, and $CoSi_2$ is formed peritectically at 1277°C.

Co_2Si is orthorhombic, with $a = 0.7109$ nm, $b = 0.4918$ nm and $c = 0.3738$ nm. CoSi is isotypic with FeSi with $a = 0.4447$ nm [13]. $CoSi_2$ is isotypic with CaF_2 whose lattice constants were reported as $a = 0.5364$ nm [11], $a = 0.5356$ nm [12], $a = 0.5367$ or 0.5376 nm [13]. Crystal parameters of Co and Co silicides and volumetric changes during Co silicide formation are shown in TABLE 4.1 [11] and TABLE 4.2 [14], respectively.

4.3 FABRICATION TECHNOLOGY

4.3.1 Introduction

Cobalt salicide processes have been developed for 0.25/0.18 μm technology nodes [3–13] and applied to the advanced technology nodes because the Co salicide process is less sensitive to the scaling of lateral feature sizes of CMOS transistors than Ti salicide processes. However, the Co salicide process is sensitive to contamination from the ambient. Therefore, several formation processes of controlling ambient contamination have been developed.

4.3.2 Co salicide process

A cross-sectional schematic drawing of a Co salicide CMOS process is shown in FIGURE 4.6. A gate oxide is formed on the Si substrates. A poly-Si film is deposited by low-pressure chemical vapour deposition (LPCVD). The poly-Si gates are patterned by photolithography and reactive ion etching. After sidewall spacers of the gate electrodes are formed, the gate electrodes and the source/drain regions of NMOS transistors are doped by arsenic (As) ion implantation. Those of PMOS transistors are doped by BF_2 ion implantation [5,6]. The doped impurities are activated by 1000°C rapid thermal annealing (RTA) for 10 s in nitrogen (N_2) ambient. After removing the thermal oxide from the Si surface of source/drain regions and gate poly-Si electrodes with a diluted hydrofluoric acid (DHF) as shown in FIGURE 4.6(a), a blanket Co thin film is deposited by direct current (DC) magnetron sputtering on the wafer as shown in FIGURE 4.6(b), where the normalised thicknesses and depth of Co and its silicides are illustrated in the inserted figures [14].

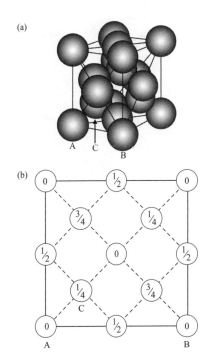

FIGURE 4.3 Schematic diagram of a Si crystal structure.
(a) A perspective packing drawing of Si. (b) The atomic positions in the unit cell of the diamond arrangement projected on a cube face (after [12]).

The Co salicide process has two RTA steps. In the first RTA, the deposited Co thin film is thermally reacted with Si in a N_2 ambient at temperatures in the 450–600°C range to start the self-aligned reaction between Co and isolated source/drain Si as well as gate poly-Si regions, resulting in CoSi or Co_2Si formation as shown in FIGURE 4.6(c). Co silicide is not formed on the gate sidewall spacers and field isolation oxide regions. Co silicide is formed predominantly by Co atom diffusion so that no special annealing ambient is required to suppress the lateral overgrowth of Co silicide over oxide regions.

A chemical selective wet etching of the unreacted Co and metal compounds on the field and sidewall spacer SiO_2 is carried out by hydrochloric acid and hydrogen peroxide solution [8] as shown in FIGURE 4.6(d). Then, the second RTA is carried out to transform CoSi to $CoSi_2$. Growth of the low resistivity $CoSi_2$ phase is achieved at temperatures in the 750–900°C range. High temperature RTA promotes uniform diffusion kinetics and suppresses non-uniform grain boundary diffusion, resulting in smooth silicide/Si interface as shown in FIGURE 4.7. $CoSi_2$ formation from CoSi does not yield nucleation problems like $TiSi_2$ formation does with its nucleation of the C54 phase from the previously formed C49 phase [3].

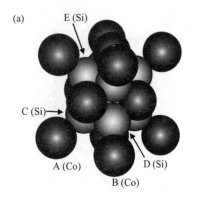
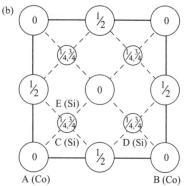

FIGURE 4.4 Schematic diagram of a $CoSi_2$ crystal structure. (a) A perspective packing drawing of $CoSi_2$. (b) The position of the atoms in the unit cell of $CoSi_2$ projected on a cube face (after [12]).

4.3.3 Co single layer deposition

The simplest method to form a Co salicide structure is to deposit a Co single layer on both source/drain Si regions and gate poly-Si electrodes. However, the conventional Co single layer deposition process results in a high resistance after silicide formation. This is due to the oxidation of the sputtered Co film in the atmosphere or during RTA. Secondary ion mass spectroscopy (SIMS) analysis of the conventional single layer Co silicide formation process indicates diffusion of oxygen from the surface into the film before and after RTA as shown in FIGURE 4.8 [15]. Therefore, it is necessary to prevent oxidation of Co during RTA. One of the solutions is high temperature sputtering of Co and subsequent *in situ* vacuum annealing [5, 6].

Cobalt is sputter-deposited at an elevated temperature and at 0.93 Pa Ar pressure, promoting an initial reaction of Co/Si during deposition. Since Co does not reduce any interfacial oxide, sputter-etching of Si surface before Co sputter-deposition is also necessary for cleaning the surface of Si and removing surface contaminants. An *in situ* vacuum annealing after Co sputtering is carried out to suppress oxidation of the Co film at 450°C at 0.93 Pa Ar ambient for a duration from 0 to 10 min in the same sputtering chamber. The first RTA is performed at the temperature range from 450 to 600°C in N_2 ambient. After selective wet etching of unreacted Co layers, the second RTA is carried out at the temperature range between 600 and 850°C in N_2 ambient.

FIGURE 4.9 shows transmission electron diffraction (TED) patterns of the Co silicide films on arsenic (As) doped Si [6]. Single layer Co films are deposited at room temperature (RT) and at 400°C, respectively. Annealing conditions of the first and second RTA are for 60 s at 450 and 600°C, respectively.

The diffracted rings from Co_2Si and $CoSi$ layers showed azimuthal textures. This is attributed to the silicide formation by high temperature sputtering of Co at 400°C and annealed in vacuum at 450°C for 60 s. The TED of the Co silicide film sputtered at 400°C shows strong (020)-type diffracted beams originated from epitaxial $CoSi_2$. On the other hand, the TED of the RT sputtered Co silicide shows preferred orientation for neither Co_2Si and $CoSi$ grains nor epitaxial $CoSi_2$ grains. FIGURE 4.10 shows diffraction patterns and plan-view TEM images of Co film sputtered at RT and 400°C after RTA at 400°C for 120 s and *in situ* vacuum annealing of 5 nm thick Co film on BF_2-doped Si [6]. In addition to diffraction spots associated with Co_2Si and $CoSi$, diffraction spots from Co_3O_4 are also observed for the RT sputtering and subsequent N_2 RTA. The dark field image of $CoSi_2$ film after second RTA shows that 70–80% of the $CoSi_2$

Cobalt silicide technology

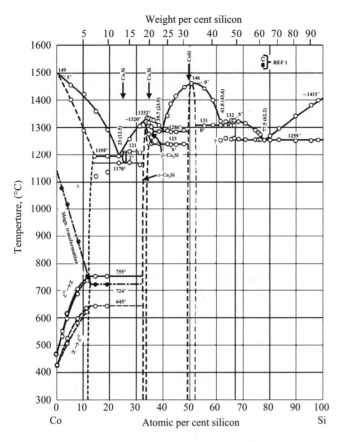

FIGURE 4.5 Phase diagram of Co/Si alloy (after [13]).

TABLE 4.1 Crystal parameters of Co and Co silicides (after [11]).

Compound	Crystal system	Lattice constants (nm)			Density (g/cm^3)
		a	b	c	
α-Co	Cubic	0.35446			8.789
ε-Co	Hexagonal	0.25071		0.40695	8.871
α-Co$_3$Si	Tetragonal	0.8420	0.3738	0.5810	
α-Co$_2$Si	Orthorhombic	0.4918	0.3737	0.7109	7.42
CoSi	Cubic	0.4447			6.65
CoSi$_2$	Cubic	0.53640			4.95

occupied the epitaxial {100} orientation on the BF$_2$ ion-implanted Si. One hundred per cent {100} and {221} epitaxial grains are found in the CoSi$_2$ on the As ion-implanted Si.

Consequently, the oxidation of Co and Co silicide films has been suppressed by the *in situ* vacuum annealing in conjunction with 400°C high temperature sputtering.

Cobalt silicide technology

TABLE 4.2 Volume changes during Co silicide formation (after [14]).

Metal		Silicide			
Phase	Number density of metal (10^{22}/cm^3)	Phase	Consumed silicon thickness normalised to metal thickness	Thickness of silicide normalised to metal thickness	Number density of silicide (10^{22}/cm^3)
α-Co	9.03	Co$_2$Si	0.90	1.47	3.062
β-Co	8.85	CoSi	1.81	1.98	4.562
		CoSi$_2$	3.61	3.49	2.590

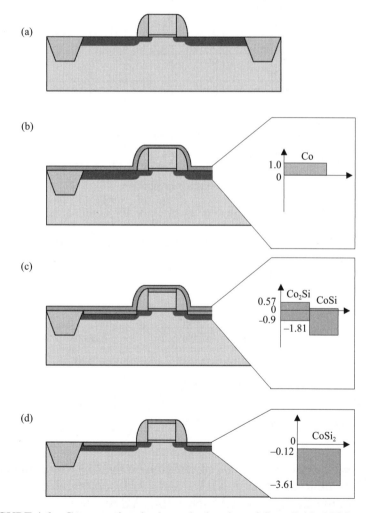

FIGURE 4.6 Cross-sectional schematic drawing of Co salicide MOS transistor. (a) Selective SiO$_2$ removal on Si source/drain and poly-Si regions. (b) Co deposition. (c) First rapid thermal annealing (RTA). (d) Selective wet etching and second RTA.

Cobalt silicide technology

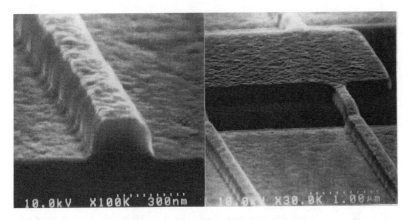

FIGURE 4.7 SEM micrographs of salicided MOS structures (after [6]).

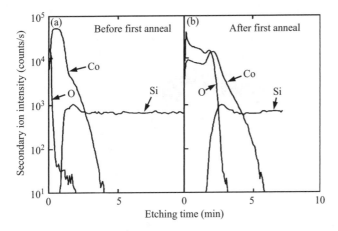

FIGURE 4.8 SIMS analysis of conventional Co salicide structure: (a) before RTA; (b) after RTA (after [7]).

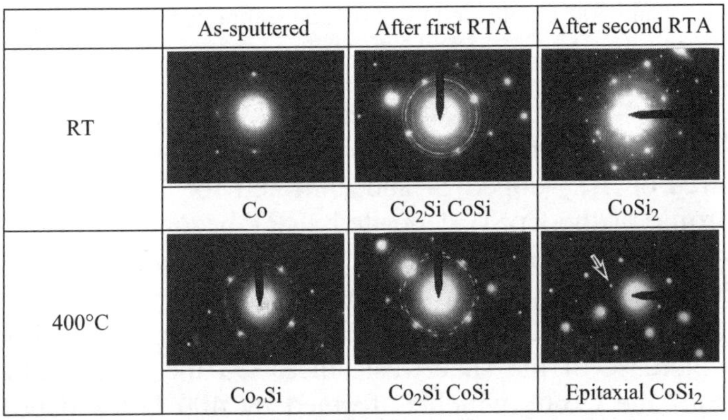

	As-sputtered	After first RTA	After second RTA
RT	Co	Co_2Si CoSi	$CoSi_2$
400°C	Co_2Si	Co_2Si CoSi	Epitaxial $CoSi_2$

FIGURE 4.9 Transmission electron diffraction (TED) patterns of Co silicide films deposited at room temperature and 400°C and annealed by the first and second RTAs on arsenic-doped Si (after [6]).

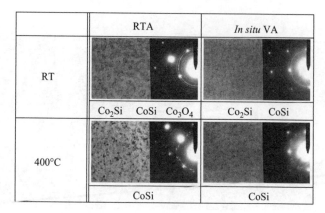

FIGURE 4.10 Diffraction patterns and plan-view TEM images of Co film sputtered at room temperature and 400°C after conventional RTA in N_2 and *in situ* vacuum annealing (after [6]).

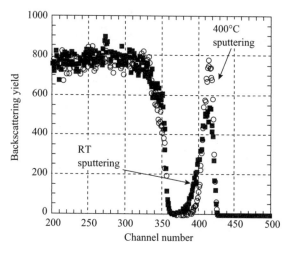

FIGURE 4.11 RBS spectra of the $CoSi_2$ deposited by sputtering at room temperature and 400°C followed by 800°C annealing for 1 h (after [6]).

FIGURE 4.11 shows Rutherford backscattering spectrometry (RBS) spectra of the $CoSi_2$, where a 5 nm thick Co film was deposited on BF_2-doped Si and annealed for 1 h at 800°C [6]. The profile of the $CoSi_2$ deposited at RT becomes broader after 800°C annealing, indicating that the conventional $CoSi_2$ film which is formed by RT Co deposition process agglomerates at high temperature. This is in good agreement with the observed sharp increase of the sheet resistance. On the other hand, the profile of the $CoSi_2$ which is formed by 400°C Co deposition process remains sharper after 800°C annealing, indicating that the high temperature sputtered Co results in high thermal stability of

Cobalt silicide technology

$CoSi_2$ film. These results demonstrate the high thermal stability of the $CoSi_2$ films epitaxially grown by use of high temperature Co sputtering and subsequent *in situ* vacuum annealing.

4.3.4 TiN/Co bilayer deposition (TiN cap)

In order to prevent oxidation of Co during RTA, an alternative process for Co salicide formation is to deposit a cap layer on the Co layer. A TiN cap layer is deposited on the Co layer as a diffusion barrier for this purpose.

A TiN-capped Co salicide process and a conventional process are shown in TABLE 4.3 [8]. After the cleaning the Si surface with DHF, 10 nm thick Co is sputtered by DC magnetron sputtering, then 20 nm thick TiN capping layer is sputtered successively without breaking the vacuum. The first RTA is carried

TABLE 4.3 Process conditions of conventional Co salicide process and TiN-capped Co salicide process (after [7]).

	Conventional Co salicide	TiN-capped Co salicide
Film thickness	Co: 18 nm	Co: 10 nm TiN: 20 nm
First RTA	700°C, 30 s	450°C, 30 min
Selective etch	HCl + H_2O_2: 3 min at RT	NH_4OH + H_2O_2 + H_2O: 90 s at 65°C HCl + H_2O_2: 3 min at RT
Second RTA	–	750°C, 30 s

RT: room temperature

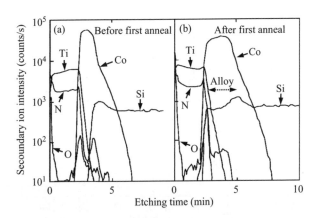

FIGURE 4.12 SIMS analysis of TiN-capped Co salicide structures: (a) before RTA; (b) after RTA (after [7]).

out at temperatures ranging from 400 to 550°C for 30 s. CoSi films are formed selectively on the source, drain and gate electrodes. The TiN and unreacted Co films are removed chemically by $NH_4OH + H_2O_2 + H_2O$ for 90 s at 65°C and $HCl + H_2O_2$ for 3 min at RT. The second RTA is carried out at temperatures ranging from 800 to 850°C for 30 s. The Co silicide is changed from CoSi to $CoSi_2$ by annealing to lower its resistivity. The optimal process conditions to reduce leakage current for the TiN capped Co salicide structure are the

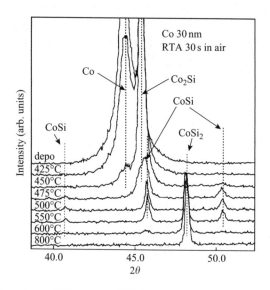

FIGURE 4.13 X-ray diffraction of TiN-capped Co silicide films annealed at various temperatures (after [8]).

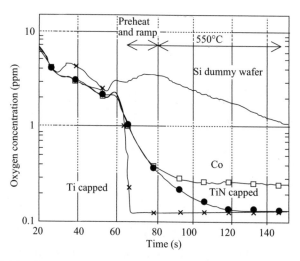

FIGURE 4.14 Oxygen concentration in N_2 ambient during RTP for Si, Co/Si, TiN (cap)/Co/Si, Ti (cap)/Co/Si substrates (after [18]).

first RTA at 550°C for 30 s and the second RTA at 850°C for 30 s.

Another TiN cap salicide process is as follows [16]. Co and TiN films are deposited subsequently by DC magnetron sputtering.

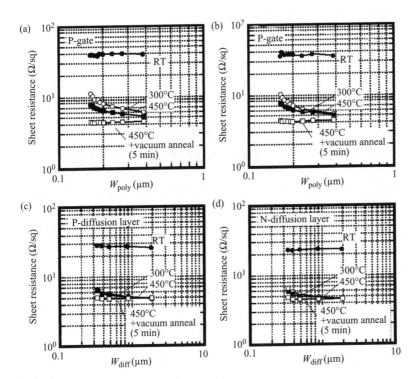

FIGURE 4.15 Sheet resistances of p- and n-type gate poly-Si electrodes and p- and n-type diffusion layers as a function of linewidth, where Co film thickness is 15 nm. (a) p-type poly-Si gate. (b) n-type poly-Si gate. (c) p-type source/drain diffusion layer. (d) n-type source/drain diffusion layer (after [6]).

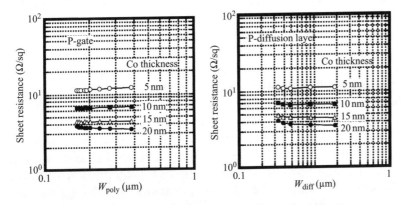

FIGURE 4.16 Sheet resistances of p-type gate electrodes and p-type diffusion layers as a function of linewidth for various Co thicknesses (after [6]).

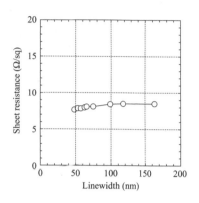

FIGURE 4.17 Sheet resistance of gate electrodes as a function of linewidths ranging from 160 to 45 nm (after [21]).

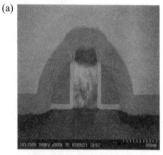

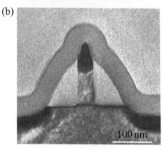

FIGURE 4.18 A cross-sectional TEM micrographs of CMOS transistors with Co salicide structure fabricated by high temperature Co single layer deposition at 400°C followed by *in situ* vacuum annealing: (a) 90 nm CMOS; (b) 65 nm CMOS (after [21]).

CoSi films are formed on the source, drain and gate electrodes selectively by annealing at 500°C for 60 s. Non-reacted Co and TiN films are removed by $H_2O_2 + H_2SO_4$ solution. Then, the Co silicide is changed from CoSi to $CoSi_2$ to lower its resistivity by annealing at 750°C for 30 s.

SIMS analysis of TiN cap Co salicide structure is shown in FIGURE 4.12 [7]. The effect of TiN cap layer on the suppression of oxidation is demonstrated in comparison with FIGURE 4.8 [8]. X-ray diffraction of TiN-capped Co silicide films is shown in FIGURE 4.13 with respect to various annealing temperatures [8].

4.3.5 Ti/Co bilayer deposition (Ti cap)

Titanium reduces the O_2 concentration in the RTA chamber instantaneously and plays an active role in reducing the native oxide of Si. The use of a Ti capping layer deposited *in situ* following Co deposition reduces the sensitivity of the Co/Si reaction to interfacial contaminants that would be present during the silicidation process [9, 17–19].

When O_2 or moisture is present during silicidation, the SiO_2 will grow thicker on top of the silicide and finally block the silicidation process. The most critical part in the Co salicide process is the $CoSi_2$ formation at the very beginning of the heat cycle of RTA. The Ti cap on top of Co can suppress H_2O release from the wafer and is effective in reducing ambient contamination by reacting with residual moisture in the RTA chamber.

A 15 nm thick blanket Co film followed by a 10 nm thick Ti film are deposited by DC magnetron sputtering on the CMOS transistors of the wafers where source/drain and poly-Si regions are isolated. Co silicide with a film thickness of 25 nm is formed by two-step silicidation. The transformation temperature for CoSi is reduced so that the broad temperature window for the first RTA is obtained; 510–600°C [9, 17–19]. The most critical part in the $CoSi_2$ process is the Co_2Si formation at the very beginning of the heat cycle. Since Co is the main moving species, it is necessary to overcome any SiO_2 formation on top of the growing silicide. Desorption of impurities can occur from the field oxide areas, the spacer areas, the Si areas, and the backside of the Si wafer. A typical *in situ* degas before deposition is insufficient to remove all water absorbed in the wafer so that the deposition of a capping layer on top of the Co allows to achieve the control on the ambient pollution. The Ti cap on top of Co suppresses any water release of the wafer. The Ti cap is efficient in reducing the oxygen and residual moisture concentrations in the chamber below 0.1 ppm as shown in FIGURE 4.14 [18]. In the case of no capping

Cobalt silicide technology

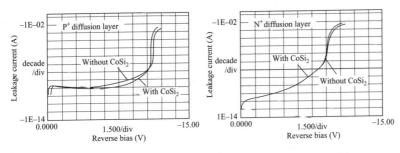

FIGURE 4.19 Junction leakage current as a function of reverse bias voltages for 100 nm junction depth and 0.25 mm² junction area: (a) PMOS; (b) NMOS (after [6]).

layer the Co ambient is contaminated by water present on top of the oxide areas. In the case of a diffusion barrier capping layer, such as TiN, the Co is shielded from the ambient contamination, but not from the water present on top of the oxide. The unreacted metal is selectively removed after the first RTA step. The temperature of the second RTA step of silicidation is varied between 800 and 850°C.

4.3.6 Co/Ti bilayer deposition (Ti at interface)

The reactivity of Ti and its ability to reduce native oxides incorporate Ti at the deposition stage. The Co/Ti with Ti at the interface has been introduced to alleviate the requirements of cleaning and reduction of interfacial native oxide, resulting in formation of $CoSi_2$ epitaxial layer [20]. After DHF dip the SiO_2 is covered with Si–OH groups, which desorb during high temperature RTA cycle. Therefore, *in situ* degas before Co deposition is not sufficient to remove all the H_2O from the wafer.

A 10 nm thick Ti film followed by 25 nm thick Co films are deposited by DC magnetron sputtering and annealed at 750°C in nitrogen atmosphere to obtain a bilayer of $CoSi_2$ covered with TiN on a p-type (100)Si wafer. This is followed by two-step wet etching to remove unreacted Ti and Co films. A second anneal is carried out at 900°C to obtain a $CoSi_2$/TiN bilayer on Si in nitrogen ambient [21]. Co diffuses from the top into the Ti and reacts with Si forming $CoSi_2$, and Ti migrates to the top to form TiN in a nitrogen environment. A Ti–Co layer remains on top of the $CoSi_2$ layer.

Formation of epitaxial $CoSi_2$ layers on (100)Si is achieved by Co/Ti multilayer (Ti at interface). A main reaction takes place between Co and Si but the absorption of interfacial impurities takes place by the thin Ti layer. The interface between $CoSi_2$ and

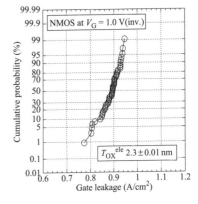

FIGURE 4.20 Distribution of gate leakage current of NMOS with T_{OX} of 2.3 nm at 1 V inverse bias (after [21]).

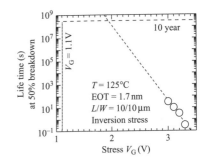

FIGURE 4.21 Time-dependent dielectric breakdown characteristics of gate oxide with EOT of 1.7 nm (after [21]).

Cobalt silicide technology

Si is atomically flat. CoSi$_2$ layers on Si have their facets along the (111) planes.

4.4 ELECTRICAL CHARACTERISTICS

4.4.1 Introduction

Co salicide processes have been developed for 180 nm ULSI technology node as a substitute for Ti salicide processes [5, 6] and extended to 65 nm technology nodes [21–23]. This is attributed to

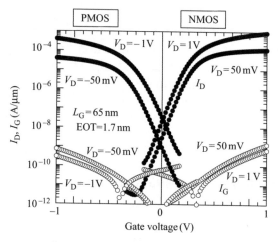

FIGURE 4.22 Sub-threshold characteristics of I_D and I_G versus gate voltage for 65 nm Co salicide CMOS with EOT of 1.7 nm (after [21]).

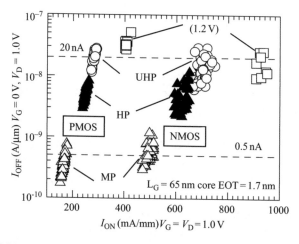

FIGURE 4.23 I_{ON}–I_{OFF} characteristics of Co salicide CMOS transistors, where MP, HP and UHP stand for medium-speed performance, high-speed performance and ultra-high-speed performance, respectively (after [21]).

TABLE 4.4 Characteristics of 65 nm Co salicide CMOS transistors (after [21]).

V_D (V) (type)	1.0 (UHP)	1.2 (UHP)	1.0 (HP)	1.0 (MP)	2.5 (I/O)
L_G (nm)	65	65	65	65	260
NMOSFET					
T_{OX}inv (nm)	2.3	2.3	2.3	2.3	5.0
I_{ON} (μA/μm)	740	940	640	520	600
I_{OFF} (A/μm)	20n	20n	5n	0.5n	10p
I_G (nA/μm)	0.4	0.4	0.4	0.4	–
PMOSFET					
T_{OX}inv (nm)	2.5	2.5	2.5	2.5	5.0
I_{ON} (μA/μm)	285	405	260	185	420
I_{OFF} (A/μm)	20n	30n	5n	0.5n	10p
I_G (nA/μm)	0.2	0.2	0.2	0.2	–

UHP, ultra-high-speed performance; HP, high-speed performance; MP, medium-speed performance; I/O, input/output

the superior scaling characteristics of the Co salicide structure than Ti salicide structure. In this section, the electrical characteristics of the Co salicide structure are described in terms of the scaling.

4.4.2 Effect of scaling

Effect of linewidth on the sheet resistances of the Co salicide structure on n- and p-type gate poly-Si electrodes and source/drain diffusion layers after the first and second RTAs for various sputtering temperatures is shown in FIGURE 4.15 [6]. The sheet resistances of the Co salicide structure decrease with increasing sputtering temperatures. High temperature Co sputtering without vacuum annealing shows the increase of the sheet resistances with decreasing linewidth because the surfaces of Co_2Si and $CoSi$ are oxidised inhomogeneously during the RTA. *In situ* vacuum annealing at 450°C could suppress the linewidth dependence of the Co salicide sheet resistance which was deposited by 450°C sputtering.

FIGURE 4.16 shows the thickness dependence of the Co salicide sheet resistances on p-type gate electrodes and p-type diffusion layers as a function of linewidth [6]. Co is sputtered at 450°C with thicknesses ranging from 5 to 20 nm and annealed by the first and second RTA. FIGURE 4.17 shows the sheet resistance of the gate Co salicide gate electrode as a function of gate length below 180 nm down to 45 nm [21–23]. As can be seen, the $CoSi_2$ sheet resistance is independent of gate electrode linewidth down to 45 nm. A cross-sectional TEM micrographs of 90 and 65 nm

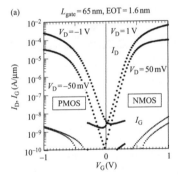

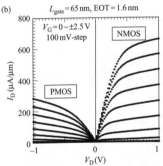

FIGURE 4.24 *I–V* characteristics of 65 nm Co salicide CMOS transistors with EOT of 1.6 nm for high-speed performance application. (a) I_D and I_G versus gate voltage. (b) I_D–V_D characteristics of CMOS transistors (after [22]).

Cobalt silicide technology

CMOS transistors with the Co salicide structure fabricated by high temperature Co single layer deposition at 400°C and *in situ* vacuum annealing are shown in FIGURE 4.18 [21].

4.4.3 Leakage current

Junction leakage current characteristics as a function of reverse bias voltage for p- and n-type diffusion layers with 100 nm junction depth and 0.25 mm² junction area are shown in FIGURE 4.19 [6]. High temperature Co deposition for the Co salicide formation process shows the same leakage current characteristics as the junction without silicide [6]. Gate leakage current characteristics of NMOS transistors with gate oxide thickness of 2.3 nm are shown in FIGURE 4.20 [21]. The gate dielectric leakage current is suppressed below 1 A/cm². Time-dependent dielectric breakdown (TDDB) characteristics of NMOS with 1.7 nm equivalent oxide thickness (EOT) indicate that the lifetime of 50% breakdown at 125°C is longer than 10 years as shown in FIGURE 4.21 [21].

4.4.4 I–V characteristics

FIGURE 4.22 shows sub-threshold characteristics of NMOS and PMOS transistors with 65 nm gate length and 1.7 nm EOT [21]. I_{ON}–I_{OFF} characteristics of CMOS transistors are shown in FIGURE 4.23 and the performance is summarised in TABLE 4.4 [21]. I_{ON} values of 740 and 285 μA/μm at 1.0 V were obtained for NMOS and PMOS high-speed performance (HP) transistors, respectively. I_{ON} and I_{OFF} values of ultra-high-speed performance (UHP) NMOS/PMOS transistors at 1.2 V overdrive voltage were 940/405 μA/μm and 20/30 nA/μm, respectively.

I_D–V_D characteristics of 65 nm CMOS with 1.6 nm gate EOT for high-speed use and 240 nm CMOS with 5.0 nm gate EOT for 2.5 V I/O use are shown in FIGURES 4.24 and 4.25, respectively [22].

4.4.5 Applicability

The Co salicide process has a potential in scaling because the dependence of the Co salicide linewidth on the resistivity is less apparent than in the Ti salicide process. Therefore, Co salicide technology has been used for mass production of CMOS ULSIs in several technology generations from 180 to 65 nm. A cross-sectional TEM micrograph of 65 nm Co salicide CMOS ULSI with Cu damascene multilevel interconnects is shown in FIGURE 4.26 [21]. A ULSI chip photograph is shown in FIGURE 4.27.

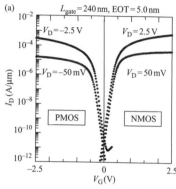
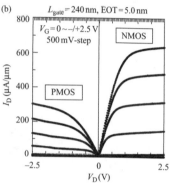

FIGURE 4.25 I–V characteristics of 240 nm Co salicide CMOS transistors with EOT of 5 nm for 2.5 V I/O application. (a) I_D and I_G versus gate voltage. (b) I_D–V_D characteristics of CMOS transistors (after [22]).

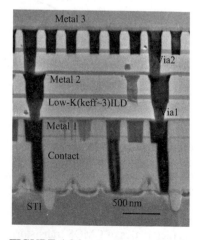

FIGURE 4.26 A cross-sectional TEM micrograph of 65 nm Co salicide CMOS ULSI (after [21]).

4.5 CONCLUSIONS

The Co salicide process technology has been applied to CMOS ULSI front-end of line fabrication as a substitute for the Ti salicide process. In order to form thin Co silicide films with low resistivity for CMOS transistor active area, several Co salicide process technologies have been developed. Since, the Co deposition process suffers from poor reproducibility, various Co salicide formation processes have been proposed such as a high temperature Co single layer sputtering followed by *in situ* vacuum annealing and the bilayer processes with Ti or TiN capping. Consequently, the Co salicide process technology has been successfully introduced to mass production of CMOS ULSIs in several technology generations from 180 to 65 nm.

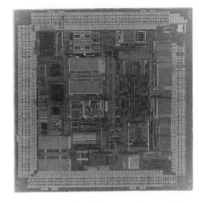

FIGURE 4.27 A photograph of a Co salicide CMOS ULSI chip (courtesy of NEC Corp).

REFERENCES

[1] R.H. Dennard, F.H. Gaensslen, H.-N. Yu, V.L. Rideout, E. Bassous, A.R. LeBlanc [*IEEE J. Solid-State Circuits (USA)* vol.SC-9 (1974) p.256–68]
[2] J.A. Kittel, W.T. Shiau, Q.Z. Hong, D. Miles [*Microelectron. Engng. (UK)* vol.50 (2000) p.87–101]
[3] K. Fujii, K. Kikuta, T. Kikkawa [*Symposium on VLSI Technology Digest of Technical Papers* (IEEE, USA, 1995) p.57–8]
[4] J.A. Kittel, Q.Z. Hong [*Thin Solid Films (UK)* vol.320 (1998) p.110–21]
[5] K. Inoue, K. Mikagi, H. Abiko, T. Kikkawa [*International Electron Devices Meeting Technical Digest* (IEEE, USA, 1995) p.445–8]
[6] K. Inoue, K. Mikagi, S. Chikaki, T. Kikkawa [*IEEE Trans. Electron Devices (USA)* vol.45 (1998) p.2312–18]
[7] T. Yamazaki, K. Goto, T. Fukano, Y. Nara, T. Sugii, T. Ito [*International Electron Devices Meeting Technical Digests* (IEEE, USA, 1993) p.906–8]
[8] K. Goto, A. Fushida, J. Watanabe, et al. [*IEEE Trans. Electron Devices (USA)* vol.46 (1999) p.117–23]
[9] K. Maex, A. Lauwers, P. Besser, E. Kondoh, M. dePotter, A. Steegen [*IEEE Trans. Electron Devices (USA)* vol.46 (1999) p.1545–50]
[10] T. Ohguro, M. Saito, E. Morifuji, et al. [*IEEE Trans. Electron Devices (USA)* vol.47 (2000) p.2208–13]
[11] K. Maex, M. Van Rossum, A. Reader [Crystal structure of TM silicides, in *Properties of Metal Silicides* Ed. K. Maex, M. Van Rossum, EMIS Datareviews Series No. 14 (IEE, UK, 1995)]
[12] R.W.G. Wyckoff [*Crystal Structures*, 2nd Ed. vol.1 (Robert E. Krieger Publishing Company, USA, 1982)]
[13] M. Hansen, K. Anderko [*Constitution of Binary Alloys*, 2nd Ed. (McGraw-Hill Book Company, New York, USA, 1985)]
[14] M. Ostling, C. Zaring [Mechanical properties of TM silicides, in *Properties of Metal Silicides*, Ed. K. Maex, M. Van Rossum, EMIS Data Reviews Series No. 14 (IEE, UK, 1995) p.18]
[15] K. Goto, T. Yamazaki, A. Fushida, S. Inagaki, H. Yagi [*Symposium on VLSI Technology Digest of Technical Papers* (IEEE, USA, 1994) p.119–20]

[16] J.A. Kittel, Q.Z. Hong, Y. Young, N. Yu, S.B. Samavedam, M.A. Gribelyuk [*Thin Solid Films (UK)* vol.332 (1998) p.404]
[17] Q.F. Wong, K. Maex, S. Kubicek, et al. [*Symposium on VLSI Technology Digest of Technical Papers* (IEEE, USA, 1995) p.17–18]
[18] A. Lauwers, P. Besser, M. de Potter, et al. [*Proceedings of International Interconnect Technology Conference* (IEEE, USA, 1998) p.99–101]
[19] A. Lauwers, M. de Potter, O. Chamirian, et al. [*Microelectron. Engng. (UK)* vol.64 (2002) p.131–42]
[20] M.L.A. Dass, D.B. Fraser, C.S. Wei [*Appl. Phys. Lett. (USA)* vol.58 (1991) p.1308–10]
[21] K. Fukasaku, A. Ono, T. Hirai, et al. [*Symposium on VLSI Technology Digest of Technical Papers* (IEEE, USA, 2002) p.64–5]
[22] A. Ono, K. Fukasaku, T. Hirai, et al. [*International Electron Devices Meeting Technical Digest* (IEEE, USA, 2001) p.511–14]
[23] N. Nakamura, T. Fukai, M. Togo, et al. [*International Electron Devices Meeting Technical Digest* (IEEE, USA, 2003) p.281–4]

Chapter 5

Nickel silicide technology

C. Lavoie, C. Detavernier and P. Besser

5.1 SCOPE OF THE CHAPTER

In this chapter, we present nickel silicide as the next material for contact to microelectronic CMOS devices. After briefly introducing the main advantages and challenges, we will describe the material properties of NiSi with an emphasis on characteristics that are different from the prior $CoSi_2$ and $TiSi_2$ contacts. Section 5.4 covers the *in situ* measurements of the phase formation sequence, putting emphasis on the presence of multiple metal-rich phases, on the very low formation temperature and on formation mechanisms. The following two sections (5.5 and 5.6) cover interesting properties of Ni monosilicide that are either new or hardly known in the microelectronics field, namely the large anisotropy in the thermal expansion and the unexpected texture in NiSi films formed on single crystal silicon. It is anticipated that the negative thermal expansion along one crystal axis together with the crystal alignments during NiSi formation (axiotaxy) will be of importance for the degradation of the film at high temperature. Section 5.7 covers thin film degradation at high temperature either through the formation of $NiSi_2$ or through grain grooving and agglomeration. We also present several material improvements that have been reported to delay the degradation of films at higher temperature. The last section (5.8) contains some examples of electrical and physical characterisation from state-of-the-art devices.

5.2 INTRODUCTION

The reactions of Ni and Si for possible use in microelectronic manufacturing have been extensively studied starting in the mid-1970s [1–13]. Multiple studies followed that further determined the electrical and mechanical properties of Ni silicide phases, the mechanisms for their formation in thinner films and its dependence

Scope of the chapter p.95

Introduction p.95
  Limitations of $CoSi_2$ p.96
  Advantages and limitations of NiSi p.98

Characteristics and properties of Ni silicide phases p.99
  Phase diagrams p.99
  Crystal structures and volumetric changes p.100
  Properties of Ni silicide phases p.102
  Reduction in silicon consumption p.104

Phase formation p.104
  Thermal budget p.104
  Complex phase sequence p.106
  Formation mechanism p.114
  Disadvantages of low temperature diffusion p.117

Thermal expansion of NiSi and related stress effects p.118
  Unit cell dimensions and CTE for bulk samples p.119
  Unit cell dimensions and CTE for thin films p.120
  Consequences of CTE anisotropy for stress in thin films of NiSi p.122

Texture of NiSi films on single crystal Si p.124
  Pole figures and orientation distribution functions p.124
  Standard classification of texture in thin films p.125

Pole figures for NiSi films on Si(001) p.125
Axiotaxy: a new type of texture p.128

High temperature limitations p.129
Formation of NiSi$_2$ p.129
Morphological stability p.130
High temperature stabilisation of NiSi films p.134

Device characterisation p.136
Addressing the device roadmap p.137
Shallow junctions p.139
NiSi$_2$ and resistivity issues p.141
Dopant segregation p.142

Conclusions p.143

Acknowledgments p.144

References p.145

on various impurities [1, 14–52]. Recently, the interest in the low resistivity NiSi increased significantly because of the projected near-term use as contact to the source, drain and gates of CMOS devices. Recent publications from microelectronic companies on NiSi contacts [53–68] point out again that the practical advantages of this material reside in a low thermal budget for formation, a low resistivity in narrow dimensions and a low device leakage. While these device data have shown the feasibility of a NiSi process, the actual use of this new material depends to a great extent on the performance and limitations of the current CoSi$_2$ contacts as well as on improvements in yield for the NiSi process. As many properties of NiSi are very different from those of its CoSi$_2$ counterpart, a good understanding of the advantages and limitations of each material is required for NiSi to become the contact material in coming generations of microelectronic devices.

5.2.1 Limitations of CoSi$_2$

In the production of transistors with gate lengths significantly shorter than 50 nm, the formation of cobalt silicide contacts becomes increasingly difficult. There are three main factors that limit the continued use of this material in future devices:

1. The rise in resistance in very narrow lines.
2. The reduction in available Si for the reaction as the silicon-on-insulator becomes very thin.
3. The introduction of SiGe substrates.

The first concern is reminiscent of the problem that was faced in the 1990s with TiSi$_2$. In this case, the low nucleation density of the desired C54-TiSi$_2$ phase was at the origin of a resistance increase in lines narrower than 350 nm [69, 70]. While the nucleation density could be increased with the addition of transition elements and allow for low resistance TiSi$_2$ down to about 200 nm [71], the further decrease in gate lengths required the introduction of CoSi$_2$. This low resistivity material did not show formation problems in the smallest achievable dimensions at the time of implementation (~100 nm). Recent work [54, 55, 59, 61, 65, 72] shows, however, that the resistance of CoSi$_2$ lines increases dramatically with a further decrease in line width (see also FIGURE 5.22). The width at which the resistance increases depends on the process itself and the test site (line length and geometry) and has been reported to be linked to the presence of infrequent voids in narrow silicide lines [72]. The origin of these voids is not clear and may depend on the presence of impurities, on early and non uniform agglomeration in smaller dimensions, on local stresses, or even on the mechanisms of formation (diffusion or nucleation controlled/diffusing species). Most likely, one is confronted with a combination of these factors.

The second factor limiting $CoSi_2$ is the use of thin Si on insulating (SOI) substrates. The sheet resistance requirements in current devices are such that the thickness of $CoSi_2$ must be in the range of 20–30 nm. From the crystal structures and atomic volumes, one easily determines that the consumption of Si necessary to form this low resistivity layer is 3% thicker than the silicide itself [73]. To the silicide layer thickness, one must also add the peak to peak roughness of the interface. Indeed, the $CoSi_2/Si$ interface is inherently rough because of the nucleation controlled formation of $CoSi_2$ from CoSi [9, 14, 22, 74, 75]. While this roughness can be controlled through an optimised cleaning procedure and alloying [76, 77], it cannot be eliminated and the typical maximum layer thickness can locally be 20–30% larger than the average $CoSi_2$ thickness. When the thickness of the SOI layer reaches about 40 nm, at least part of the silicide film will touch the underlying oxide layer causing degradation in contact resistance and device properties.

The last factor restricting the extended use of $CoSi_2$ is the current trend of adding Ge to the Si substrate to modify the stress in the surface layer and thereby increase the carrier mobility and the device switching speed. The formation of $CoSi_2$ in the presence of Ge is extremely arduous [77–79]. Germanium is soluble in CoSi and immiscible in $CoSi_2$. As a result, the formation of $CoSi_2$ from CoSi requires that the Ge be expelled from a growing $CoSi_2$ grain. From *in situ* measurements of phase formation, we have determined that not only the growth of the phase is retarded but its nucleation is also elevated to much higher temperatures. From classical nucleation theory [22], it can be shown that the change in entropy of mixing from a solution to a mixture of phases raises the barrier for nucleation [22, 51, 77, 79]. Thus the addition of Ge results in an increase in the nucleation temperature of $CoSi_2$ from about 600°C to above 800°C. This processing temperature is not only too high for manufacturing of advanced devices but also for the integrity of the silicide itself as the high temperature prevents the formation of a continuous silicide with a low resistance. Indeed, the presence of the low melting point Ge in the film reduces the agglomeration temperature to the point where the process window between silicide formation and agglomeration becomes non existent.

All the concerns above can be somewhat alleviated either through the optimisation of the material by alloying [77] or through the selective addition of epitaxial Si to the source and drain of transistors (raised source/drain process – RSD). The possibility of a NiSi process remains extremely interesting in terms of reduction of both the cost and the process complexity involved.

5.2.2 Advantages and limitations of NiSi

The front up candidate for the replacement of $CoSi_2$ shows improved performance with respect to the three limitations presented above. The main advantages of NiSi can be separated in the four categories listed below:

1. reduced thermal budget;
2. lower resistivity and reduced Si consumption;
3. formation controlled by diffusion of Ni;
4. low resistivity phase formation possible on SiGe.

These advantages will be covered in detail in the following sections. Each leads to specific improvements in device characteristics. The lower formation temperature allows for a reduction in the thermal budget and limits dopant deactivation in shallow junctions and in poly-Si gates. Because of the reduced resistivity and of the reduced Si density in the silicide compound, the Si consumption can be decreased by about 30%. Furthermore, the fact that the formation of NiSi is controlled by diffusion results in two significant advantages, one being the limitation of bridging and the second being the formation of silicide films with much smoother surfaces and interfaces compared to films of $TiSi_2$ and $CoSi_2$ for which the formation is nucleation controlled. This improvement of interface roughness leads to an important reduction in device leakage. Finally, NiSi is also advantageous for the formation of low resistance contacts on SiGe devices. While the process window for Ni/SiGe is smaller than for Ni/Si, this window is not affected to the point where it becomes impracticable. The formation of contacts to SiGe will be covered in Chapter 6.

As expected with any material change or optimisation, the advantages are matched with a set of challenges for integration. Four of the main challenges can be described as follows:

1. to understand phase formation at low temperature;
2. to control and limit diffusion of Ni in Si;
3. to avoid the formation of $NiSi_2$;
4. to increase the morphological stability of NiSi.

The interaction between Ni and Si is more complex than that between Co and Si. For the reaction of relatively thick Ni films ($>100\,nm$) with Si substrates, the phase formation sequence has been reported as similar to the one observed in the Co/Si system (Co_2Si, $CoSi$, $CoSi_2$). For films of thickness and preparation conditions relevant for contacts to current microelectronic devices, we observe dramatic modifications to the phase formation sequence. Several metal-rich phases form sequentially at low temperature, prior to the appearance of the low resistivity NiSi phase [53].

Recent reports also suggest that epitaxial $NiSi_2$ can form locally before NiSi formation [80–82]. The formation sequence and its dependence on process parameters such as dopant concentration, substrate type, substrate preparation and annealing conditions need to be understood in order to build a process that is reliable and reproducible. Since Ni diffuses readily in Si at very low temperatures and causes trapping of carriers in actual devices, it is important to limit the possible reaction of NiSi close to the edges of gates and source and drain areas and to form a contact of uniform thickness.

Two phenomena affect the stability of NiSi films at high temperature. One of these is the formation of $NiSi_2$. Since NiSi is not the phase that is in thermal equilibrium with the Si substrate, the formation of the higher resistivity $NiSi_2$ is expected upon heating. This formation could be driven alone by the increased diffusion at high temperature and also enhanced by the reduction in interfacial energies as the $NiSi_2$ forms epitaxially with the Si substrate. The formation of $NiSi_2$ is to be avoided since it consumes twice as much Si and is three times as resistive as NiSi. The epitaxial formation of $NiSi_2$ also causes some faceting at the interface and on the edges of contacts. Second, the degradation of the NiSi films can occur through thermal grooving and agglomeration. We will show in Section 5.7 that this is the main degradation factor for the electrical contacts. For film thickness and processing conditions relevant to the microelectronics industry, agglomeration occurs before $NiSi_2$ formation.

We note here that, as the thermal expansion of NiSi is extremely anisotropic (Section 5.5), the thermal degradation of NiSi should depend on grain orientation. As such, the texture of NiSi was carefully studied and we will show in Section 5.6 that thin NiSi films show an interesting new type of off-normal fibre texture that may affect agglomeration kinetics in narrow dimensions.

5.3 CHARACTERISTICS AND PROPERTIES OF Ni SILICIDE PHASES

5.3.1 Phase diagrams

The Ni/Si phase diagram is much more complex than that of Co/Si as shown in FIGURE 5.1. Whereas only three phases are found at room temperature in the Co/Si system (Co_2Si, CoSi and $CoSi_2$), one can count up to 11 phases in the Ni/Si diagram, six of which being stable at room temperature (Ni_3Si, $Ni_{31}Si_{12}$ or Ni_5Si_2, Ni_2Si, Ni_3Si_2, NiSi and $NiSi_2$) [83]. This considerably increases the complexity of the phase formation sequence and its possible dependence on processing parameters and substrate

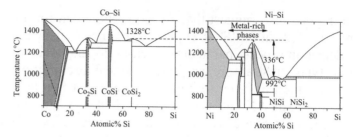

FIGURE 5.1 Comparison of phase diagrams for Co/Si and Ni/Si. Note the increased complexity in number of stable phases at room temperature and the lower temperature for melting.

variations (dopant type and concentration, cleaning conditions, substrate type). Even if two of the metal-rich phases appear to have a very narrow composition range ($Ni_{31}Si_{12}$, Ni_2Si), the phase diagram shows larger limits of composition for the other two metal-rich phases (Ni_3Si and Ni_3Si_2). Another noticeable difference is the large variation in the melting points of the compounds of interest. The low resistivity phases are the $CoSi_2$ in the Co/Si system and the $NiSi$ in the Ni/Si system. Their melting points are about 1330°C for $CoSi_2$ and 990°C for $NiSi$. This dramatically affects the temperature at which each phase becomes rougher and agglomerates. Morphological stability will be discussed in Section 5.7. From the phase diagrams, it is also interesting to note that about 10 at.% Si can be dissolved in either Co or Ni at 800°C under equilibrium, whereas little of either metal dissolves in the Si itself. Co is orders of magnitude less soluble than Ni in Si [19].

5.3.2 Crystal structures and volumetric changes

Changes in volume and consumption of Si during silicide formation can easily be calculated once the crystal structure and unit cell dimensions of each possible silicide are known. The simple procedure for these calculations can be followed as in TABLE 5.1 which first lists the crystal structures as extracted from the Joint Committee on Powder Diffraction Standard Database (JCPDS) published by the International Centre for Diffraction Data (card numbers: 04-0850, 32-0699, 24-0254, 03-943, 17-0881, 38-0844). The crystal structures, space groups and unit cell dimensions are given for the six nickel silicide phases that are stable at room temperature as well as for pure Ni and pure Si. From the lattice dimensions and structure, one easily calculates the volume of a unit cell (unit volume). From the number of "formula units" per unit volume (i.e. the 3 atom unit Ni_2Si appears 16 times in the unit volume) and the mass of Ni (58.7) and Si (28.1), the bulk density of each of the phases can then be calculated. The following two columns present the average volume per Ni atom and per Si

TABLE 5.1 Comparison of crystalline properties of Ni, Ni/Si phases and Si.

Phase	Crystal structure/ space group	Lattice constants (Å)	Unit volume (Å^3)	Formula unit per unit V	Density (g/cm^3)	V per Ni (Å^3)	V per Si (Å^3)	Silicide t/ Ni t	Silicon consumed/ Ni t
Ni	Cubic/$Fm3m$	3.5238	43.76	4	8.91	10.94	–	1	0
Ni$_3$Si	Cubic/$Pm3m$	3.5056	43.08	1	7.87	14.36	43.08	1.31	0.61
Ni$_{31}$Si$_{12}$	Hex./$p3_21$	6.671/12.288	1421	3	7.56	15.28	39.46	1.40	0.71
Ni$_2$Si	Orthor./	7.39/9.90/7.03	514.3	16	7.51	16.07	32.15	1.47	0.91
Ni$_3$Si$_2$	Orthor./$Cmc21$	12.29/10.805/6.924	919.5	16	6.71	19.16	28.73	1.75	1.22
NiSi	Orthor./$Pnma$	5.233/3.258/5.659	96.48	4	5.97	24.12	24.12	2.20	1.83
NiSi$_2$	Cubic/$Fm3m$	5.416	158.0	4	4.80	39.50	19.75	3.61	3.66
Si	Cubic/$Fd-3m$	5.4309	160.1	8	2.33	–	20.01	–	–

atom for each of the crystal structures. This is also the inverse of the atomic density for each element in the compound and leads to the ratio of the silicide thickness to the deposited Ni thickness. The amount of Si consumed by the formation of each of the silicides can be obtained by multiplying by the ratio of the volume per Si atom in the pure Si and in the silicide. All the numbers given here are extracted directly from the lattice constants provided by the JCPDS. These usually represent data on relatively thick films or powders and are therefore more representative of bulk samples. In thin films one can expect variations depending on grain boundaries, defects, impurities and strains. The density in thin films can be slightly smaller than in bulk samples and may depend on the microstructure of the original Ni film and the type of substrate as well as on the quality and cleanliness of the interface. As the grain size in the film can change during formation, the impurities can segregate, some defects can be either generated or annihilated by the phase formation and the strain can vary. Thus, the ratio of the film thickness to the Ni thickness could be either larger or smaller. However, from our experience, this ratio is usually close to the expected ratio from bulk samples.

5.3.3 Properties of Ni silicide phases

Some physical properties of the Ni silicide phases are presented in TABLE 5.2. Values for resistivity, melting points or transformation temperatures, average Young's modulus, average

TABLE 5.2 Comparison of physical properties of Ni, Si and Ni/Si phases. Note that the CTE and the Young's modulus represent an average in polycrystalline thin films. These quantities vary significantly with crystalline direction.

Phase	Resistivity ($\mu\Omega$ cm)	Melting or transformation/ melting (°C)	Avg. Young's modulus (GPa)	Avg. CTE for thin film ($\times 10^{-6}$ K^{-1})	Enthalpy of formation (kJ/mol)
Ni	7–10	1455	200	13.4	–
Ni$_3$Si	80–90	1035/1170	139	9.0	149
Ni$_{31}$Si$_{12}$	90–150	1242	177	–	1850
Ni$_2$Si	24–30	1255/1306	161	16.5	132–143
Ni$_3$Si$_2$	60–70	830/845	167	–	224–232
NiSi	10.5–18	992	132	~12	85–90
NiSi$_2$	34–50	981/993			87–94
Si	Dopant dependent	1414	130–187	2.60	–

coefficient of thermal expansion (CTE) and enthalpy of formation are presented for each of the six Ni/Si phases that are stable at room temperature.

The lowest resistivity phase (NiSi) is the desired phase for contacts to devices. The melting points of the silicides provide an indication of their morphological stability. The lower the melting point, the lower the temperature at which the diffusion of the components is expected, leading to degradation of the silicide film at lower temperature. Note that the metal-rich phases show much higher melting points than either NiSi or $NiSi_2$. Only the Ni_3Si_2 shows lower transformation temperatures with first a polymorphic transformation at about 830°C and a peritectoid transformation to NiSi and the high temperature θ phase. Since these temperatures do not refer to the presence of liquid, the morphological stability of $Ni_{31}Si_{12}$ may not be much lower than that of other metal-rich phases.

The elastic constants together with the CTE (both are thin film averages) allow for the determination of average strain and stresses in thin Ni silicide films. While it is considered that NiSi should be advantageous over $CoSi_2$ or $TiSi_2$ in terms of stress reduction in devices [84, 85], we will show in Section 5.5 that the anisotropy in the CTE is such that the stress in individual NiSi grains will vary drastically from tensile to compressive as the grain orientation changes. These large variations in local stresses may be detrimental to device properties and device reliability. The values given here are extracted from thin film measurements and therefore represent an average of the elastic properties. It is expected that the elastic characteristics of films will vary with film texture as the properties of a crystalline material may vary significantly along different crystal directions. The Young's modulus for any direction in a crystal as well as the shear stress can be extracted from the second order stiffness tensor that relates strain and stress [86, 87]. In a cubic material such as Si, the Young's modulus varies from 130 GPa up to 187 GPa depending on crystal orientation. One can therefore expect the values given in TABLE 5.2 to be very sensitive to film texture and vary from one reference to the next.

Enthalpies of formation are listed in kilojoules per mole of compound (formula unit). The higher number for $Ni_{31}Si_{12}$ is a consequence of the presence of a larger number of atoms in one formula unit (43). Some energy values are presented as a range covering available values from the literature. This should be a reminder of the relative accuracy of these measurements. The enthalpy of formation is useful in evaluating the magnitude of the driving force for a reaction. As will be discussed later, from the difference in enthalpies, one can evaluate if nucleation may be a controlling mechanism in the formation of a specific compound.

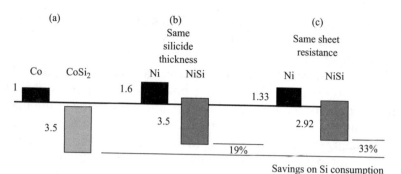

FIGURE 5.2 Schematic illustration of the reduction in Si consumption when using NiSi instead of $CoSi_2$.

5.3.4 Reduction in silicon consumption

Two factors contribute to the reduction in Si consumption when moving from $CoSi_2$ to NiSi contacts. The first one relates to the lower resistivity of NiSi. Such a film can show the same sheet resistance and yet be thinner than a $CoSi_2$ film, therefore consuming less Si during its formation. The second one is the lower density of Si in NiSi compared to $CoSi_2$. The combination of these factors is schematically shown in FIGURE 5.2. The evaluations of thickness ratios and Si consumption originate from the last two columns of TABLE 5.1. FIGURE 5.2(a) shows the formation of $CoSi_2$ from a Co film of thickness unity. The formation of a NiSi film of the same thickness as $CoSi_2$ is shown in FIGURE 5.2(b). The original Ni thickness must be 60% thicker and the Si consumption is only 81% of the amount needed for the $CoSi_2$ formation. Combining this with a conservative decrease in resistivity from $18\,\mu\Omega\,cm$ for $CoSi_2$ to $15\,\mu\Omega\,cm$ for NiSi, one can deposit a thinner Ni film and achieve the same sheet resistance leading to a reduction of Si consumption of more than 30% as shown in FIGURE 5.2(c). Note that the effective decrease in resistivity from $CoSi_2$ to NiSi depends on film thickness, microstructure as well as dopant types and their concentrations.

5.4 PHASE FORMATION

5.4.1 Thermal budget

The thermal budget for the formation of the low resistivity NiSi is much lower compared to its $CoSi_2$ counterpart. An advantage is that once the low resistivity NiSi is formed, one does not have to anneal further and reach the $NiSi_2$ phase. Even when comparing the formation of the monosilicides, it is found that NiSi is much easier to form than CoSi. In FIGURE 5.3, we show a comparison

Nickel silicide technology

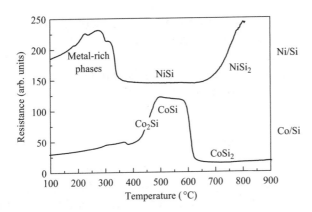

FIGURE 5.3 Sheet resistance versus annealing temperature measured *in situ* while heating Co/Si and Ni/Si films at 3°C/s. Note that the lower resistive NiSi phase forms at considerably lower temperature than $CoSi_2$. However, the NiSi film also degrades at lower temperature.

of the process windows for forming the low resistance phases in both the Ni/Si and Co/Si systems using *in situ* resistance measurements. The traces shown in FIGURE 5.3 were measured from thin films of Ni and Co deposited on undoped poly-Si while the samples were annealed at 3°C/s in nitrogen. The resistance trace for the Ni/Si reaction has been shifted up for clarity. As the temperature increases, the film resistance is modified by the successive formation of silicide phases. From the graph, it is clear that the low resistance $CoSi_2$ phase forms slightly above 600°C for these anneal conditions, whereas the formation of the low resistivity NiSi is achieved at 350°C. Note that at the temperature of formation of the $CoSi_2$ phase, the resistance of NiSi is already starting to increase, showing the importance of reducing the thermal budget for the process after NiSi formation. A material change from $CoSi_2$ to NiSi thus not only allows for a reduction in the thermal budget but it also requires it. The high temperature degradation of NiSi layers will be discussed in greater detail below.

The lower temperature formation of NiSi compared to CoSi may reside partly in the fact that the concentration and diffusivity of Ni is higher than that of Co. The ratio in solubilities for Ni and Co atoms in the silicon substrate is very large [19, 88, 89]. At 900°C, for example, the solubility of interstitial Ni is about four orders of magnitude larger than that of Co. As the temperature decreases, the ratio increases rapidly. By 500°C, the solubility of Ni in Si is close to six orders of magnitude larger than that of Co. Ni also has a considerably larger diffusivity in Si than does Co at low temperatures [87]. Although Ni is both more soluble and more mobile than Co in Si, lower temperature formation of phases is by no means guaranteed. The diffusivity in the silicide phases themselves is also of importance to determine which phase will be measured since Ni diffuses through some forming silicide extremely rapidly. This may become a determining factor as Ni diffuses even faster in some metal-rich silicides than in Si. The determination of the first nucleating phase is then further complicated [20].

5.4.2 Complex phase sequence

As shown in FIGURE 5.1, the Ni/Si phase diagram is more complex than that for Co/Si. In spite of the complexity of the Ni/Si system, traditional studies of the interaction of Ni films with Si substrates carried out using relatively thick films (~100 nm Ni) and isothermal anneals followed by *ex situ* characterisation have revealed only the sequential formation of Ni_2Si, $NiSi$ and $NiSi_2$, remarkably similar to the Co/Si reaction [2–4, 15, 18, 23, 24, 29, 37, 47, 81, 87, 90–96]. We show here that for samples prepared by standard device manufacturing processes and for film thicknesses relevant to the state of the art CMOS technology, several phases appear in sequence during rapid thermal anneals before NiSi formation [53]. Recent works also suggests that some epitaxial $NiSi_2$ forms early in the phase sequence and is consumed by Ni_2Si or NiSi [81, 82]. This formation would be driven by a reduction in interfacial energy as the unit cell dimension of $NiSi_2$ at room temperature (5.41 Å) is similar to that of Si (5.43 Å). Moreover, this small difference will be reduced as the temperature is increased since the thermal expansion of $NiSi_2$ is larger than that of Si [73, 87]. We note here that this phenomenon may be related to oxide mediated epitaxy [10, 97–102] in which a very thin oxide would slow diffusion enough to help in forming an epitaxial disilicide layer. In our work, $NiSi_2$ inverted pyramids were not detected at low temperature with X-ray diffraction (XRD) or transmission electron microscopy (TEM). Since their formation might be dependent on the presence of oxygen at the interface, a reduction in oxygen contamination before Ni deposition may be sufficient to eliminate them. Note that p-doped areas of a substrate would be more prone to the formation of $NiSi_2$ pyramids since the chemical oxide is easier to form (unlike thermal oxidation) and is thus thicker on these substrates [103].

5.4.2.1 In situ *measurements of Ni/Si reactions*

The increased complexity of the Ni/Si reaction compared with Co/Si can already be observed from the low temperature part of the resistance traces in FIGURE 5.3. In contrast to the Co/Si sample which shows a relatively simple evolution before CoSi formation, the resistance of the Ni/Si film shows multiple maxima before formation of the low resistance phase. This is a signature for the formation of multiple metal-rich phases and of the complexity of the phase sequence. This complex sequence can be better observed when the resistance measurements are combined with *in situ* X-ray and light scattering measurements during annealing.

The annealing experiments were performed at the National Synchrotron Light Source (NSLS) X20C beamline at Brookhaven National Laboratory (NY, USA) where we follow *in situ* the formation of different phases via the development of their characteristic diffraction patterns [77, 104–106]. Annealing is performed in a purified atmosphere and the temperature measurements, calibrated using metal–silicon eutectics, are precise to $\pm 3°C$. For XRD, we selected an energy of 6.9–7 keV ($\lambda = 0.177$–0.180 nm) for which the X-ray flux is greater than 10^{13} photons/s. Unless otherwise specified, the X-ray beam was incident at an angle of 27° from the plane of the sample. This incidence angle was selected since it allows a near Bragg–Brentano geometry for characteristic X-ray diffraction peaks from all important phases in the Ni/Si system within the 2θ window of the linear detector. In the current configuration, a diffraction curve (covering a range of 2θ angles of up to 14°) from a ∼10 nm metal film can be acquired in less than 100 ms. The experimental apparatus is designed [105, 107] so that one can simultaneously monitor both the film resistance using a four-point probe setup and the surface roughness via the elastic scattering of monochromatic light. Roughness measurements [108] are made using a He/Ne laser light with a wavelength of 633 nm, at an incident angle of 65° and scattering angles of −20° and 52° (with two detectors) providing roughness information at lateral length scales of about 0.5 and 5 μm, respectively. This technique is used to determine degradation of the film at high temperatures. The detection of low intensity scattered light during high temperature anneals requires optical filtering and lock-in detection in order to eliminate intense black body radiation from the sample.

The results from the reaction of a 15 nm film of Ni with p-doped polycrystalline Si are displayed in FIGURE 5.4, as measured using time-resolved XRD during an anneal at 3°C/s in purified nitrogen. Two peaks are present at low temperature in the selected 2θ window (50–60°). The first one near 52° is the 111 diffraction peak of Ni. The second, slightly above 55°, is the 220 diffraction line from the poly-Si substrate. As the temperature increases to 300°C, the Ni peak starts to decrease in intensity while another peak appears at about the same position as the Si(220). Considering the metal-rich phases alone, this peak position matches the (350) spacing of Ni_3Si_2 [9]. When the temperature reaches 400°C, the peak

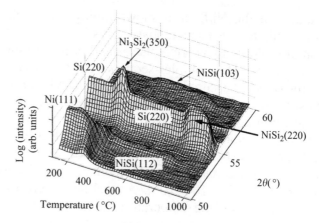

FIGURE 5.4 Three-dimensional rendering of the *in situ* X-ray diffraction measurements during annealing of a 15 nm Ni film deposited on p-doped poly-Si (3°C/s).

from the metal-rich phase has disappeared and two low intensity peaks from the monosilicide phase can be observed: NiSi(103) just below 60° and NiSi(112) at about 53°. Note that there is a clear decrease of the Si(220) peak intensity: a consequence of Si consumption during Ni silicide formation. At about 800°C, the two weak NiSi peaks disappear as the $NiSi_2(220)$ peak intensity increases. This phase forms relatively quickly and is stable over a temperature range of less than 200°C as both the poly-Si and $NiSi_2$ intensities decrease drastically upon reaching the eutectic melting temperature of 966°C. The decrease in $NiSi_2$ intensity is a direct consequence of melting. However, the disappearance of the poly-Si peak is more surprising and could be the result of texturing in the poly-Si layer in the presence of the eutectic liquid.

The X-ray data are replotted in a top view (FIGURE 5.5(b)), together with the simultaneous resistance and light scattering data (FIGURE 5.5(a)). The X-ray intensity is presented both as grey logarithmic scale (white is high intensity) and as intensity contours. In this way, it becomes clear that the high photon flux available from the synchrotron ring allows the detection of significant changes to the Ni layer starting just above 200°C for these annealing conditions. The peak identification is difficult because of the large number of phases over a narrow temperature range and because of the presence of the poly-Si(220) peak, which overlaps many peaks from the metal-rich phases. Thermal expansion and strain relaxation during annealing also complicate the phase identification. It is, however, clear that multiple metal-rich phases form at low temperature before the appearance of the low resistivity NiSi phase. Note that the temperature

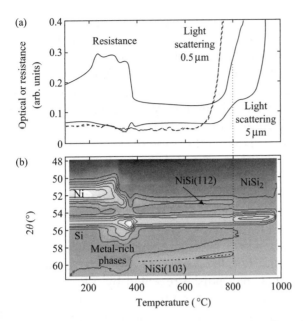

FIGURE 5.5 (a) Resistance and light scattering from 0.5 to 5 μm length scales together with (b) X-ray diffraction measurements performed *in situ* during annealing (3°C/s) of a 15 nm Ni film deposited on p-doped poly-Si.

windows over which each of the metal-rich phases exist do overlap significantly, and as a result, for many temperatures, multiple metal-rich phases coexist in the film. This can also be observed in FIGURE 5.7.

Although the phase formation sequence at low temperature is relatively complex, one can already learn a great deal from this narrow *in situ* 2θ XRD window. Starting around 200°C, the Ni(111) peak already shifts towards larger 2θ values (lower part of the Ni peak on FIGURE 5.5(b)). This shift could be the result of some strain relief related to grain growth combined with the possible appearance of some Ni_3Si. Because this phase has a high symmetry (cubic, Cu_3Au) only one diffraction line is present in the current 2θ window at ~52.2°. From 250 to 300°C, the increases in intensity on each side of the Ni peak as well as the appearance of the peak below the Si(220) correspond well to the 024 and 122 (49.9° and 51.1°), 115 (53.3°) and 205 (56.6°) diffraction lines of the $Ni_{31}Si_{12}$ phase [10]. At about 340°C, the clear peak at 53.0° may be the 301 and 121 lines (same d spacing) of orthorhombic Ni_2Si, with the extra shoulder at about 57° corresponding to the 002 line of the same phase [11] (will be referred to only as Ni_2Si below). Although there is a clear shift of the peaks around 330°C that suggests a phase change to Ni_2Si, we want to emphasise that this shift could also originate from a stress relief in an originally strained Ni_2Si layer. The very intense peak seen around 370°C at about 55.6° can be indexed as Ni_3Si_2 (350 line at 55.4°) [9]. From 390 to 420°C, the peak at about 53.5° which remains up to 750°C, corresponds to NiSi(210). Additional details regarding the identification of these metal-rich phases were published earlier and supported with limited TEM and other XRD data [53, 104]. Note that the phase sequence presented here relies mainly on XRD data and on JCPDS identification of unstrained films or powders. As mentioned above, some of the peak shifts observed may be associated with stress modifications in a given metal-rich phase and not with a phase change. In particular, the identification of $Ni_{31}Si_{12}$ is tentative and we believe that this layer could also be a strained Ni_2Si layer.

The simultaneous measurements of resistivity and scattered light shown in FIGURE 5.5(a) are useful in revealing film transformations that may not be revealed by XRD. The resistance is particularly sensitive to the formation of the metal-rich phases. At low temperature, the sheet resistance of the Ni film follows the standard monotonic increase as the phonon population increases with temperature. It first deviates upwards from this regular behavior around 250°C as atoms start to interdiffuse and then stabilises at a lower value as the first metal-rich phase, crystallises. The following variations in resistance correspond to the resistivities listed in TABLE 5.2. The second local minimum occurs when the lower resistivity Ni_2Si forms (or relaxes) at about 340°C, while the last increase before NiSi formation correlates with the formation of the high resistivity Ni_3Si_2 at about 370°C. As NiSi forms, the resistance abruptly drops and then remains low for a temperature window of about 400°C. Above 800°C, the resistance increases sharply as the film agglomerates, the higher resistivity $NiSi_2$ phase forms and the film finally melts.

During the formation sequence, the light scattering signals also show very distinct signatures. The shorter length scale light scattering signal (0.5 μm) remains low throughout the formation of all Ni rich and that of the monosilicide phases, up to a temperature higher than 600°C. In contrast, the 5 μm signal changes significantly slightly below 400°C. On this length scale, the signal first goes through a maximum that corresponds exactly to the sharp decrease in resistance and then stabilises at a higher intensity. This increase could result from different optical constants for laterally non-uniform phases. It could also be correlated with either surface pits appearing as NiSi starts to form or alternatively be related to the reported low temperature formation of $NiSi_2$ inverted pyramids [81, 82]. The large increase in surface roughness at high temperature begins below 700°C for the 0.5 μm signal and at about 750°C for the 5 μm one. For both length scales, the roughening clearly starts before the transformation to the high resistivity disilicide as observed by XRD. This suggests that for the processing conditions used here, the agglomeration of the thin NiSi film precedes the disilicide formation and should be considered the limiting factor for integration of nickel silicide in microelectronics. This will be discussed further in Section 5.7.

5.4.2.2 First metal-rich phases: low temperature formation

During the phase formation of Ni silicides, the Ni_3Si phase may be the first silicide to form, but it is not easily detected using XRD since it would only appear in the current XRD window in the form of a very slight shoulder above the Ni(111) peak (Ni_3Si and Ni are both cubic with similar lattice constants). The first crystalline phase formed that can be clearly identified by XRD is either the $Ni_{31}Si_{12}$ phase (also referred to as Ni_5Si_2) or a strained Ni_2Si phase. Note that the diffusion coefficients of Ni are reported to be the highest in $Ni_{31}Si_{12}$ phase [20, 109]. Isothermal anneals at low temperature can be used not only to determine the lowest temperature at which Ni silicide phases appear but can also help in understanding the formation mechanisms and formation kinetics of this first phase. In FIGURE 5.6, data from such isothermal heat treatments show the intensity variations of the diffraction peak at about $2\theta = 57°$ for different annealing temperatures. The formation mechanisms will be discussed later. Here, we will briefly address the variation in diffracted intensity with time. In the case of a diffusion-controlled formation, one would expect the volume of the new phase to be proportional to the square root of time [56]. The shape of the XRD intensity with time is rather complex and does not support a simple formation of crystalline layer controlled solely by diffusion through the layer. The formation could occur as follows. As the diffusion of Ni in the Si begins, an amorphous Ni/Si layer is forming at the interface [110]. This layer thickens with time and is only detected by XRD when is crystallises at a given thickness (sudden increase in intensity). The XRD then further increases but at a rate that is significantly slower than what is expected from diffusion ($<t^{1/2}$). This could be explained by significant grain growth through the crystallised layer. Beside showing the advantage of *in situ* detection for the study of formation kinetics, this experiment clearly shows the low temperature diffusion of Ni since a significant amount of silicide forms during a 3 h anneal at temperatures as low as 160°C. By 225°C, the silicide can form in times that are relevant to the microelectronics industry. Such a fact must be taken into consideration since most microelectronics tooling systems go through preheat cycles that will dramatically affect the formation of phases.

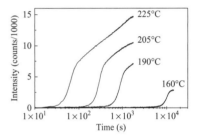

FIGURE 5.6 Evolution of the X-ray intensity for the first crystalline metal-rich phase during isothermal annealing at low temperature. Note the peculiar shape of the intensity increase that can even occur at extremely low temperature (160°C).

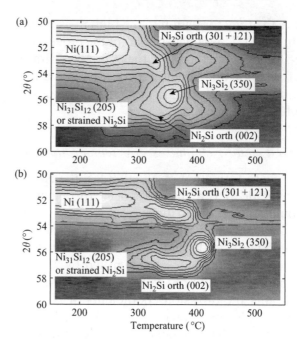

FIGURE 5.7 *In situ* XRD measurement during annealing of a 15 nm Ni film deposited on (a) p-doped and (b) n-doped SOI substrate.

5.4.2.3 Effect of dopants on phase formation

The complexity of the low temperature phase formation and its dependence on dopant type is illustrated in FIGURE 5.7 using *in situ* XRD during annealing of 15 nm Ni films deposited on n- and p-doped SOI substrates. These substrates are built of a thin single crystal Si layer on a thicker buried oxide and are currently used in the semiconductor industry. For the n-type substrate, phosphorus was implanted at a concentration of 5×10^{15} cm^{-2} and an energy of 12 keV while for the p-doped substrates, boron was implanted at a concentration of 3.5×10^{15} cm^{-2} and an energy of 8 keV. The anneals were performed at 3°C/s in purified He. The sequence of phases described earlier for a poly-Si substrate is presented directly on the figure and can be seen more clearly because of the absence of the poly-Si(220) peak. Importantly for processing devices, it is clear that for different dopant types, the phase formation is shifted by more than 50°C as phases are formed at lower temperature on the boron-doped substrate. The consequence of this is that very low temperature anneals will lead to different phase formation on p- and n-doped parts of CMOS structures. This in itself may not be a problem for a later formation of a uniform NiSi layer but may lead to different types of problems on different locations of a wafer during processing.

It is not clear at this point how the dopant influences silicide formation. For the case of Ti and Co silicide formation, ternary phase diagrams and diffusion data show that there are compounds forming with Ti and B, with Ti and P as well as with Ti and As. In this case, some of the compounds, namely TiAs, $TiAs_2$, TiB_2 and TiP can be in thermal equilibrium with the Si substrate, and if they precipitate during silicide formation, one can expect the diffusion properties at interfaces to be affected. In the case of Co, only CoAs or CoP_3 could be present at the interface as none of the Co/B compounds can be in thermal equilibrium with Si.

The ternary phase diagrams for Ni, Si and possible dopants are not well established. A first estimation of the diagrams is given in a chapter by Madar in Reference 73. The structure of the Ni/Si/B ternary phase diagram is very similar to the Co diagram suggesting that Ni-B compounds should not be in thermal equilibrium with the Si. While the ternary phase diagrams containing P or As are very tentative, we do expect the Ni to behave similarly to Co. This would suggest that Ni arsenide or Ni phosphide compounds would be more likely to affect the formation by comparison to an undoped or B-doped sample. Consistently, compared to undoped poly-Si, larger variations have been observed here for the n-doped samples.

Multiple studies of dopant redistribution during silicide formation show that both phosphorus and arsenic tend to partially segregate at the interface NiSi/Si. This is particularly important in affecting the concentration of dopant at the interface as a high dopant concentration ensures a low contact resistance [111]. Although few studies compare the contact resistance of $TiSi_2$ or $CoSi_2$ and NiSi in similar conditions, most of them point toward a lower contact resistance for the NiSi contacts [26, 73, 84, 112–114]. Consistent with the increased concentration of P or As at the interface, the contact resistance of silicide on n^+-doped Si is much lower than the contact resistance on p^+-doped substrates.

Another important aspect of dopant diffusion is that the silicide formation itself can affect the diffusion of dopants in the Si below. It was found in the past that formation of either $CoSi_2$ or $TiSi_2$ injects point defects in the crystalline substrate [115–117]. For both silicides, the vacancy concentration increased during formation while the interstitial concentration decreased. Consequently, the diffusion of B and P is reduced, because their movement depends on the concentration of interstitials. In contrast, the diffusion of As or Sb is increased, because they move mainly through vacancies [118]. Although this injection of point defects was shown to be independent of diffusion species for Co and Ti disilicide formation [115], we believe that it could be different for Ni silicide formation because all phases in this system form through Ni diffusion as will be discussed in the next section. As such, the number of vacancies could be reduced in the Si substrate and the final junction profile could be significantly different than that for $CoSi_2$ formation.

5.4.3 Formation mechanism

5.4.3.1 Diffusion-controlled formation: limited nucleation effects

Not only does the formation of NiSi consume considerably less Si from the substrate as compared to $CoSi_2$ formation, it also results in a much smoother surface and interface. As a result, barring the eventual formation of $NiSi_2$ pyramids, the NiSi/Si interface can be closer to the buried oxide of a SOI wafer without risking contact with this underlying oxide. The formation of NiSi has been reported as being diffusion controlled [2, 23]. Contrary to a nucleation-controlled reaction (low resistivity Ti and Co silicides), where the reaction evolves rapidly, non-uniformly and leads to some characteristic roughness, a diffusion-controlled reaction is characterised by growth fronts of new phases that are ideally planar and move uniformly following the standard relation where the thickness is proportional to the square root of time [56].

The mechanism for the formation of a new phase depends strongly on the amplitude of the driving force for the reaction, which is controlled by the variation in free energy. While multiple factors contribute to the free energy, to first order, one can get an idea of the type of reaction by following the variations in the enthalpy of formation. If the difference in enthalpy of formation between the product and the reactants is large, the drive for the reaction to occur is large and the elements will react as soon as they are in close proximity. The diffusion of the reactants through the growing phase is then the limiting factor and the phase grows very uniformly. On the other hand, if the difference in enthalpy of formation is small, the nucleation of the phase can be very difficult. For small nuclei, the energy of the new interfaces is larger than that of the enthalpy reduction in the volume. In this case, the formation is controlled by nucleation and the phase can grow laterally from these nucleation centres generating some characteristic roughness [22]. In the formation of $CoSi_2$ from CoSi and Si, nucleation is important and film roughness typically represents 20–30% of the film thickness. In this case, the variation in free energy from the variation in enthalpy of formation is only about -2.5 kJ/mol. For Ni silicide phases, it has been reported that nucleation is only important for the formation of $NiSi_2$. These conclusions need to be revisited since the formation sequence is different in current thin films. For the prior phase sequence, only three phases were observed. The sequence and the energy difference at each step are estimated to be:

Reactants	Products	ΔH (kJ/mol of Ni atoms)	ΔH (kJ/cm^3 of products)
$2Ni + Si$	Ni_2Si	−66	−6.8
$Ni_2Si + Si$	$2NiSi$	−19	−1.3
$NiSi + Si$	$NiSi_2$	−2	−0.08

For one of the phase sequence currently suggested, we get:

Reactants	Products	ΔH (kJ/mol of Ni atoms)	ΔH (kJ/cm^3 of products)
$31Ni + 12Si$	$Ni_{31}Si_{12}$	−60	−6.5
$2Ni_{31}Si_{12} + 7Si$	$31Ni_2Si$	−6	−0.62
$3Ni_2Si + Si$	$2Ni_3Si_2$	−9	−0.78
$Ni_3Si_2 + Si$	$3NiSi$	−10	−0.69
$NiSi + Si$	$NiSi_2$	−2	−0.08

We want to bring attention to three important points regarding the data above. First, the precision of these enthalpy measurements is typically about 2–4 kJ/mol, possibly leading to values that could be considerably different, especially when the ΔH is small. One is only certain that all values are negative since the phases are present in equilibrium diagrams. Second, the enthalpy differences will slightly depend on temperature through the heat capacity of each of the phases present [119, 120]. While this may slightly modify the numbers presented here, it does not change the basic idea that with more phases present, nucleation may become more important. Third, we choose to show the data in kilojoules per mole of Ni atoms and in kilojoules per cubic centimetre. This choice of normalisation, even if helpful in the comparison of each of the reactions, does not give any indication of the sequence of phase formation. It is an indication that the reaction can occur, not that it will (for a given temperature and a finite anneal time). In microelectronics, since we are typically in a situation where Si is in excess, it is true that one normally observes a phase sequence ending with the phase that is the most rich in Si ($NiSi_2$). However, in an environment where Ni would be in excess, $NiSi_2$ would

Nickel silicide technology

necessarily transform to phases that are more rich in Ni. The table above could then be rewritten from such reactions as $NiSi_2$ and Ni and the ΔH could be given in kilojoules per mole of Si atoms. The ΔH would not only still be negative but would be much larger, as in a Ni rich environment, one $NiSi_2$ can form two NiSi for a ΔH in excess of -40 kJ/mol of Si atoms.

From the two tables given earlier, it is clear that the driving force for the formation of each of the phases (after the first phase) is considerably lower when more phases appear in the sequence of formation. Potentially, the nucleation could become an important issue in these layers. Since thin NiSi films are extremely smooth by comparison to Ti, Co or Ni disilicides which form through nucleation, the formation of NiSi in the current films is still clearly controlled by diffusion. However, other factors such as important increases in dopant concentration, changes in the density of defects or variation in mixing entropy in the presence of alloying elements may reduce the free energy of formation to the point where nucleation may become an issue and the film could become significantly rougher.

5.4.3.2 Dominant diffusing species: reduction of bridging and reduction of Kirkendall voiding

Not only is the formation of metal-rich phases controlled by diffusion but it is important to point out that Ni is the dominant diffusing species. This leads to two considerable advantages with regard to the process. The first one is that in the self-aligned process, if the temperature of formation is low enough so that Si is not significantly mobile, the possibility of bridging is practically eliminated. With a low diffusivity for Si atoms, the formation of a silicide above sidewalls or oxide areas that would resist the selective etch is not possible.

The second advantage of Ni diffusion is that vacancies generated by the diffusion itself are mainly located in the metal layer (see FIGURE 5.8). The consequences of this, although not obvious at first, are of critical importance in small dimensions. In current

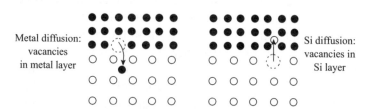

FIGURE 5.8 Schematic illustration of the importance of the diffusing species in determining where void formation is likely to occur when reacting a metal film with a thin Si layer (e.g. on SOI wafers).

devices, the formation of CoSi is controlled by Si diffusion leading to the generation of vacancies in the Si. As the diffusivity of vacancies is very high, in bulk substrates, they normally get distributed throughout the wafer thickness. However, in many cases of bulk diffusion couples the vacancies can result in the development of diffusional porosity also called Kirkendall voiding [121–124]. In SOI wafers and poly-Si gates, the vacancies are limited to a Si volume that is barely larger than the volume of the vacancies generated. Under such conditions, one can imagine that voids may be much easier to form. Note that vacancies diffuse readily in single crystal substrates and it is expected that void formation will be worse on poly-Si substrates where the grain boundaries act as sinks for the vacancies. In the case of metal diffusion, the vacancies are generated in the metal layer. The morphology of this layer after reaction is of little interest since it is removed by selective etching. We believe this to be an important reason why $CoSi_2$ shows limitations in small poly-Si dimensions whereas NiSi does not (see also FIGURE 5.22).

5.4.4 Disadvantages of low temperature diffusion

The very fast diffusion of Ni has led to unexpected resistance variations in narrow dimensions [59, 61, 64, 65]. For Ti and Co silicide contacts, when the linewidth of conductive lines was reduced the resistance was found to increase. In the case of Ni, when the annealing temperature is such that Ni is allowed to diffuse over distances longer than the thickness of the film, the resistance in smaller dimension actually decreases. By comparison to the effect seen in Ti and Co, this behavior has been termed the "reversed fine line effect". This decrease in resistance is explained through an increase in the reaction volume of Ni and Si close to line edges and is schematically shown in FIGURE 5.9. Because of the geometry, more Ni is available on the edge of either a poly-Si gate or a source/drain area. The extra Ni atoms above the spacer or on the STI that are within the diffusion distance can increase the amount of reaction close to the boundaries. This is a small percentage of

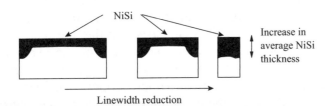

FIGURE 5.9 Schematic illustration of the inverse linewidth effect, whereby a thicker NiSi layer is formed for thinner poly-Si gates.

Nickel silicide technology

large area structures, but it becomes very significant in narrow dimensions.

While it is clear that a low anneal temperature may be necessary to limit this narrow line effect, a relatively high temperature anneal may also be needed to ensure that all Ni atoms have reacted with Si and have become a part of the silicide contact. In Si, interstitial Ni atoms diffuse extremely rapidly and would be detrimental to device performance. Literature reports suggest that annealing at a temperature as high as 500°C may be necessary to eliminate carrier trapping by Ni atoms as measured by deep-level transient spectroscopy (DLTS) [125, 126]. The source of this effect is not clear at present. While the Ni could simply move to a different position within the Si crystal structure itself (interstitial versus substitutional), the high diffusivity could also ensure that the Ni has moved to a location outside of the probing area of the DLTS measurement. At 500°C, Ni atoms can quickly cover relatively large distances (of the order of 1 mm in minutes) and any reaction or segregation would become a sink for these atoms. Not only is a careful optimisation of the process necessary to limit these effects, but the reproducibility at low temperatures is also critical as any deviation from wafer to wafer at very low temperature can cause significant variation in material properties and large variations in yield.

Once a good-quality NiSi film can be formed, the next challenge is to keep its integrity as the film must withstand temperature exposure during further processing. Before we discuss the degradation of the film at high temperature, we will cover two important and unexpected characteristics of the NiSi film itself. The first one is the very large anisotropy in thermal expansion and the second one is the unexpected texture seen in NiSi films on single crystal Si substrates. Both these properties/characteristics will be important for the evolution of the film morphology at high temperature.

5.5 THERMAL EXPANSION OF NiSi AND RELATED STRESS EFFECTS

In FIGURE 5.5, one observes a striking difference in the way that the NiSi(112) and (103) diffraction peaks shift as a function of increasing temperature. This suggests that the thermal expansion of NiSi is strongly anisotropic. The value of the average CTE determined from bending measurements and given in TABLE 5.2 does not represent the picture on a microscopic scale. In the following section, we discuss the magnitude of the thermal expansion of NiSi in bulk and in thin films, and its implications on the occurrence of local stress in thin NiSi layers.

5.5.1 Unit cell dimensions and CTE for bulk samples

NiSi has the orthorhombic MnP structure, a slightly deformed hexagonal NiAs structure. The unit cell dimensions at room temperature were reported by Rabadanov and Ataev [127] as $a = 0.51752$ nm, $b = 0.3321$ nm, $c = 0.56094$ nm from XRD studies using NiSi single crystals. Wilson and Cavin [128] studied polycrystalline bulk samples, and obtained unit cell dimensions of $a = 0.5177$ nm, $b = 0.3325$ nm and $c = 0.5616$ nm. The small difference between the two sets of dimensions is within the experimental errors.

Wilson and Cavin [128] measured the CTE for bulk samples by using XRD to measure the lattice spacing of various crystallographic planes as a function of temperature. They obtained a polynomial expression for the dimensions of the orthorhombic unit cell as a function of the temperature T (expressed in Kelvin):

$$a = 0.5187 - 1.807 \times 10^{-5}T + 6.026 \times 10^{-8}T^2$$
$$- 2.709 \times 10^{-11}T^3$$
$$b = 0.3286 + 2.977 \times 10^{-5}T - 6.557 \times 10^{-8}T^2$$
$$+ 2.915 \times 10^{-11}T^3$$
$$c = 0.5626 - 1.779 \times 10^{-5}T + 5.727 \times 10^{-8}T^2$$
$$- 2.546 \times 10^{-11}T^3$$

Both Wilson and Cavin [128] and Rabadanov and Ataev [127] reported the surprising observation that the shortest axis of the NiSi unit cell actually contracts for increasing temperature. The usual increase in unit cell dimensions for increasing temperature is attributed to the asymmetry (with respect to the equilibrium interatomic spacing) of the cohesive energy versus atomic distance curve. Normally, an increase in temperature implies an increase in entropy, increased disorder and consequently increased volume. However, there is no fundamental reason, of thermodynamic origin or otherwise, that excludes a decrease in lattice parameter with increasing temperature. "Thermal contraction" is nevertheless extremely rare; in pure elements encountered only in one phase of U and in Zn and Si at low temperatures. Other materials for which negative thermal expansion coefficients have been observed along one crystallographic axis are calcite ($CaCO_3$), beryl (beryllium aluminium silicate) and silver iodide [129]. It should be noted that although the b axis of the NiSi unit cell contracts upon heating, the total volume of the unit cell actually increases for increasing temperature.

5.5.2 Unit cell dimensions and CTE for thin films

Unit cell dimensions for thin NiSi films have been reported by d'Heurle et al. [15, 130] as $a = 0.5233$ nm, $b = 0.3258$ nm and $c = 0.5659$ nm. These values are significantly different from the values for bulk samples quoted above, suggesting that thin NiSi films are in fact strained. Detavernier et al. [131] studied the CTE in thin films. Samples of 10 nm Ni deposited on poly-Si were annealed to 500°C to form NiSi and subsequently quenched to room temperature. The NiSi layers thus formed were then re-annealed. While ramping up the temperature from 50 to 750°C at 1°C/s, the NiSi and poly-Si peaks were measured. The peak positions of all observed NiSi peaks were determined as a function of temperature. On the basis of the peak positions, the dimensions of the unit cell could be calculated as a function of temperature. In FIGURE 5.10, we present the unit cell value as a function of temperature for the thin NiSi film together with the values for bulk NiSi. The good match between bulk data and thin film behavior at high temperature (above 500°C) implies that the high temperature data for thin films correspond to a properly relaxed equilibrium

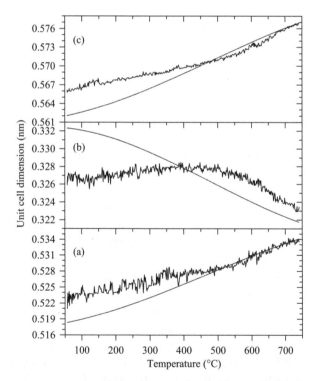

FIGURE 5.10 The evolution of the unit cell dimensions for NiSi (measured for a 20 nm NiSi film) during heating at a constant ramp rate of 1°C/s. The solid lines were calculated based on the published expansion data for bulk samples [128].

sample. However, the low temperature thin film data were obtained on a strained sample.

The room temperature dimensions of the a and c axes in thin films on Si substrates are significantly larger than the reported bulk values, whereas the b dimension is slightly shorter in thin films than in bulk samples. During the first heat treatment, aimed at forming the NiSi phase, the NiSi film will be relaxed at temperatures above 500°C, but the difference in thermal expansion coefficients between film and substrate will cause stress buildup during cooling to room temperature. In view of the apparently anisotropic thermal expansion of NiSi, the stress–strain state in the NiSi film at room temperature will depend on the grain orientation, the plane parallel to the NiSi/Si interface, the expansion coefficients in this plane and the stress–strain tensor. Moreover, inter-grain stresses originating from the mechanical interaction between neighbouring grains may give rise to important second-order effects [132]. A truly quantitative interpretation would require a precise knowledge of the strain–stress tensor for this material with undoubtedly peculiar elastic properties, but qualitatively these results can be understood as follows. To begin with, it is clear that the a and c axes have large expansion coefficients, averaging to about 40 ppm/°C, considerably greater than the value of 2.6 ppm/°C that is known for Si. For the b axis, contraction follows a sigmoidal curve as a function of increasing temperature, quite far from linear. Because of the diffraction geometry, the dimension along the b axis derived from the present experiments is determined for grains for which a plane that contains the a and c axes is parallel to the substrate. Supposing that the grains are relaxed at 500°C (after forming the NiSi phase during the first anneal), the big difference in thermal expansion between Si and NiSi will cause strong tensile stresses in these grains in the plane of the film during sample cooling to room temperature. Hence, the Poisson ratio effect implies a compressive strain in the direction perpendicular to the interface. Therefore, the b unit cell dimension in thin films should be smaller than the equilibrium value in the direction perpendicular to the film and larger for all directions within the plane of the film, the more so at lower temperature. When reheating the NiSi films (to determine the unit cell dimensions as a function of temperature), the thermal stress in the film decreases, and the unit cell dimensions converge towards their equilibrium values. Above a "relaxation temperature", which is not a fixed point but according to FIGURE 5.10 can be estimated to be in the vicinity of 500°C for a heating rate of 1°C/s, the thin film relaxes and the data are in good agreement with bulk data. For the a or c axes, the stress–strain conditions are more complex, since in this case the measurements correspond to directions normal to b and c, and a and b, respectively. The observation that the room temperature length of the *a* and *c* axes in thin films in the direction perpendicular to the surface is greater than their equilibrium values implies through the Poisson ratio that there are compressive stresses in the plane of the film for these grains. According to the thermal expansion coefficients, an ideal unit cell that is embedded in a Si substrate and relaxed at 500°C, would be in tension in one direction (*c* and *a*) and in compression in the other (*b*) after cooling to room temperature. The experiments indicate that in the unknown stress–strain tensor the effect of compression along the *b* axis dominates.

A consequence of these considerations is that the unit cell dimensions in the JCPDS record [130] for NiSi, however faithfully they match room temperature measurements on thin films, are not equilibrium values. They correspond to films on Si substrates that suffer both from considerable local tensile or compressive stresses depending on grain orientation. Valid unit cell dimensions are found in bulk studies [127, 128].

5.5.3 Consequences of CTE anisotropy for stress in thin films of NiSi

If one approximates the thermal expansion of NiSi as being linear as a function of temperature, one obtains values for the CTE of 42, −43 and 34 ppm/°C along the *a*, *b* and *c* axes, respectively. As long as the crystal symmetry remains the same, the CTE along any direction (x, y, z) in the crystal can be estimated using

$$\text{CTE} = \begin{bmatrix} x & y & z \end{bmatrix} \cdot \begin{bmatrix} 42 & 0 & 0 \\ 0 & -43 & 0 \\ 0 & 0 & 34 \end{bmatrix} \cdot \begin{bmatrix} x \\ y \\ z \end{bmatrix}$$

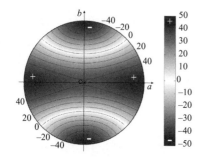

FIGURE 5.11 Calculated CTE along different directions in the NiSi unit cell. The *a* and *b* axes of the unit cell coincide with the horizontal and vertical axes on the figure, while the *c* axis extends out of the page. The pole-figure style representation shows the calculated thermal expansion along any direction in the unit cell using a grayscale code.

FIGURE 5.11 shows a graphical representation of the CTE as calculated for different directions in the NiSi unit cell. The direction of the *a* and *b* axes of the unit cell are indicated on the figure, while the *c* axis is orthogonal to the figure. Every possible direction in the crystal can be represented by a unit vector, which is then projected orthogonally onto the plane formed by the *a* and *b* axes (i.e. the plane of the paper). The colour of the projected point is determined by the calculated CTE along that direction. FIGURE 5.11 illustrates that the CTE along certain directions is actually zero, indicating that the *d*-spacing for the planes normal to these directions is independent of temperature.

When comparing the CTE values for the axes of the NiSi unit cell to known values for other silicides (typically 10 ppm/°C) and to the CTE for the Si substrate (2.6 ppm/°C), it is clear that the thermal expansion in NiSi is significantly larger. This seems in apparent contradiction with the relatively low stress that has been reported in NiSi films using wafer curvature methods [74, 85]. The low average stress (as measured by wafer curvature) at room temperature is first of all related to the low stress relaxation temperature for NiSi. For most silicides, this stress relaxation temperature is similar to the typical growth temperature at which the phase forms in a reasonably short time in thin film experiments. Since the stress relaxation temperature for NiSi is much lower than that of $CoSi_2$ or $TiSi_2$, less thermal stress can build up while cooling the sample back to room temperature. Second, the rate at which thermal stress builds up during cooling of the sample is determined by the product of the elastic modulus and the thermal expansion mismatch. The bulk modulus for NiSi has been reported as 132 GPa in the literature [133]. However, this value should be used with caution, since the elastic properties of NiSi are undoubtedly also strongly anisotropic. The bulk moduli are expected to be approximately inversely proportional to the thermal expansion coefficient, given that both quantities are related to the attractive part of the atomic potential within the solid: when the bulk modulus is large for a certain material, it is more difficult to increase the separation between atoms, and hence one expects a small tendency for the bond length to increase upon heating. This may explain why the slope of the increase in thermal stress when the NiSi layer is cooled down after reaction is similar to the slopes observed for other silicides, in spite of the significantly larger mismatch in thermal expansion for NiSi.

Although the average stress in NiSi films is low, the strong anisotropy of the CTE will cause significant localised stresses between neighbouring grains. Indeed, XRD observations indicate that while some grains are under tension, other grains within the same film are experiencing compressive stress. The strain energy density within a grain is determined by the product of σ (stress) and ε (strain). Since $\sigma = M\varepsilon$ (with M the modulus of the material), and $\varepsilon = \Delta\alpha\Delta T$ for thermal stress (with $\Delta\alpha$ the difference in CTE between film and substrate, and ΔT the difference between the current temperature and the stress relaxation temperature), the strain energy density is given by $M(\Delta\alpha\Delta T)^2/2$. In the case of anisotropic materials, the modulus M and/or the CTE α are actually tensors. As a consequence, the strain energy density within each individual grain of an anisotropic material is dependent on the orientation of the grain with respect to the substrate. For metal films (e.g. Cu [134, 135]), anisotropy in the modulus M has been proposed as a possible texture selection mechanism since differences in strain energy between neighbouring grains may provide a driving force for well-oriented grains to grow faster, thereby influencing the texture of the film after annealing. In view of the strong anisotropy for the CTE for NiSi, one could expect that a similar mechanism will influence the texture of the NiSi layer. Although the next paragraphs will discuss a peculiar type of texture that was observed for NiSi films on a single crystal Si substrate, similar experiments on poly-crystalline Si or amorphous Si did not provide evidence of strong preferential orientation, suggesting that the differences in strain energy between neighbouring grains do not provide a sufficient driving force to significantly influence the measured texture of the NiSi film.

5.6 TEXTURE OF NiSi FILMS ON SINGLE CRYSTAL Si

When performing *in situ* diffraction experiments to study the formation kinetics of nickel silicide on Si(001), it is surprising to observe the relatively low intensity of the NiSi peaks (see FIGURES 5.4 and 5.5). However, when moving away from the Bragg–Brentano condition, either by rocking θ or χ, diffraction conditions can be found for which the NiSi(103) peak is significantly more intense. This shows that a NiSi film on a Si(001) substrate does not have a random (powder-like) distribution of grain orientations, but that there is a strong tendency for the grains to align in a certain way with respect to the substrate, that is, NiSi films are "textured" [136].

5.6.1 Pole figures and orientation distribution functions

"Texture" is the statistical distribution of the orientation of grains in a polycrystalline material [137–140]. When positioning a grain on a substrate, there are three rotational degrees of freedom to choose the orientation of the grain, and three translational degrees of freedom to choose its location. The orientation distribution function (ODF) provides the complete description of the percentage of grains in the material that are oriented in a certain way. It basically consists of a three-dimensional histogram as a function of three variables describing the orientation of the grains (e.g. Euler angles). Because of its intrinsic three-dimensional nature, an ODF is not easily interpreted visually. An alternative, though less complete, representation of the texture of a material is provided by pole figures.

A pole figure for a (hkl) plane depicts the statistical angular distribution of the direction of the normal to this plane (pole). One can imagine placing the sample in the centre of an imaginary hemisphere and marking the intersection between the normal to the chosen (hkl) plane and the sphere for each grain in the film. If one projects the density of marks on the sphere onto a planar surface, one obtains the *(hkl)* pole figure. This procedure is illustrated in FIGURE 5.12. The angles χ and ϕ are defined as spherical coordinates (azimuth and elevation) in the hemisphere.

Texture is frequently studied using XRD. The sample is mounted on a four-circle diffractometer. By fixing the sample and detector at a given θ and 2θ angle, one fixes the d-spacing of the crystallographic plane in the film for which diffraction will be detected. The pole figure is then obtained by rotating the sample around the axis that is normal to its surface (ϕ scan) and around the axis that is formed by the intersection of the sample surface and the plane defined by the X-ray beam and detector (χ scan).

Texture measurements are very time-consuming when using a standard scintillation detector and a laboratory based X-ray source. Especially for the interesting case of thin films (with a film thickness less than 40 nm for present-day silicides), it can take several days to measure a single pole figure. Measuring a detailed pole figure (e.g. with steps of 0.5° in χ and ϕ) is impractical. One way to deal with this experimental difficulty is the use of a two-dimensional area detector and software to translate the intensity differences that are observed along the length of a Debye ring into an orientation distribution function and hence into pole figures for different *(hkl)* planes. Another approach is to use a standard scintillation detector, but to improve the measurement time by using high intensity synchrotron radiation.

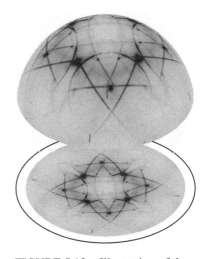

FIGURE 5.12 Illustration of the projection method used to obtain a planar representation of the NiSi(112) pole figure for a 60 nm NiSi film on Si(001).

5.6.2 Standard classification of texture in thin films

In literature, texture in thin films has been classified into three different categories: random, fibre and in-plane texture [141, 142]. For the case of random texture, no restrictions are imposed on grain orientation, and all pole figures have a homogeneous intensity. For fibre texture, an *(hkl)* plane is parallel to the surface (with its normal fixed at $\chi = 0°$), leaving a single degree of freedom to rotate the grains around the fibre axis when positioning them on the substrate. This single degree of freedom results in circles centred around the middle of the pole figure. For in-plane alignment, the orientation of the grain is completely determined by the substrate, resulting in well-defined spots at certain locations on the pole figure.

5.6.3 Pole figures for NiSi films on Si(001)

Although complete pole figures have been measured for $CoSi_2$ [143] and C54-$TiSi_2$ [144], no pole figures have been reported in the literature for NiSi. However, a considerable amount of work

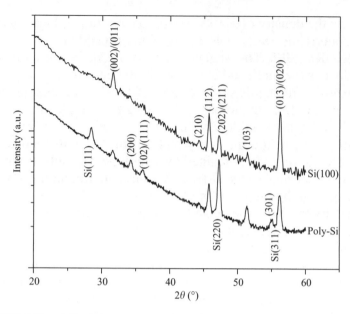

FIGURE 5.13 θ/2θ XRD measurements for NiSi films on Si(001) and poly-Si substrates. The peaks are indexed according to orthorhombic NiSi (JCPDS# 73-1843).

has been reported on the effect of alloying elements like Pt and Pd on improving the thermal stability of the NiSi phase. It has been claimed that the improved thermal stability that is achieved when adding these alloying elements is related to the strongly textured growth of the NiSi layer [145]. Other recent work reports on textured growth of NiSi on SiGe substrates [146–148].

One can usually quickly determine whether there is a strong preferential orientation in a film by performing a standard θ/2θ XRD measurement. Any significant discrepancy between the XRD peak intensity ratios from a powder sample and from the thin film under study constitutes a tell-tale sign of a textured film. FIGURE 5.13 shows θ/2θ measurements for NiSi formed on poly-crystalline Si, for which no in-plane alignment with the substrate is possible, and single-crystalline Si(001). At first sight, one does not detect a strong texture in the NiSi thin films, since in both cases most of the expected XRD peaks for a powder sample are present. However, their relative intensity is not exactly the same, indicating some degree of preferential alignment.

The (112) and (103) pole figures for a film of 60 nm NiSi on a Si(001) substrate are shown in FIGURE 5.14. As mentioned in the previous section, pole figures for thin films typically consist of simple geometric features like spots or circles. The pole figures for NiSi consist of a surprisingly detailed set of complex though symmetrical line patterns. From the complexity of these pole figures, it is immediately clear that the texture of NiSi films cannot be

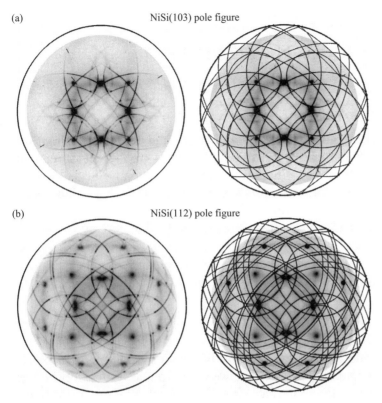

FIGURE 5.14 (a) NiSi(103) and (b) NiSi(112) pole figures measured for a 60 nm NiSi film formed on Si(001). The left image shows the measured data, while in the right image calculated patterns for NiSi⟨202⟩//Si⟨220⟩ and NiSi⟨211⟩//Si⟨220⟩ were overlaid on top of the measurement.

categorised within the standard classification of random, fibre or in-plane texture. The circular features observed on the spherical representation of the (103) pole figure (FIGURE 5.14) suggest that the lines that are observed in the planar representation of the pole figures are related to a fibre-like texture with the fibre axes at $\chi = 45°$ (instead of normal to the substrate, that is, at $\chi = 0°$ as is expected for a standard fibre texture). Indeed, it was found that the complex patterns on the pole figures are generated by

1. an off-normal NiSi(101) fibre texture with the fibre axis at $\chi = 45°, \phi = 45°$;
2. an off-normal NiSi(211) fibre texture with the fibre axis at $\chi = 45°, \phi = 45°$;
3. a standard in-plane texture, with the NiSi(200) plane nearly parallel to Si(010) and the NiSi(014) plane nearly parallel to Si(001).

Apart from these three most intense components, two weaker features could also be observed when plotting the data using a logarithmic grayscale:

4. an off-normal NiSi(103) fibre texture with the fibre axis at $\chi \sim 41°$, $\phi = 45°$;
5. an off-normal NiSi(112) fibre texture with the fibre axis at $\chi \sim 47°$, $\phi = 45°$.

The calculated pole figures for the three most intense texture components are overlaid on top of the measured data in FIGURE 5.14, illustrating the good fit between calculated and measured patterns. It should be noted that the calculations should not be considered as fitting, since they are purely based on geometry, without invoking any adjustable parameters.

5.6.4 Axiotaxy: a new type of texture

The mechanisms that have been reported to act as a driving force for texture formation include minimisation of surface, interface and strain energy. However, none of these mechanisms seem to provide a valid explanation for our observations, since they all require that a fixed crystallographic plane in the film be parallel to the surface/interface. For the case of off-normal fibre texture, the fact that the fibre axis is not normal to the substrate means that the surface and interface consist of different crystallographic planes for different rotations around the fibre axis. An important clue to explain why the NiSi grains have this strange texture is provided by our understanding of the crystallographic orientation of the textured grains: it was shown that the NiSi(202) and (211) planes are parallel to Si(220) planes. The distance between NiSi(202) planes is 1.921 Å, while the distance between NiSi(211) planes is 1.919 Å. Both values are within 0.06% of the spacing of 1.9201 Å between Si(220) planes. The almost identical d-spacing between planes in the film and substrate results in the formation of an interface which has a periodic structure in one dimension. A periodic arrangement at the interface (although only along a single direction) may allow interface reconstruction to occur, thus decreasing the interface energy and providing the driving force for the grains in the film to align the NiSi(202) or (211) plane parallel to the Si(220) plane (see FIGURE 5.15).

Since the texture that causes the pattern of lines on the pole figures is characterised by a periodic arrangement of atoms at the interface in only one direction or along one axis, we proposed the terminology "axiotaxy" [136]. The non-uniform diffracted intensity along the circular features on the pole figures is related to a varying degree of lattice matching along the other direction.

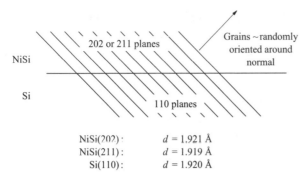

NiSi⟨202⟩: $d = 1.921$ Å
NiSi⟨211⟩: $d = 1.919$ Å
Si⟨110⟩: $d = 1.920$ Å

FIGURE 5.15 Illustration of the principle of axiotaxy. The pattern of lines on the pole figures is created by grains for which the NiSi⟨211⟩ or ⟨101⟩ planes are preferentially aligned parallel to Si⟨110⟩. As a consequence of the constraint that a set of planes in the film is preferentially parallel to a set of planes in the substrate, the texture manifests itself as though it were a fibre texture, with the fibre axis normal to Si⟨110⟩ (i.e. at 45° from the surface normal). Since the spacing between the NiSi planes is almost identical to the value of 0.192 nm for Si(220) planes, the resulting interface structure is periodic along a Si(100)-type direction.

5.7 HIGH TEMPERATURE LIMITATIONS

The initial sections of this chapter described the properties, the formation mechanisms and characteristics of Ni silicide phases that are stable at room temperature. As the low resistivity monosilicide phase is the relevant one for microelectronics applications, the unexpectedly large anisotropy in the thermal "expansion" and the texture of films were carefully analysed in Sections 5.5 and 5.6. The current section deals with the degradation of NiSi films at "high" temperature. It is crucial for the film to retain its integrity and low resistance to as high a temperature as possible to allow for further anneals after silicide formation. There are two mechanisms by which the NiSi film can degrade at high temperature. In the first one, the NiSi film transforms to $NiSi_2$. In the second one, the film suffers morphological degradation through grain boundary grooving and agglomeration.

5.7.1 Formation of $NiSi_2$

As can be seen on the Ni/Si phase diagram (FIGURE 5.1), the NiSi phase is not at equilibrium with Si. At high temperature, it is expected that the NiSi film will react with Si to form $NiSi_2$. Fortunately, the ΔH for the reaction of NiSi and Si is very small resulting in a high nucleation barrier for the formation of $NiSi_2$. This formation is observed with XRD in FIGURES 5.4 and 5.5 through the appearance of the $NiSi_2(220)$ diffraction line at about 800°C. The sudden appearance of the phase at high temperature is a typical signature of a nucleation-controlled formation. There are three main disadvantages to the appearance of this phase. The first one is that the phase is three times as resistive as NiSi. The second one is that it consumes twice as much Si, and it is much rougher as it forms through a nucleation-controlled reaction. Because of high Si consumption and roughness, the $NiSi_2$ may come in contact with the buried oxide of the SOI substrate leading to a reduction of the interfacial area between the silicide and the Si effectively reducing the contact area for the device. Note here that the resistance of the film itself only increases by 50% when $NiSi_2$ forms since the threefold increase in resistivity is counterbalanced by a twofold increase in thickness. The third disadvantage is that any part of the film transforming to $NiSi_2$ will become epitaxial with the underlying Si substrate. The crystal structures and dimensions of the unit cells for $NiSi_2$ and Si are very similar. Both crystals are cubic and their lattice constants at room temperature are 5.431 and 5.416 Å (see TABLE 5.1) for Si and $NiSi_2$, respectively. The difference in lattice constant becomes even smaller with increasing temperature since the thermal expansion of Si is smaller than that of $NiSi_2$.

In the case of epitaxial formation, the growth of these grains will not be limited to the open contact areas but will tend to simply follow the crystal directions of the underlying substrate leading, for example, to encroachment under sidewall spacers or to the formation of epitaxial facets that increase interface roughness. This was initially observed for epitaxial grains of $CoSi_2$ [76, 77] and was one of the reasons why the industry shied away from the otherwise promising epitaxial silicides [97, 149, 150]. If any $NiSi_2$ forms on Si, one can expect the epitaxy problem to be more serious than for $CoSi_2$ since the lattice mismatch between $NiSi_2$ and Si is smaller (0.4%) than that of $CoSi_2$ and Si (1.3%).

5.7.2 Morphological stability

Beside formation of $NiSi_2$, the NiSi film can also degrade at high temperature through grain boundary grooving and agglomeration. The minimisation of total energy at high temperature leads to separation of the film into more spherical islands for which there is an overall decrease in surface or interface areas for the same volume of NiSi and therefore a decrease in surface and interface energies. Although this morphological instability was not expected to be dominant at first because the early studies were performed with thicker Ni films [52, 114, 151–155], for thinner films relevant to the current microelectronic devices, the agglomeration of NiSi becomes the main degradation mechanism at high temperatures [66, 81, 156]. As discussed earlier, the tendency to agglomerate is worsened because NiSi layers are thinner than their $CoSi_2$ counterparts (for the same sheet resistance) and because of the lower melting point of NiSi. The first evidence of this degradation was observed in FIGURE 5.5 for a 15 nm Ni film deposited on a poly-Si substrate and annealed at 3°C/s in purified He. Even though $NiSi_2$ is forming at about 800°C, it is clear from the two light scattering signals and the resistance measurement that the film has already started to roughen by 700°C. In the next two subsections, we cover the dependence on film thickness and annealing ramp rate of both the agglomeration of NiSi and the formation of $NiSi_2$.

5.7.2.1 Dependence on film thickness

In order to determine which of these degradation mechanisms occurs first, we studied the stability of NiSi films as a function of silicide layer thickness. FIGURE 5.16 shows the *in situ* XRD data for annealing 5, 10, 15 and 30 nm Ni films on poly-Si at a constant heating rate of 3°C/s. For all thicknesses, the NiSi(103) peak (near $2\theta = 60°$) is clearly visible. From the XRD data, one can easily determine the formation of $NiSi_2$, since the NiSi peaks disappear and the $NiSi_2$(220) peak appears on top of the Si(220) peak from the poly-Si substrate (near $2\theta = 55°$). While the change from the NiSi to the $NiSi_2$ phase occurs over a narrow temperature range for the thicker Ni films (15 and 30 nm), the phase formation is much more gradual and occurs over a temperature range larger than 100°C for the two thinner films. Solid black markers have been superimposed on top of the XRD data to roughly indicate the temperature at which $NiSi_2$ is observed to form.

Agglomeration of the NiSi is most easily observed using the *in situ* sheet resistance data (FIGURE 5.17). As expected, thinner films agglomerate much faster than thicker ones. For this work, the agglomeration temperature was arbitrarily selected as the temperature at which the sheet resistance has doubled from its minimal value. The dotted markers included on the XRD data of FIGURE 5.16 indicate this agglomeration temperature. It is immediately clear that for Ni layers with a thickness below 15 nm, the agglomeration of the NiSi will occur long before any $NiSi_2$ nucleates. From the viewpoint of CMOS applications, agglomeration is the failure mechanism that defines the upper limit of the temperature range in which a NiSi layer may be annealed after formation.

According to the data in FIGURE 5.16, the film thickness not only affects the agglomeration temperature, but also strongly influences $NiSi_2$ formation. It is known that if the formation of a phase is controlled by nucleation, the formation temperature is dependent on the geometry and thickness of the film into which

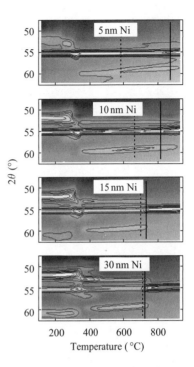

FIGURE 5.16 *In situ* XRD data for 5, 10, 15 and 30 nm Ni on poly-Si for ramping at 3°C/s. The solid black markers indicate $NiSi_2$ formation, while the dotted markers indicate the agglomeration temperature (i.e. at which the sheet resistance has doubled from its minimal value, see FIGURE 5.3).

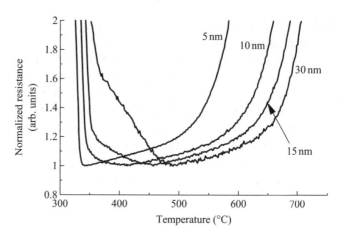

FIGURE 5.17 Normalised sheet resistance versus annealing temperature for 5, 10, 15 and 30 nm Ni on poly-Si. XRD data for the same samples are shown in FIGURE 5.16.

the phase nucleates. For example, if the density of nuclei is constant per unit volume of NiSi, as the film becomes thinner, the density of nuclei per unit area decreases and the nucleation temperature increases. A well-known example is the formation of C54-TiSi$_2$ from C49-TiSi$_2$, where the nucleation temperature increases for both thin films and narrow lines. Therefore, an increase of the nucleation temperature of NiSi$_2$ for decreasing layer thickness would not be such a surprise. However, the slow increase in the NiSi$_2$(220) intensity and slow decrease in the NiSi(103) peak intensity for the 5 and 10 nm Ni films in FIGURE 5.16 is unexpected a priori and suggests that the rise in NiSi$_2$ formation temperature with film thickness has a different origin than a decrease in nucleation density. For a nucleation controlled reaction, one expects a very fast growth rate after nucleation. This is especially true for the case of NiSi$_2$, since the melting point of 992°C is close to the nucleation temperature. This apparently slow formation of the NiSi$_2$ in thin films is related to the agglomeration. Indeed, for thin films, by the time nucleation of NiSi$_2$ becomes possible, the NiSi layer is severely agglomerated. The two limiting cases of NiSi$_2$ growth are schematically described in FIGURE 5.18. While the nucleation in a uniform NiSi film can lead to the rapid growth of NiSi$_2$, the growth of NiSi$_2$ in an agglomerated NiSi film is limited to the NiSi island. To transform, each island must contain a nucleation centre. As the dimensions of the islands are of the order of the film thickness, a complete formation of NiSi$_2$ requires a variety of nucleation sites to become active and this can only occur gradually and at much higher temperature.

Nickel silicide technology

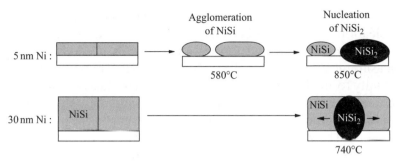

FIGURE 5.18 Schematic overview of nucleation of NiSi$_2$ in a thick NiSi film (where lateral growth of the NiSi$_2$ nucleus may transform the surrounding NiSi grains) as compared to a thin NiSi film, where the film agglomerates below 600°C, and a nucleus of NiSi$_2$ has to be formed in each separate NiSi island.

5.7.2.2 Dependence on annealing ramp rate

To explore the kinetics of a solid-state reaction, isothermal anneals can be performed to look at the time dependence of the formation at different temperatures. A time-efficient alternative is the Kissinger analysis, for which the temperature is ramped at a constant rate, and a certain physical property is measured *in situ* (e.g. sheet resistance or XRD peak intensity). If one performs this type of ramping experiments for a wide range of ramp rates, it is possible to extract activation energies [93, 157]. From the data of FIGURE 5.19, the Kissinger analysis can be performed to determine the apparent activation energy for the formation of NiSi. It is clear that the temperature at which the NiSi(103) peak starts to appear increases for increasing ramp rate. This can be clearly seen using the vertical dashed line across FIGURE 5.19. The appearance of the (103) peak in FIGURE 5.19(d) for the highest ramp rate occurs with a delay of about 50°C by comparison with the slowest temperature ramp rate. This is expected, since for higher ramp rates, the total thermal budget to which the sample has been exposed to reach a certain temperature is smaller. Kissinger type analysis for NiSi formation results in an activation energy of about 1.5 eV, in agreement with values reported previously in the literature [32]. These measurements of activation energies are only meaningful when the formation mechanisms are well identified. In the case of NiSi$_2$ formation for example, the formation temperature decreases with increasing ramp rate as can be observed from the vertical solid lines in FIGURE 5.19. Attempting the same analysis to determine the activation energy will lead to a negative apparent activation energy, a result which is clearly unphysical. The explanation is related to FIGURE 5.18. When ramping up the temperature slowly, the film is allowed to agglomerate before any NiSi$_2$ nucleation is possible.

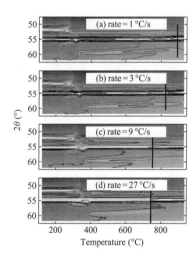

FIGURE 5.19 *In situ* XRD data for a 10 nm Ni film on poly-Si, annealed at 1, 3, 9 and 27°C/s. The markers indicate NiSi$_2$ formation.

As described above, once the film is agglomerated, a single nucleation event will only lead to NiSi$_2$ formation in a small agglomerated island and further nucleation is needed to have a measurable proportion of NiSi$_2$. Hence, for slow ramping, NiSi$_2$ is actually formed at a higher temperature than for fast ramping (solid vertical line in FIGURE 5.19) since in the latter case the film was not fully agglomerated. These observations emphasise the fact that one should be very cautious with activation energy determinations. Without a precise knowledge of the controlling mechanism of the observed phenomenon, an activation energy determination is often meaningless.

Results on film agglomeration presented in this chapter concerned NiSi films on poly-Si. The agglomeration temperature on this substrate is strongly dependent on substrate preparation. There are important effects of dopant type and concentration as well as of poly-Si grain sizes. Grain growth in the poly-Si layer is known to lead to early agglomeration of silicide films [158–160]. Since the NiSi$_2$ formation depends strongly on the agglomeration of the NiSi film, we expect to see variations in formation temperature from one substrate to the next, if prepared under different conditions. One example of this can be seen when comparing the XRD data for the 15 nm films of FIGURES 5.5 and 5.16. For these cases, only the poly-Si grain structure was nominally different with both substrates being undoped (not intentionally doped). One can see a difference of about 80°C in NiSi$_2$ formation, showing the importance of film preparation and microstructure. A poly-Si substrate with larger initial grains will retard agglomeration of NiSi and favour the formation of NiSi$_2$. Although not shown in this chapter, it is interesting to note that the agglomeration of NiSi on SOI substrates is peculiar. One would expect the NiSi films on SOI substrates to be more stable against agglomeration since the substrate is single crystal (equivalent to one extremely large grain). Although this is clearly the case for NiSi films formed from Ni films thicker than 15 nm, a different behavior is observed for very thin films. For these thinner films (<15 nm Ni), the agglomeration rate on SOI is actually faster than on poly-Si substrates. This difference may be related to the texture of the NiSi; more work is needed to investigate this in detail.

5.7.3 High temperature stabilisation of NiSi films

In the past five years, a fair amount of work was published on stabilising NiSi films against transformation to NiSi$_2$. Among the methods specified, the addition of either Pd or Pt was shown to push the formation of NiSi$_2$ to the highest temperatures [161–166]. Other techniques have been suggested, such as the implantation of nitrogen [17, 167–170], hydrogen [153] or BF$_2$ [171, 172] and the use of capping layers [173], which are proven techniques to retard the agglomeration of films. Very recently, it was shown that fluorine introduced through the implantation of BF$_2$ segregates to the NiSi/Si interface and retards agglomeration significantly [174–176]. We will here show two examples of stabilisation: the addition of Pt to the Ni/Si system and the effect of increasing the dose of implanted BF$_2$ on the agglomeration of the NiSi film.

5.7.3.1 Alloying

In FIGURE 5.20, we present the effect of Pt addition to the Ni/Si system as measured using the time resolved XRD system for alloys of Ni/Pt deposited on poly-Si substrates. Ramp anneals of (a) a pure Ni film, (b) a Ni (5 at.% Pt) film and (c) a Ni (10 at.% Pt) film were performed at 3°C/s in purified He. The formation of NiSi$_2$ is marked on each of the X-ray graphs as a large vertical bar. The first noticeable difference is the clear rise in NiSi$_2$ formation temperature as the Pt concentration in the deposited Ni-Pt film increases from 0, to 5 and 10 at.%. With the addition of Pt, it is clear that the nucleation of the disilicide phase is retarded to higher temperatures. This phenomenon can be attributed to a change in the entropy of mixing of solutions as the system transforms from the monosilicide to the disilicide [50, 51, 77, 79, 161]. The fact that Pt is miscible in NiSi and not soluble in NiSi$_2$, requires that the Pt be expelled from the growing NiSi$_2$ grain. One can describe this phenomenon with equilibrium thermodynamic arguments using entropy of mixing and classical nucleation theory. Because a solution with complete solubility possesses higher entropy, the phase change from Ni(Pt)Si to NiSi$_2$+PtSi results in a decrease in the total entropy of the system. The driving force for the disilicide formation originates from the change in free energy that depends on both the difference in enthalpy of formation and on the entropy changes ($\Delta G = \Delta H - T\Delta S$). With smaller $|\Delta G|$, the higher the nucleation barrier becomes [9, 14, 177] because of the smaller drive for the reaction to occur. Since $\Delta S < 0$ (reduction of mixing entropy), the addition of Pt results in a decrease in the absolute value of ΔG and therefore causes a reduction in the driving force to nucleate NiSi$_2$. For a more thorough description, see References 22, 50, 51, 77, 79 and 161.

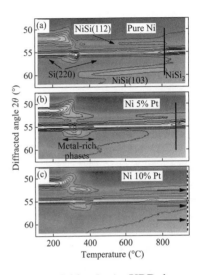

FIGURE 5.20 *In situ* XRD data for the reaction of 10 nm films of pure Ni (top), Ni(5 at.% Pt) (middle) and Ni(10 at.% Pt) (bottom). The markers indicate NiSi$_2$ formation.

In addition to raising the formation temperature of the disilicide, Pt raises and widens the temperature window over which the metal-rich silicide phases are present. With pure Ni/Si, the metal-rich phases are present from about 250 to 350°C. The addition of Pt raises the window to about 300–400°C. These relatively large temperature variations would dramatically affect the fabrication process if the first silicide anneal is performed at low temperature. Another variation observed with the addition of Pt is the change in texture in the monosilicide phase. The NiSi(112) and NiSi(103) peaks, clearly visible in FIGURE 5.20(a) for a silicide prepared from pure Ni, gradually decrease in intensity as Pt is added to the Ni film. The disappearance of these peaks shows that the statistical distribution of grain orientation is modified by the presence of Pt. As we mentioned earlier, we do expect the agglomeration of the film to be dependent on its texture. In the case of the Pt addition, we detected an improvement of the morphological stability, although this improvement was not as significant as the increase in the formation temperature of the $NiSi_2$. Moreover, recent results also show that the presence of Pt may be detrimental to device performance as measured using DLTS [178].

5.7.3.2 BF$_2$ implantation

An unusual way to stabilise NiSi against agglomeration is to use BF$_2$ as the source of p dopant, as was shown recently by Wong et al. [174]. They observed that the segregation of the fluorine at the NiSi/Si interface retards agglomeration significantly. This stabilisation can be illustrated using *in situ* resistance measurements. FIGURE 5.21 shows four resistance traces as a function of temperature for increasing doses of implanted BF$_2$. Although the resistance already starts to increase at about 600°C for the undoped poly-Si sample, for implantation concentrations of BF$_2$ at 1×10^{15} cm^{-2}, the start of agglomeration is pushed up by about 100°C. Increasing the concentration of BF$_2$ to 4×10^{15} cm^{-2} delays the start of thermal degradation to about 800°C. This 200°C rise in process window is extremely large and shows that either diffusion at the NiSi/Si interface or interfacial energy itself can be very well controlled. By modifying the silicide through addition of elements or impurities, the properties can be adjusted to withstand the temperature exposure necessary for the processing of contacts and interconnections.

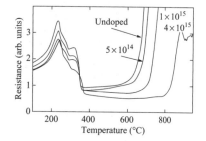

FIGURE 5.21 Resistance traces as a function of temperature for undoped and three doses of implanted BF$_2$. Increases in the concentration of BF$_2$ delay the start of NiSi thermal degradation.

5.8 DEVICE CHARACTERISATION

In this section, we highlight the advantages and challenges of a NiSi process using both electrical and physical characterisation of devices in narrow dimensions. As mentioned above, the inherent advantages of NiSi over its predecessors, $TiSi_2$ and $CoSi_2$, include the lower resistivity, lower Si consumption, lower thermal budget, lower contact resistance and lack of increase in sheet resistance with narrower poly-Si linewidths. One should not infer from this that NiSi is easily inserted in place of CoSi$_2$ since multiple challenges are associated with its use in device fabrication, ranging from the undesired agglomeration of NiSi films and the formation of NiSi$_2$ at low temperatures, to some excessive Ni diffusion either on narrow Si(100) areas or poly-Si lines leading to junction leakages that may become abnormally high [59, 82].

5.8.1 Addressing the device roadmap

In the device development community, the use of NiSi is viewed in terms of the technology node at which it can be inserted and in terms of a product upon which it can be qualified. As the device community continues to scale down the devices, the requirements for silicide contacts become more stringent. The integration challenges for the silicide are best described by the International Technology Roadmap for Semiconductors (ITRS) [179] and are presented in TABLE 5.3. This table highlights important device requirements (first column) related to the device contacts for the 130, 90 and 65 nm technology nodes (defined here as the half-pitch of the microprocessor unit – MPU), which have reached or will reach volume manufacturing in 2002, 2004 and 2007, respectively. Integrated circuit manufacturers will target NiSi for a specific node, plan the development accordingly and design the product with the properties of this new material in mind. Since the 130 nm node is already in volume manufacturing, the next technology node is that at 90 nm, and several prominent manufacturers of integrated circuits have suggested that NiSi could be part of production at that point [54, 55, 59, 61, 64, 180]. Other device manufacturers are investigating the insertion of NiSi at the 65 nm node [58, 63, 65].

TABLE 5.3 Comparison of the key characteristics affecting silicide formation for the current and the upcoming two technology nodes.

Key characteristics	Technology node		
	130 nm	90 nm	65 nm
Gate length (nm)	53	37	25
Silicide thickness (nm)	29.2	20.4	13.8
Max. Si consumption (nm)	19–38	13–26	9–18
Silicide sheet resistance (Ω/sq)	5.1	7.4	10.9
Max. contact resistivity ($\Omega\,cm^2$)	3.2×10^{-7}	2.1×10^{-7}	1.1×10^{-7}
Junction depth (nm)	39–78	27–45	18–37
Active doping (cm^{-3})	9.2×10^{19}	1.5×10^{20}	1.87×10^{20}

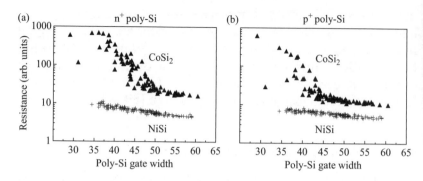

FIGURE 5.22 Resistance of patterned poly-Si lines as a function of linewidth for $CoSi_2$ and NiSi. $CoSi_2$ on n-doped (*a*) and p-doped (*b*) poly-Si resistors exhibit a resistance increase at gate dimensions less than 45 nm while NiSi resistance is independent of width.

Insertion of NiSi at a particular technology node implies that NiSi must be compatible with products and meet performance requirements (TABLE 5.3). One of the most daunting challenges to NiSi may be the aggressive shrinking of the poly-Si gate length. It was shown recently that the performance and yield of NiSi on narrow poly-Si gate is by far superior to $CoSi_2$ [54, 59, 61, 65, 181, 182]. This is shown in FIGURE 5.22, using data from Reference 182, in which the resistance of silicided poly-Si gates is plotted against the physical gate length. For $CoSi_2$, the resistance is stable for larger poly-Si lines, while in narrower dimensions, the process leads to much higher resistance in a relatively noisy fashion (flyers). The dimension at which the $CoSi_2$ process becomes limited depends on preparation conditions and on the type of structures measured. In the current case, it is about 50 nm for the n-doped poly-Si, while it reaches down to 45 nm for the p-doped structures. As discussed earlier, the resistance increase is linked with the presence of voids or openings in the lines, possibly originating from agglomeration, stress effects or Kirkendall voiding (see Section 5.4.3.2). It is clear that for the NiSi process, no degradation in resistance is observed, suggesting that gate patterning is not limiting the silicide formation. Note that the gate widths in FIGURE 5.22 are measured close to the gate oxide since this is the important value for devices. Given the typical 10–20 nm gate tapering originating from the gate patterning and etch process, the silicided area of the polysilicon lines are actually smaller than 35 nm at the top of the gate. This dimension is close to the expected gate dimension of 25 nm at the 65 nm technology node. Note that the advantage of using NiSi on such narrow poly-Si gate may become irrelevant if the poly-Si gate material is replaced by a combination of metals or conductive nitrides (predicted for 2007 [179]). It is worth mentioning that another process leading to more

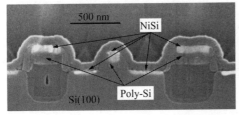

FIGURE 5.23 An example of the "reverse linewidth effect". Large poly-Si features and wide active Si areas show NiSi of expected thickness while the NiSi is twice as thick on narrow poly-Si gates. The enlargement of the edge of a poly-Si area shows that the rapid diffusion of Ni from the reservoir of Ni over the spacer can enhance the formation of NiSi along the vertical edge of the gate.

conductive gates consists of the complete silicidation of the poly-Si structures. In this way, the silicide becomes in contact with the gate oxide or the high dielectric constant material. Feasibility of both fully silicided $CoSi_2$ and NiSi gate conductors has been demonstrated [183–185].

The following three requirements of TABLE 5.3 relate to silicide thickness, Si consumption and sheet resistance. Although a NiSi film of equivalent resistance consumes significantly less Si than $CoSi_2$, narrow dimensions may behave differently because of the possible "reverse linewidth effect" described earlier in Section 5.4.4 and in FIGURE 5.9. An example of this effect is shown in FIGURE 5.23. Excessively thick NiSi was found on narrower poly-Si gates or other isolated narrow Si features where excess Ni could diffuse through the NiSi into the Si from the surrounding area. In the current figure, the NiSi film on the narrow poly-Si gate is close to twice a thick as the film on the larger poly-Si area in the vicinity. The enlargement of the edge of a poly-Si area in the same figure also shows that the rapid diffusion of Ni from the reservoir of Ni over the spacer can enhance the formation of NiSi along the vertical edge of the polysilicon gate.

5.8.2 Shallow junctions

A clear example of the difference in the formation of NiSi on poly crystalline or single crystalline Si is illustrated in the device cross-sectional image in FIGURE 5.23. The relative thickness of NiSi on Si(100) areas is barely half of the thinnest NiSi films on poly-Si regions. As more grain boundaries are present in NiSi films formed on poly-Si substrates, we expect a much faster diffusion of Ni during the phase formation for this substrate type (see diffusion coefficients in TABLE 5.2) and a difference in formation temperature for each of the phases. As a result, the judicious optimisation

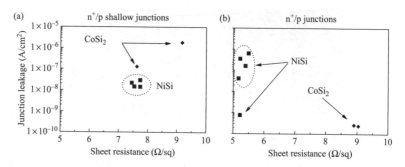

FIGURE 5.24 A comparison of leakage data for $CoSi_2$ and NiSi on shallow junctions. For a similar silicide sheet resistance (a), using NiSi leads to at least an order of magnitude decrease in leakage compared to the best $CoSi_2$ contact, for these ultra-shallow junctions [54, 181]. During process optimisation for the reduction of leakage (b) [181], the sheet resistance of NiSi films is more than 30% lower than the $CoSi_2$ controls suggesting that the NiSi film is thicker and consequently more subject to junction leakage.

of a process may lead to thicker silicide on the poly-Si areas and less silicide on active junctions. Such a process would be a clear advantage for devices built on SOI substrates or very shallow junctions as the limited Si consumption is critical for the source and drain areas of transistors.

The very thin and smooth NiSi film on active areas should lead to lower junction leakage in shallow junction devices. A comparison of leakage data for $CoSi_2$ and NiSi on shallow junctions is presented in FIGURE 5.24(a). This data originates from the work of Kittl et al. [54, 181] and has been presented recently. It is clear that for a similar sheet resistance, the use of NiSi leads to at least an order of magnitude decrease in leakage compared to the best $CoSi_2$ contact. The high leakages of $CoSi_2$ films are not a signature of typical leakage in state-of-the-art devices since the junctions are made extremely shallow for the purpose of evaluating the leakage of NiSi films.

The high diffusivity of Ni at low temperature that leads to thicker NiSi films in narrow dimensions also can lead to large increases in device leakage. Reported results [182] on process optimisation for the reduction of leakage are reproduced in FIGURE 5.24(b). For this set of data, the sheet resistance of NiSi films is more than 30% lower than the $CoSi_2$ controls suggesting that the NiSi film is thicker and consequently more subject to junction leakage. Even in these conditions, the best NiSi films compared well with the optimised $CoSi_2$ process as both films showed leakage currents smaller than 10^{-9} A/cm^2. With a process that is not optimised some films show as much as three orders of magnitude larger leakage currents. One of the possible explanations for these increases is the excessive formation of NiSi on the edge of active areas. This phenomenon

Nickel silicide technology

is illustrated in the TEM image presented in FIGURE 5.25. The sample has been decoratively etched to reveal the junction depth. While both the junction and the NiSi reach deeper at the edge of the shallow trench isolation (STI), the NiSi comparatively reaches further so that it is getting extremely close to touching the bottom of the junction. As the NiSi tends to form preferentially along the edge of the STI, this excessive Ni diffusion leads to junction leakage.

5.8.3 NiSi$_2$ and resistivity issues

The limitations on silicide thickness, silicon consumption and sheet resistance also require that NiSi$_2$ does not form either globally or locally in narrow structures. Although the nucleation of NiSi$_2$ does not normally occur at temperatures below 800°C in blanket films, it is presumed that the stresses that are generated by the device structures can locally lead to some formation of the disilicide. FIGURE 5.26 represents an example of such formation. The silicide film was formed by annealing at 550°C for 60 s. Using TEM and selective area electron diffraction (SAED), it was confirmed that most of the film was NiSi, while at the end of one of the structures some disilicide (cubic structure) had formed locally. This result stresses further the need for a reduction of thermal budget both during and after NiSi formation.

Another challenge related to the silicide thickness and sheet resistance given in TABLE 5.3, resides in the resistivity of the NiSi

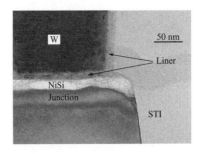

FIGURE 5.25 This cross-sectional TEM sample has been decoratively etched to reveal the junction depth. Even though the depth of both the junction and the NiSi increase at the edge of the shallow trench isolation (STI), NiSi tends to form preferentially along the edge of the STI, leading to junction leakage.

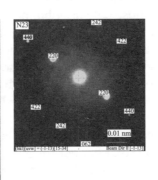

FIGURE 5.26 This cross-sectional TEM and corresponding micro-diffraction analysis reveals NiSi$_2$ forms on narrow poly-Si lines after an anneal to 550°C for only 60 s.

Nickel silicide technology

thin films. It is clear that the material is expected to maintain a resistivity of around $15\,\mu\Omega\,\text{cm}$ (product of thickness and sheet resistance) as the device scaling continues. Maintaining the resistivity in much thinner film is a challenge as the scattering from surfaces and small grains becomes important. Efforts will need to focus on the roughness of surfaces and interfaces as well as the grain structure and texture of the silicides in order to limit charge carrier scattering and maintain a low resistivity.

5.8.4 Dopant segregation

According to the ITRS roadmap, NiSi will be introduced in NMOS and PMOS devices which will exhibit considerably higher levels of doping. We mentioned earlier (Section 5.4.2.2 and [26, 73, 84, 111–114]) that the levels of dopants at the interface are critical for establishing a low contact resistance and that for NiSi the contact resistance was considerably higher for B-doped substrates. Recent results suggest that the contact resistance of NiSi on p-doped substrates is lower than that of Co silicided junctions. Such an example is presented in FIGURE 5.27 from the work of Kittl et al. [54, 181]. In this context, contact resistance refers to the resistance of the silicide to Si interface. This resistance contributes to the total source to drain parasitic resistance of the transistor. Measurement of the silicide contact resistance is not trivial and requires special test structures to distinguish the contribution of the unsilicided active resistance from the silicide to active contact resistance. These data are obtained from transmission line structures (or contact front resistance test structures) [111] (Section 5.3.4.2) and are given as the product of the contact resistance (R_{co}) and the contact width (W). The dependence of the contact resistance of NiSi and $CoSi_2$ on PMOS structure is presented as a function of formation temperature. NiSi is clearly advantageous compared to $CoSi_2$. Origins of this advantage may reside in the material characteristics themselves but are most likely the result of the lower thermal budget for NiSi formation. As the annealing temperature is reduced, the dopants at the interface do not have the possibility to deactivate.

The segregation of the dopants at the surface and interface is believed to be a significant advantage in lowering the contact resistance. An example of As segregation during NiSi formation is shown in FIGURE 5.28 using Auger depth profiling. The higher As concentration at the interface is undoubtedly an advantage. However, we have observed that the accumulation of As at the surface may be detrimental in certain conditions. Elevated levels of As at the surface accelerate the oxidation of NiSi during processing

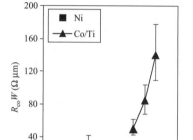

FIGURE 5.27 Contact resistance as a function of RTP temperature for $CoSi_2$ and NiSi, from the work of Kittl et al. [54, 181]. In this figure, contact resistance refers to the resistance of the silicide to Si interface. The contact resistance of p-doped substrates with NiSi is lower than those with $CoSi_2$.

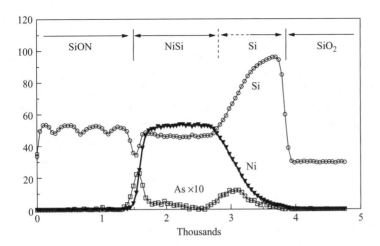

FIGURE 5.28 Auger depth profiles for a NiSi film formed on a Si substrate with As doping. The As concentration is higher at the NiSi/Si interface.

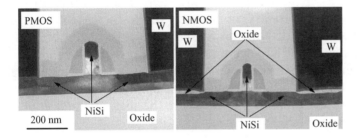

FIGURE 5.29 Cross-sectional TEM images of a PMOS (left) and NMOS (right) transistor fabricated with NiSi. Anomalously thick oxides can be found at the bottom of contact landings on As-doped Si (right) due to enhanced oxidation of the NiSi in the presence of As.

after silicide formation. As a result, without an optimised cleaning process, contact chain structures on As-doped NiSi are prone to displaying unacceptably high contact resistance. As shown in FIGURE 5.29, anomalously thick oxides can be found at the bottom of contact landings on As-doped Si, because of enhanced oxidation of the NiSi in the presence of As.

5.9 CONCLUSIONS

The use of NiSi as contacts to source, drain and gate regions of CMOS devices depends first on the possible extendibility of $CoSi_2$ as the device dimensions continue to decrease, as the available Si thickness for the formation of the silicide is further limited and as the substrate material may be modified with the addition of Ge.

The limitations on the current $CoSi_2$ process associated with these dimensional and material constraints were described. Although some clever modifications to the process or slight adjustment of the material may extend the use of $CoSi_2$ contacts for a limited number of generations, fundamental material limitations will demand the introduction of either a raised-source–drain process or the implementation of a new contact material such as NiSi.

Among the advantages of NiSi contacts are the lower thermal budget, the lower resistivity and lower Si consumption, the formation being controlled by diffusion of Ni instead of nucleation, and the possible formation of a low resistivity phase on SiGe substrates. Special attention was drawn both to the anisotropy in thermal expansion of NiSi and to the unexpected texture in films on single crystal Si. As many properties of NiSi differ from its $CoSi_2$ counterpart, there are serious challenges to the use of this material. Such challenges reside in the more complex phase formation sequence at low temperatures, in the control and limitation of Ni diffusion into the Si, in the avoidance of high temperature formation of $NiSi_2$ and most importantly, in the necessity to increase the morphological stability of NiSi on Si substrates. The addition of Pt and implantation of BF_2 were reviewed as possible ways to increase the temperature process window for NiSi. These examples demonstrate that a good understanding of the material set is required to increase the yield to a level where the implementation of a NiSi process in products becomes a reality.

ACKNOWLEDGMENTS

The authors thank Cyril Cabral, Roy Carruthers, Robert Purtell, Patricia O'Neil, Cedrik Coia, Martin Tremblay, Simon Gaudet, Patrick Desjardins and Jean Jordan-Sweet for contributions to the experiments and stimulating discussions. We are very grateful to François d'Heurle and Jim Harper for sound scientific guidance and unfailing support. Thanks to François d'Heurle again and to Prof. Katy Barmak for thorough reviews of this manuscript that led to major modifications and surely deeper understanding. Special acknowledgments are given to Jorge Kittl of Texas Instruments for stimulating discussions and for the access to his electrical results for the device characterisation section. The synchrotron XRD experiments were conducted under DOE contract DE-AC02-76CH-00016. C. Detavernier acknowledges the Fonds Voor Wetenschappelijk Onderzoek-Vlaanderen (FWO) for financial support.

The authors would also like to thank the following collaborators from AMD for valuable technical discussions and data input: Simon Chan, Witek Maszara, Thorsten Kammler and David Brown. We also acknowledge the following collaborators for providing analytical results and interpretation: Paul King (AES), Holger Saage (TEM), Hans-Juergen Engelmann (TEM), Liliana Thompson (TEM), Heiko Stegmann (TEM and micro-diffraction) and Max Sidorov (TEM).

REFERENCES

[1] G. Ottaviani, K.N. Tu, J.W. Mayer [*Phys. Rev. B (USA)* vol.24 (1981) p.3354]
[2] T.G. Finstad, J.W. Mayer, M.-A. Nicolet [*Thin Solid Films (Netherlands)* vol.51 (1978) p.391]
[3] K.N. Tu [*J. Appl. Phys. (USA)* vol.48 (1977) p.3379]
[4] R. Pretorius, C.L. Ramiller, S.S. Lau, M.-A. Nicolet [*Appl. Phys. Lett. (USA)* vol.30 (1977) p.501]
[5] K. Nakamura, J.O. Olowolafe, S.S. Lau, et al. [*J. Appl. Phys. (USA)* vol.47 (1976) p.1278]
[6] S. Petersson, E. Mgbenu, P.A. Tove [*Phys. Stat. Sol. A (Germany)* vol.36 (1976) p.217]
[7] N.F. Podgrushko, N.F. Selivanova, L.A. Dvorina [*Izvestiya Akademii Nauk SSSR, Neorganicheskie Materialy (Russia)* vol.10 (1974) p.150]
[8] K.E. Sundstrom, S. Petersson, P.A. Tove [*Phys. Stat. Sol. A (Germany)* vol.20 (1973) p.653]
[9] R. Anderson, J. Baglin, J. Dempsey, et al. [*Appl. Phys. Lett. (USA)* vol.35 (1979) p.285]
[10] M.W. Kleinschmit, M. Yeadon, J.M. Gibson [*Appl. Phys. Lett. (USA)* vol.75 (1999) p.3288]
[11] C. Canali, F. Catellani, G. Ottaviani, M. Prudenziati [*Appl. Phys. Lett. (USA)* vol.33 (1978) p.187]
[12] J.O. Olowolafe, M.-A. Nicolet, J.W. Mayer [*Thin Solid Films (Netherlands)* vol.38 (1976) p.143]
[13] V. Koos, U. Beck, H.-G. Neumann [*Phys. Stat. Sol. A (Germany)* vol.37 (1976) p.193]
[14] F.M. d'Heurle [*Thin Films and Interfaces II. Materials Research Society Symposium—Proceedings* (1984) p.3]
[15] F.M. d'Heurle, C.S. Petersson, J.E.E. Baglin, S.J. La Placa, C.Y. Wong [*J. Appl. Phys. (USA)* vol.55 (1984) p.4208]
[16] F.M. d'Heurle [*Thin Solid Films (Netherlands)* vol.105 (1983) p.285]
[17] F.M. d'Heurle, S. Petersson, L. Stolt, B. Strizker [*J. Appl. Phys. (USA)* vol.53 (1982) p.5678]
[18] J.E.E. Baglin, H.A. Atwater, D. Gupta, F.M. d'Heurle [*Thin Solid Films (Netherlands)* vol.93 (1981) p.255]
[19] E.R. Weber [*Appl. Phys. A (Solids Surfaces) (USA)* vol.30 (1983) p.1]
[20] J. Gulpen [*Reactive Phase Formation in the Ni–Si System. PhD Thesis* (Eindhoven University of Technology, Eindhoven, 1985)]
[21] P. Gas, F.M. d'Heurle, F.K. LeGoues, S.J. La Placa [*J. Appl. Phys. (USA)* vol.59 (1986) p.3458]
[22] F.M. d'Heurle [*J. Mater. Res. (USA)* vol.3 (1988) p.167]
[23] T.G. Finstad [*Phys. Stat. Sol. A (Germany)* vol.63 (1981) p.223]

[24] N.W. Cheung, P.J. Grunthaner, F.J. Grunthaner, J.W. Mayer, B.M. Ullrich [*J. Vac. Sci. Technol. (USA)* vol.18 (1980) p.917]
[25] Y.-J. Chang, J.L. Erskine [*Phys. Rev. B (USA)* vol.28 (1983) p.5766]
[26] M. Finetti, S. Guerri, P. Negrini, A. Scorzoni, I. Suni [*Thin Solid Films (Netherlands)* vol.130 (1985) p.37]
[27] A. Witzmann, A. Dittmar, K. Gartner, G. Gotz [*Phys. Stat. Sol. A (Germany)* vol.91 (1985) p.439]
[28] G. Majni, M. Costato, F. Panini [*Thin Solid Films (Netherlands)* vol.125 (1985) p.71]
[29] E.J. van Loenen, J.F. van der Veen, F.K. LeGoues [*Surf. Sci. (USA)* vol.157 (1985) p.1]
[30] I. Ohdomari, M. Akiyama, T. Maeda, et al. [*J. Appl. Phys. (USA)* vol.56 (1984) p.2725]
[31] S. Valeri, U. Del Pennino, P. Lomellini, P. Sassaroli [*Surf. Sci. (USA)* vol.145 (1984) p.371]
[32] G. Majni, M. Costato, F.D. Valle [*Il Nuovo Cimento (Italy)* vol.4D (1984) p.27]
[33] G. Majni, F. Della Valle, C. Nobili [*J. Phys. D (UK)* vol.17 (1984) p.L77]
[34] S. Valeri, U. Del Pennino, P. Sassaroli [*Surf. Sci. (USA)* vol.134 (1983) p.L537]
[35] F. Comin, J.E. Rowe, P.H. Citrin [*Phys. Rev. Lett. (USA)* vol.51 (1983) p.2402]
[36] A. Humbert, A. Cros [*J. Physique Lett. (France)* vol.44 (1983) p.L929]
[37] Y.-J. Chang, J.L. Erskine [*J. Vac. Sci. Technol. A (USA)* vol.1 (1983) p.1193]
[38] K.N. Tu, G. Ottavania, U. Gosele, H. Foll [*J. Appl. Phys. (USA)* vol.54 (1983) p.758]
[39] Y.J. Chang, J.L. Erskine [*Phys. Rev. B (USA)* vol.26 (1982) p.4766]
[40] L.S. Wielunski, C.-D. Lien, B.X. Liu, M.-A. Nicolet [*J. Vac. Sci. Technol. (USA)* vol.20 (1982) p.182]
[41] P.J. Grunthaner, F.J. Grunthaner, D.M. Scott, M.-A. Nicolet, J.W. Mayer [*J. Vac. Sci. Technol. (USA)* vol.19 (1981) p.641]
[42] P.J. Grunthaner, F.J. Grunthaner, A. Madhukar, J.W. Mayer [*J. Vac. Sci. Technol. (USA)* vol.19 (1981) p.649]
[43] D.M. Scott, M.-A. Nicolet [*Phys. Stat. Sol. A (Germany)* vol.66 (1981) p.773]
[44] N.W. Cheung, P.J. Grunthaner, F.J. Grunthaner, J.W. Mayer, B.M. Ullrich [*J. Vac. Sci. Technol. (USA)* vol.18 (1981) p.917]
[45] D.M. Scott, M.-A. Nicolet [*Nucl. Instrum. Methods (USA)* vol.182–183 (1981) p.655]
[46] L. Wielunski, D.M. Scott, M.-A. Nicolet, H. von Seefeld [*Appl. Phys. Lett. (USA)* vol.38 (1981) p.106]
[47] E.G. Colgan, M. Maenpaa, M. Finetti, M.-A. Nicolet [*J. Electron. Mater. (USA)* vol.12 (1983) p.413]
[48] R.M. Boulet, A.E. Dunsworth, J.-P. Jan, H.L. Skriver [*J. Phys. F (UK)* vol.10 (1980) p.2197]
[49] P.S. Ho, M. Liehr, P.E. Schmid, et al. [*Surf. Sci. (USA)* vol.168 (1986) p.184]
[50] T.G. Finstad, D.D. Anfiteatro, V.R. Deline, et al. [*Thin Solid Films (Netherlands)* vol.135 (1986) p.229]
[51] F.M. d'Heurle, D.D. Anfiteatro, V.R. Deline, T.G. Finstad [*Thin Solid Films (Netherlands)* vol.128 (1985) p.107]

[52] V.V. Tokarev, A.I. Demchenko, A.I. Ivanov, V.E. Borisenko [*Appl. Surf. Sci. (USA)* vol.44 (1990) p.241]
[53] C. Lavoie, R. Purtell, C. Coïa, et al. [*Electrochem. Soc. Symp. Proc.* vol.2002-11 (2002) p.455]
[54] J.A. Kittl, A. Lauwers, O. Chamirian, et al. [*Microelecron. Engng. (UK)* vol.70 (2003) p.158]
[55] C. Lavoie, F.M. d'Heurle, C. Detavernier, C. Cabral [*Microelectron. Engng. (UK)* vol.70 (2003) p.144]
[56] F.M. d'Heurle [*J. Electron. Mater. (USA)* vol.27 (1998) p.1138]
[57] T. Ohguro, S. Nakamura, M. Koike, et al. [*IEEE Trans. Electron Devices (USA)* vol.41 (1994) p.2305]
[58] E. Morifuji, M. Kanda, N. Yanagiya, et al. [*IEEE International Electron Devices Meeting* (2002) p.655]
[59] J.P. Lu, D. Miles, J. Zhao, et al. [*IEEE International Electron Devices Meeting* (2002) p.371]
[60] K. Ohuchi, K. Adachi, A. Hokazono, Y. Toyoshima [*Mater. Res. Soc. Symp. Proc. (USA)* vol.717 (2002) p.77]
[61] Q. Xiang, C. Woo, E. Paton, et al. [*Symposium on VLSI Technology. Digest of Technical Papers* (2000) p.76]
[62] X.W. Lin, N. Ibrahim, L. Topete, D. Pramanik [*Mater. Res. Soc. Symp. Proc. (USA)* vol.514 (1998) p.179]
[63] A. Hokazono, K. Ohuchi, M. Takayanagi, et al. [*IEEE International Electron Devices Meeting* (2002) p.639]
[64] R. Chau, J. Kavalieros, B. Doyle, et al. [*International Electron Devices Meeting. Technical Digest* 29.1.1 (2001)]
[65] R. Chau, J. Kavalieros, B. Roberds, et al. [*International Electron Devices Meeting. Technical Digest. IEDM* (2000) p.45]
[66] R. Mukai, S. Ozawa, H. Yagi [*Thin Solid Films (Netherlands)* vol.270 (1995) p.567]
[67] J.-S. Maa, D.J. Tweet, Y. Ono, L. Stecker, S.T. Hsu [*Mater. Res. Soc. Symp. Proc. (USA)* vol.670 K6.9.1 (2002)]
[68] J.-S. Maa, Y. Ono, D.J. Tweet, F. Zhang, S.T. Hsu [*J. Vac. Sci. Technol. A (USA)* vol.19 (2001) p.1595]
[69] R.W. Mann, L.A. Clevenger, P.D. Agnello, F.R. White [*IBM J. Res. Dev. (USA)* vol.39 (1995) p.403]
[70] R.W. Mann, L.A. Clevenger [*J. Electrochem. Soc. (USA)* vol.141 (1994) p.1347]
[71] J.M.E. Harper, C. Cabral, C. Lavoie [*Annu. Rev. Mater. Sci.(USA)* vol.30 (2000) p.523]
[72] J.H. Chung, J.E. Lee, J.S. Park, et al. [*Advanced Metallisation Conference—Proceedings* (2000) p.495]
[73] K. Maex, M. van Rossum [(INSPEC, 1995)]
[74] F.M. d'Heurle, C.S. Petersson [*Thin Solid Films (Netherlands)* vol.128 (1985) p.283]
[75] J.P. Gambino, E.G. Colgan [*Mater. Chem. Phys. (USA)* vol.52 (1998) p.99]
[76] P.D. Agnello, S. Brodsky, E. Crabbe, et al. [*Electrochem. Soc. Symp. Proc. (USA)* vol.99–100 (1999) p.217]
[77] C. Lavoie, C. Cabral Jr., F.M. d'Heurle, J.L. Jordan-Sweet, J.M.E. Harper [*J. Electron. Mater. (USA)* vol.31 (2002) p.597]
[78] C. Detavernier, R.L. Van Meirhaeghe, F. Cardon, K. Maex [*Thin Solid Films (Netherlands)* vol.384 (2001) p.243]
[79] C. Detavernier, R.L. Van Meirhaeghe, F. Cardon, K. Maex [*Phys. Rev. B (USA)* vol.62 (2000) p.12045]

[80] A. Lauwers, M. de Potter, O. Chamirian, et al. [*Microelectron. Engng. (UK)* vol.64 (2002) p.131]
[81] A. Lauwers, A. Steegen, M. de Potter, et al. [*J. Vac. Sci. Technol. B (USA)* vol.19 (2001) p.2026]
[82] V. Teodorescu, L. Nistor, H. Bender, et al. [*J. Appl. Phys. (USA)* vol.90 (2001) p.167]
[83] [*Handbook of Binary Alloy Phase Diagrams (CD Version 1.0)* (ASM International, ISBN PC-0-87170-562-1, 1996)]
[84] H. Iwai, T. Ohguro, S.-I. Ohmi [*Microelectron. Engng. (UK)* vol.60 (2002) p.157]
[85] A. Steegen, K. Maex [*Mater. Sci. Engng. R: Reports (Netherlands)* vol.R38 (2002) p.1]
[86] J.J. Wortman, R.A. Evans [*J. Appl. Phys. (USA)* vol.36 (1965) p.153]
[87] O. Madelung [Semiconductors Basic Data (Springer, Berlin, 1996)]
[88] M.Y. Lee, P.A. Bennett [*Phys. Rev. Lett. (USA)* vol.75 (1995) p.4460]
[89] C.A. Londos, K. Eftaxias, V. Hadjicontis [*Phys. Stat. Sol. A (Germany)* vol.118 (1990) p.K13]
[90] M.-A. Nicolet, S.S. Lau [*Microstructure Science*, vol.6 (Academic Press, NY, 1983)]
[91] R. Byers, R. Sinclair [*J. Appl. Phys. (USA)* vol.57 (1985) p.5240]
[92] M. Tinani, A. Mueller, Y. Gao, et al. [*J. Vac. Sci. Technol. B (USA)* vol.19 (2001) p.376]
[93] E.G. Colgan, F.M. d'Heurle [*J. Appl. Phys. (USA)* vol.79 (1996) p.4087]
[94] H. Foll, P.S. Ho, K.N. Tu [*Philos. Mag. A (UK)* vol.45 (1982) p.31]
[95] M. Costato [*Lettere Al Nuovo Cimento (Italy)* vol.32 (1981) p.219]
[96] W.J. Schaffer, R.W. Bene, R.M. Walser [*J. Vac. Sci. Technol. (USA)* vol.15 (1978) p.1325]
[97] R.T. Tung [*Mater. Chem. Phys. (USA)* vol.32 (1992) p.107]
[98] R.T. Tung [*Appl. Phys. Lett. (USA)* vol.68 (1996) p.3461]
[99] C. Detavernier, R.L. Van Meirhaeghe, F. Cardon, K. Maex [*Thin Solid Films (Netherlands)* vol.386 (2001) p.19]
[100] R.T. Tung, S. Ohmi [*Thin Solid Films (Netherlands)* vol.369 (2000) p.233]
[101] C. Detavernier, R.L. Van Meihaeghe, F. Cardon, R.A. Donaton, K. Maex [*Appl. Phys. Lett. (USA)* vol.74 (1999) p.2930]
[102] A.R. Londergan, G. Nuesca, C. Goldberg, et al. [*J. Electrochem. Soc. (USA)* vol.148 (2001) p.C21]
[103] H. Okorn-Schmidt [Private communication, Unpublished work]
[104] C. Lavoie, C. Cabral, F.M. d'Heurle, J.M.E. Harper [*Defect Diffusion Forum* vol.194–199 (2001) p.1477]
[105] C. Lavoie, C. Cabral Jr., L.A. Clevenger, et al. [*Mater. Res. Soc. Symp. Proc. (USA)* vol.406 (1996) p.163]
[106] G.B. Stephenson, K.F. Ludwig, J.L.B.S. Jordan-Sweet, et al. [*Rev. Sci. Instrum. (USA)* vol.60 (1989) p.1537]
[107] C. Lavoie, R. Martel, C.J. Cabral, L.A. Clevenger, J.M.E. Harper [*Mater. Res. Soc. Symp. Proc. (USA)* vol.440 (1997) p.389]
[108] J.M. Bennett, L.M. Mattsson [*Introduction to Surface Roughness and Scattering* (Opt. Soc. Am., Washington, 1989)]
[109] J. Gulpen, A.A. Kodentsov, F.J.J. van Loo [*Z. Metallk (Germany)* vol.86 (1995) p.530]
[110] L.J. Chen [*Mater. Sci. Engng. Rep. (Netherlands)* vol.29 (2000) p.115]
[111] D.K. Schroder [*Semiconductor Material and Device Characterisation* (John Wiley & Sons, New York, 1998)]

[112] Y. Tsuchiya, A. Tobioka, O. Nakatsuka, et al. [*Jpn. J. Appl. Phys. Part 1 (Japan)* vol.41 (2002) p.2450]

[113] T. Morimoto, T. Ohguro, S. Momose, et al. [*IEEE Trans. Electron Devices (USA)* vol.42 (1995) p.915]

[114] F. Deng, R.A. Johnson, P.M. Asbeck, et al. [*J. Appl. Phys. (USA)* vol.81 (1997) p.8047]

[115] S.B. Herner, K.S. Jones, H.-J. Gossmann, et al. [*Appl. Phys. Lett. (USA)* vol.68 (1996) p.2870]

[116] S.B. Herner, H.-J. Gossmann, R.T. Tung [*Appl. Phys. Lett. (USA)* vol.72 (1998) p.2289]

[117] A.K. Tyagi, L. Kappius, U. Breuer, et al. [*J. Appl. Phys. (USA)* vol.85 (1999) p.7639]

[118] P.M. Fahey, P.B. Griffin, J.D. Plummer [*Rev. Mod. Phys. (USA)* vol.61 (1989) p.289]

[119] O. Kubaschewski, C.B. Alcock, P.J. Spencer [*Material Thermo-Chemistry*, 6th Ed. Revised (Pergamon Press, Oxford, 1993)]

[120] K. Wark [*Thermodynamics*, 3rd Ed. (McGraw-Hill, New York, 1977)]

[121] A.D. Smigelkas, E.O. Kirkendall [*Trans. AIME (USA)* vol.171 (1947) p.130]

[122] R.S. Barnes [*Nature (UK)* vol.166 (1950) p.1032]

[123] L.C.C. da Silva, R.F. Mehl [*Trans. AIME (USA)* vol.191 (1951) p.155]

[124] E. Fitzer [*Z. Metallk. (Germany)* vol.44 (1953) p.462]

[125] B.O. Kolbesen, H. Cerva [*Phys. Stat. Sol. B (Germany)* vol.222 (2000) p.303]

[126] Y. Tian, Y.-L. Jiang, Y. Chen, F. Lu, B.-Z. Li [*Semicond. Sci. Technol. (UK)* vol.17 (2002) p.83]

[127] M.Kh. Rabadanov, M.B. Ataev [*Inorg. Mater. (USA)* vol.38 (2002) p.120]

[128] D.F. Wilson, O.B. Cavin [*Scripta Metall. Mater. (Netherlands)* vol.26 (1992) p.85]

[129] J.F. Nye [*Physical Properties of Crystals* (Oxford University Press, London, 1957)]

[130] Joint Committee on Powder Diffraction Standard (JCPDS) NiSi, p.38

[131] C. Detavernier, C. Lavoie, F.M. d'Heurle [*J. Appl. Phys. (USA)* vol.93 (2003) p.2510]

[132] U. Welzel, M. Leoni, E.J. Mittemeijer [*Philos. Mag. (UK)* vol.83 (2003) p.603]

[133] M. Qin, M.C. Poon [*J. Mater. Sci. Lett. (UK)* vol.19 (2000) p.2243]

[134] E.M. Zielinski, R.P. Vinci, J.C. Bravman [*J. Appl. Phys. (USA)* vol.76 (1994) p.4516]

[135] C.V. Thompson [*Scripta Metall. Mater. (Netherlands)* vol.28 (1993) p.167]

[136] C. Detavernier, A.S. Özcan, J.L. Jordan-Sweet, et al. [*Nature (UK)* vol.426 (2003) p.641]

[137] U.F. Kocks, C.N. Tome, H.-R. Wenk [*Texture and Anisotropy: Preferred Orientations in Polycrystals and Their Effect on Materials Properties* (Cambridge University Press, Cambridge, 1976)]

[138] H.J. Bunge [*Texture Analysis in Materials Science—Mathematical Methods*, 2nd Ed. (Butterworths, London, 1982)]

[139] V. Nokinov [*Grain Growth and Control of Microstructure and Texture in Polycrystalline Materials* (CRC Press, Boca Raton, 1997)]

[140] V. Randle, O. Engler [*Introduction to Texture Analysis: Macrotexture, Microtexture and Orientation Mapping* (Gordon and Breach Science Publishers, Amsterdam, 2000)]

[141] C.V. Thompson, R. Carel [*Mater. Sci. Engng. (Netherlands)* vol.B32 (1995) p.211]
[142] J.M.E. Harper, K.P. Rodbell, E.G. Colgan, R.H. Hammond [*J. Appl. Phys. (USA)* vol.82 (1997) p.4319]
[143] C.W.T. Bulle-Lieuwma, A.H. van Ommen, J. Hornstra, C.N.A.M. Aussems [*J. Appl. Phys. (USA)* vol.71 (1992) p.2211]
[144] A.S. Özcan, K.F. Ludwig Jr., P. Rebbi, et al. [*J. Appl. Phys. (USA)* vol.92 (2002) p.5011]
[145] D. Mangelinck, J.Y. Dai, J. Pan, S.K. Lahiri [*Appl. Phys. Lett. (USA)* vol.75 (1999) p.1736]
[146] J. Seger, S.L. Zhang [*Thin Solid Films (Netherlands)* vol.429 (2003) p.216]
[147] T. Jarmar, J. Seger, F. Ericson, et al. [*J. Appl. Phys. (USA)* vol.92 (2002) p.7193]
[148] S.-L. Zhang [*Microelectron. Engng. (UK)* vol.70 (2003) p.174]
[149] R.T. Tung [*J. Vac. Sci. Technol. A (USA)* vol. 7 (1989) p.598]
[150] R.T. Tung, A.F.J. Levi, F. Schrey, M. Anzlowar [*Evaluation of Advanced Semiconductor Materials by Electron Microscopy. Proceedings of a NATO Advanced Research Workshop* (1989) p.167]
[151] M.C. Poon, M. Chan, W.Q. Zhang, F. Deng, S.S. Lau [*Microelectron. Reliab. (USA)* vol.38 (1998) p.1499]
[152] S.C.H. Ho, M.C. Poon, M. Chan, H. Wong [*Proc. 1998 IEEE Hong Kong Electron Devices Meeting* (1998) p.105]
[153] C.-J. Choi, Y.-W. Ok, S.S. Hullavarad, et al. [*J. Electrochem. Soc. (USA)* vol.149 (2002) p.G517]
[154] M.C. Poon, F. Deng, M. Chan, W.Y. Chan, S.S. Lau [*Appl. Surf. Sci. (USA)* vol.157 (2000) p.29]
[155] M.C. Poon, C.H. Ho, F. Deng, S.S. Lau, H. Wong [*Microelectron. Reliab. (USA)* vol.38 (1998) p.1495]
[156] O. Chamirian, J.A. Kittl, A. Lauwers, et al. [*Microelectron. Engng. (UK)* vol.70 (2003) p.201]
[157] S.-L. Zhang, F.M. d'Heurle [*Thin Solid Films (Netherlands)* vol.256 (1995) p.155]
[158] S. Nygren, D. Caffin, M. Ostling, F.M. d'Heurle [*Appl. Surf. Sci. (USA)* vol.53 (1991) p.87]
[159] Q.Z. Hong, F.M. d'Heurle, J.M.E. Harper, S.Q. Hong [*Appl. Phys. Lett. (USA)* vol.62 (1993) p.2637]
[160] S. Nygren, F.M. d'Heurle [*Solid State Phenomena* vol.23-24 (1992) p.81]
[161] P.S. Lee, D. Mangelinck, K.L. Pey, et al. [*Microelectron. Engng. (UK)* vol.60 (2002) p.171]
[162] P.S. Lee, K.L. Pey, D. Mangelinck, et al. [*J. Electrochem. Soc. (USA)* vol.149 (2002) p.G331]
[163] P.S. Lee, K.L. Pey, D. Mangelinck, et al. [*IEEE Electron Device Lett. (USA)* vol.22 (2001) p.568]
[164] J.F. Liu, H.B. Chen, J.Y. Feng [*J. Crystal Growth (Netherlands)* vol.220 (2000) p.488]
[165] D. Mangelinck, J.Y. Dai, S.K. Lahiri, C.S. Ho, T. Osipowicz [*Mater. Res. Soc. Symp. Proc. (USA)* vol.564 (1999) p.163]
[166] H.L. Seng, T. Osipowicz, P.S. Lee, et al. [*Nucl. Instrum. Methods Phys. Res. Section B (USA)* vol.181 (2001) p.399]
[167] T.-S. Chao, L.-Y. Lee [*Jpn. J. Appl. Phys. Part 2 (Japan)* vol.41 (2002) p.L124]

[168] P.S. Lee, K.L. Pey, D. Mangelinck, et al. [*IEEE Electron Device Lett. (USA)* vol.21 (2000) p.566]
[169] L.J. Chen, S.L. Cheng, S.M. Chang, et al. [*Mater. Res. Soc. Symp. Proc. (USA)* vol.564 (1999) p.123]
[170] P.S. Lee, D. Mangelinck, K.L. Pey, et al. [*J. Electron. Mater. (USA)* vol.30 (2001) p.1554]
[171] P.S. Lee, K.L. Pey, D. Mangelinck, et al. [*J. Electrochem. Soc. (USA)* vol.149 (2002) p.G505]
[172] W.J. Chen, L.J. Chen [*J. Appl. Phys. (USA)* vol.70 (1991) p.2628]
[173] C.-J. Choi, Y.-W. Ok, T.-Y. Seong, H.-D. Lee [*Jpn. J. Appl. Phys. Part 1 (Japan)* vol.41 (2002) p.1969]
[174] A.S.W. Wong, D.Z. Chi, M. Loomans, et al. [*Appl. Phys. Lett. (USA)* vol.81 (2002) p.5138]
[175] S.K. Donthu, D.Z. Chi, A.S.W. Wong, S.J. Chua, S. Tripathy [*Mater. Res. Soc. Symp. Proc. (USA)* vol.716 (2002) p.465]
[176] M.-H. Juang, S.-C. Han, M.-C. Hu [*Jpn. J. Appl. Phys. Part 1 (Japan)* vol.37 (1998) p.5515]
[177] F.M. d'Heurle [*Mater. Res. Soc. Symp. Proc. (USA)* vol.402 (1996) p.3]
[178] D.Z. Chi, D. Mangelinck, A.S. Zuruzi, A.S.W. Wong, S.K. Lahiri [*J. Electron. Mater. (USA)* vol.30 (2001) p.1483]
[179] [*International Technology Roadmap for Semiconductors—2002 Update* (European Semiconductor Industry Association, Japan Electronics and Information Technology Industries Association, Korea Semiconductor Industry Association, Taiwan Semiconductor Industry Association, Semiconductor Industry Association, http://public.itrs.net/, 2002)]
[180] S. Thompson, N. Anand, M. Armstrong, et al. [*IEEE International Electron Devices Meeting* (2002) p.61]
[181] J.A. Kittl, A. Lauwers, O. Chamirian, et al. [*Mater. Res. Soc. Symp. Proc. (USA)* vol.765 (2003) p.267]
[182] P.R. Besser, S. Chan, E. Paton, et al. [*Mater. Res. Soc. Symp. Proc. (USA)* vol.766 (2003) p.59]
[183] J. Kedzierski, D. Boyd, P. Ronsheim, et al. [*International Electron Devices Meeting Technical Digest IEDM* (2003)]
[184] J. Kedzierski, D. Boyd, Y. Zhang, et al. [*International Electron Devices Meeting Technical Digest IEDM* (2003)]
[185] M. Qin, M.C. Poon, S.C.H. Ho [*J. Electrochem. Soc. (USA)* vol.148 (2001) p.271]

Chapter 6

Light-emitting iron disilicide

L.J. Chou

6.1 OVERVIEW

Metal silicides have many desirable properties such as high melting temperature, superb thermal stability and low resistivity [1]. These characteristics have made metal silicides more and more popular in VLSI process applications. For instance, WSi_2 is commonly used in poly gate to lower the contact resistance and form ohmic contact [2], whereas $TiSi_2$ is often used in IC fabrication to reduce the resistance between the source, drain and the metal layer – that is, the so-called silicide reaction [3, 4].

Silicon possesses top quality, inexpensive and easy-to-use native oxide. Nevertheless, it is an indirect bandgap material, making it difficult to produce devices high in illumination efficiency. Hence, silicon-based semiconductors have always been outperformed by III–V compound semiconductors in the field of opto-electronics. Equipped with a direct bandgap, compound semiconductors can be applied in illuminating devices such as light emitting diodes (LEDs), laser diodes (LDs) and vertical cavity surface emitting laser (VCSEL). Their exceptionally high electron mobility makes them ideal materials for high-speed devices such as high electron mobility transistor (HEMT) and heterojunction bipolar transistor (HBT). The different lattice constants of compound semiconductors and silicon have prevented the integration between illumination devices made of compound semiconductors and silicon wafers. Fortunately, findings over the past decade have offered some hope. In addition to their known qualities of high melting temperature, superb thermal stability, low resistivity and high Schottky barrier height, silicides including Cr_xSi_y, Mn_xSi_y and Fe_xSi_y have a quality that is absent in silicon-based semiconductor materials – direct bandgap [5–7]. Consequently, there is great potential for the possibility of directly forming illuminating devices on silicon-based materials in the mature silicon semiconductor industry instead of the conventional technique of integrating compound semiconductors on silicon wafers by means of wafer bonding. TABLE 6.1 lists

Overview p.153

Growth methods p.155
  Growing β-FeSi$_2$ by means of RDE p.156
  Growing β-FeSi$_2$ through IBS p.156
  Structure of the precipitate p.158

Effect of stress on the illuminating characteristics of β-FeSi$_2$ p.159
  Source of stress p.159
  Effect of stress on β-FeSi$_2$ p.161
  Using Raman optical measurement to obtain the stress value p.165
  Effect of stress on the feature of β-FeSi$_2$ p.168
  Optical properties p.169

References p.173

TABLE 6.1 The bandgap value for different silicides.

Material	Experimental energy gap (eV)	Type of gap (experimental)	Theoretical energy gap (eV)	Type of gap (theoretical)
$CrSi_2$	0.35–0.67	Indirect	0.21–0.38	Indirect
	0.5–0.9	Direct	0.37–0.47	Direct
$MnSi_x$	0.45–0.47	Indirect		
	0.78–0.83	Direct		
β-$FeSi_2$	0.83–0.89	Direct	0.8–0.82	Direct
	0.765–0.79	Indirect	0.44–0.78	Indirect
Ru_2Si_3	0.44–0.7	Direct	0.45	Direct
	0.12–0.15	Indirect	0.16	Indirect
$ReSi_2$	0.16–0.2	Direct	0.36	Direct
OsSi	0.34	(Direct)		
Os_2Si_3	2.3	(Direct)		
$OsSi_2$	1.4–1.8	(Direct)	0.95	Indirect
			1.14	Direct
Ir_3Si_5	1.2–1.57	Direct		

the metal silicides currently possessing the potential of becoming illuminating devices.

Iron disilicide ($FeSi_2$) has different phases under different annealing temperatures. The α phase (α-$FeSi_2$) is the stable phase at high temperature (>940°C). Among the seven major Bravais lattice, it has a tetragonal structure, with the lattice constant of $a = b = 2.69$ Å, $c = 5.09$ Å and the space group of $P4/mmm$. In contrast, as the annealing temperature falls below 940°C, it turns into the β phase (β-$FeSi_2$), which is, among the seven major Bravais lattice, based-centred orthorhombic, with the lattice constant of $a = 0.986$ nm, $b = 0.778$ nm, $c = 0.788$ nm. It also possesses semiconductor properties showing p-type conduction and has carrier concentrations around 1×10^{18} at room temperature [8]. At even lower temperature, it exists in the FeSi phase with a cubical structure and the lattice constant of $a = b = c = 0.448$ nm [8–10]. However, recent studies have found that thin films exist in only two phases, namely γ-$FeSi_2$ and S-$FeSi_2$, both belong to the cubic system with lattice constants of 0.5389 nm and 2.7 Å, respectively [11–14]. The following sections describe the physical properties, opto-electronic and thermo-electric characteristics of the materials used in this study.

Generally speaking, the semiconductor properties of β-$FeSi_2$ thin film are best described in terms of the doping species. For instance, the doping of p-type formation includes Ti, V, Cr and Mn, while the doping of n-type consists of Co, Ni and Pt [15].

6.2 GROWTH METHODS

There are a wide variety of methods to grow iron disilicide. The most commonly used methods include ultra high vacuum (UHV) [16], molecular beam epitaxy (MBE) [17], chemical beam epitaxy [18], ion beam synthesis (IBS) [19] and pulsed laser deposition (PLD) [20], as described below.

The basic principle is to evaporate iron on silicon-based materials using an electron beam in a UHV environment. But the thin films obtained are all polycrystalline. During the subsequent heat treatment, a layer of a-Si is grown on top of the iron to prevent oxidation. But, the thin film exhibits poor qualities in various aspects. Another growth method involves Si/Fe multilayer. Some researchers have found that an increase in Si/Fe percentage produces (β-FeSi$_2$) thin film, which changes from p-type to n-type [21]. The resulting increase in annealing time lowers the carrier concentration and raises the mobility rate. Takaura et al. reported that after 14 h of annealing at 900°C, the carrier mobility was 13,000 cm^2/V/s [22].

MBE uses a similar principle of evaporation. The difference is that MBE allows us to control the desirable content percentage. On the other hand, chemical vapour epitaxy uses gas reactants such as disilane (Si$_2$H$_6$) and pentacarbonyl (Fe(CO)$_5$) [23]. These two gases are deposited on a heated substrate through chemical vapour deposition (CVD) to form FeSi$_2$ thin films.

Metal organic chemical vapour deposition is similar to conventional CVD methods. The difference lies in the precursor. Experiment findings using *cis*-Fe(SiCl$_3$)$_2$(Co)$_4$ to grow FeSi$_2$ have concluded that different substrates produce different FeSi$_2$ [24]. The similarity of lattice constants turns most growth on Si(001) substrate into β-FeSi$_2$. However, given a Pyrex glass substrate, the growth is primarily FeSi phase. On the other hand, PLD uses pulsed laser to evaporate FeSi$_2$ target on a heated silicon substrate, which can grow β-FeSi$_2$ at a lower temperature.

IBS uses ion implantation to grow β-FeSi$_2$ thin films. The approach of implantation generates certain unwanted phenomena such as defect and channelling effect, which compromise on the thin film qualities. Nevertheless, IBS is the most commonly used method for manufacturing illuminating devices using β-FeSi$_2$ following the MBE method.

In reactive deposition epitaxy (RDE) Fe is evaporated on a heated substrate at a temperature between 400 and 500°C. The specimen then undergoes vacuum annealing at 850°C for 1 h to improve the crystal quality of β-FeSi$_2$ thin film. A 0.3 nm thick layer of p-type epi-Si (10^{16} cm^{-3}) is covered by means of MBE at 500°C to acquire the optimal illumination efficiency. The specimen is further heated to 900°C and then annealed in an Ar atmosphere for 14 h to eliminate some defects that result from the original growth or phase transformation [25, 26].

6.2.1 Growing β-FeSi$_2$ by means of RDE

A wide variety of methods for growing β-FeSi$_2$ have been developed over the past years, for instance, thermal reaction method, MBE and IBS. These methods all have their own strengths and weaknesses. It is difficult to obtain an epi-thin film using the thermal reaction method, while IBS is far from the ideal method for obtaining an even and level thin film. Thus, Suemasu et al. proposed the approach called reactive deposition epitaxy or RDE to obtain level and epitaxial β-FeSi$_2$ thin films.

The so-called RDE approach deposits Fe on a high temperature substrate. Given a high enough temperature, Fe directly reacts to form β-FeSi$_2$ after deposition on the substrate. The β-FeSi$_2$ thin film grown by this method possesses a high directionality and a considerably level surface. To identify the optimal growth temperature using RDE, Suemasu et al. experimented on different RDE temperatures and used X-ray and RHEED to determine the quality of β-FeSi$_2$ thin films [27], as shown in FIGURE 6.1.

FIGURES 6.1(a)–(e) show the β-FeSi$_2$ thin films grown using RDE at the temperatures of 400, 470, 550, 700 and 850°C, respectively. These X-ray photos indicate that epi-thin film can be obtained at 470°C. Different crystal orientations are generated when the temperature exceeds 470°C, say at 550°C. As the temperature reaches 700°C, α-FeSi$_2$ phase forms in addition to β-FeSi$_2$. These results confirm that using RDE at 470°C the best crystalline quality can be obtained. The RHEED patterns also reveal that the optimal surface levelness occurs also at 470°C. Regardless of the aspect involved, 470°C is certainly the optimal temperature for the RDE growth method.

6.2.2 Growing β-FeSi$_2$ through IBS

At present many groups adopt IBS to grow β-FeSi$_2$. Its widespread use in semiconductor processes has made it very attractive. Thus, this approach presents tremendous advantages in the manufacture of optical devices.

IBS implants high density elements into the substrate to form compounds. Then, high temperature annealing is carried out to repair the substrate and crystalline quality of the compounds. Using Fe as the element to be implanted on a Si substrate produces β-FeSi$_2$ phase. Then the crystalline quality of Si and β-FeSi$_2$ is improved through high temperature annealing. The parameters involved in IBS include the implantation density, energy and

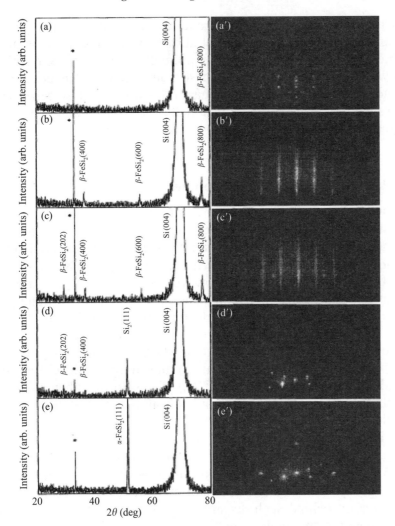

FIGURE 6.1 XRD and RHEED patterns of 200 Å thick β-FeSi$_2$ layers formed by RDE at temperatures of (a, a') 400°C, (b, b') 470°C, (c, c') 550°C, (d, d') 700°C and (e, e') 850°C. Fe deposition rate is fixed at 0.1 A/s for all samples (from Reference 27).

temperature. It is thus necessary to first define the Fe implantation density because an overly high implantation density may form a continuous β-FeSi$_2$ layer (as shown in FIGURE 6.2) [19]. However, a continuous layer is not desirable in common LED devices because dots display a better illumination effect without it. In addition, an overly high implantation density generates more defects, which cannot be eliminated even after high temperature annealing. After the implantation density is set, the optimal annealing temperature and duration must also be determined. We must rely on annealing to eliminate the defects caused by implantation. If the temperature chosen is too low, the outcome may be a poor

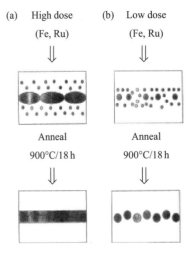

FIGURE 6.2 Schematic of the ion beam synthesis of silicides at (a) high and (b) low implant doses (from Reference 19).

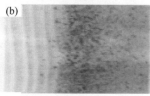
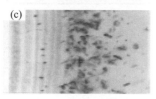
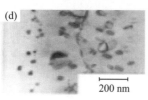

FIGURE 6.3 Cross-section TEM images of samples implanted at 250°C with 340 keV Fe 2.5×10^{15} cm^{-2} as grown (a), annealed at 700°C for 3 h (b), at 800°C for 2 h (c), and at 900°C for 2 h (d) (from Reference 28).

β-FeSi$_2$ crystal. But the temperature cannot exceed the phase transformation temperature that allows the β phase to transform into the α phase. It is extremely difficult for the α phase to fully reverse to the β phase once phase transformation has occurred. The volume reduction during phase transformation also generates structural defects. All these factors have an impact on the optical qualities of the material.

6.2.3 Structure of the precipitate

However, in practice, there are still considerable controversies instead of foregone conclusions. Homewood et al. [19] reported that they obtained the best crystalline quality by using an implantation density of 5×10^{17} cm^{-2} and an annealing temperature of 900°C for a duration of 18 h. However, Grimaldi et al. [28] warned that both an overly high implantation density and high annealing temperature generate a tremendous amount of defects. These defects form a stacking fault during the high temperature annealing process (as shown in FIGURE 6.3, where (a) shows "as grown"; (b) annealing at 700°C for 2 h; (c) annealing at 800°C for 2 h; and (d) annealing at 900°C for 2 h). To avoid this phenomenon, both the implantation density and annealing temperature must be reduced.

They proposed the appropriate conditions of an annealing temperature below 900°C and implantation density under 4.5×10^{15} cm^{-2}. The implantation densities proposed by the above two groups differ by two orders of magnitude. We speculate that the reason behind such disparity may be the different levels of implantation energy used by them.

There are two types of β-FeSi$_2$ precipitate formed by IBS. Smaller spherical precipitates with a diameter around 15–30 nm appear near the surface, while disk-like precipitates emerge near the locations registering the maximum implantation density (as shown in FIGURE 6.4). These platelets belong to the β-FeSi$_2$(101) face, which is overlapped on top of Si{111}. The platelet is around 20–30 nm thick with diameter around 100–200 nm. Thus, we know that the β-FeSi$_2$ precipitates obtained by IBS plus annealing treatment are non-directional growths.

Precipitates are formed by nucleation on the defects. When Fe is implanted on Si substrate, vacancies form near the surface while interstitial defects form inside. These defects concentrate into clusters during the annealing process. Thus, spherical precipitates are derived from nucleation on the vacancies, while platelets are formed by nucleation on the interstitial defects. Thus, these two different types of precipitates originate from different defects.

Light-emitting iron disilicide

The epi relationship between the substrate and the precipitates can be derived from the diffraction patterns. Most epitaxial growth is β-FeSi$_2$(110)||(111)Si or β-FeSi$_2$(101)||(111)Si. It is very difficult to distinguish between these two types of epi-crystal because of the limited difference between the lattice constants of b and c axes ($b = 0.779$ nm, $c = 0.783$ nm), as shown in FIGURE 6.5.

6.3 EFFECT OF STRESS ON THE ILLUMINATING CHARACTERISTICS OF β-FeSi$_2$

β-FeSi$_2$ possesses semiconducting properties. Since its bulk form is not capable of illumination, it had always been considered that β-FeSi$_2$ has indirect bandgap until the early 1980s, when β-FeSi$_2$ thin films were found capable of emitting light. The discovery initiated discussions on the material's energy band structure. A number of research teams conducted further studies on the energy band structure and its illuminating characteristics during the early 1990s and presented important findings related to the energy band simulation of β-FeSi$_2$. That is, the energy band of β-FeSi$_2$ is a quasi-direct bandgap and extremely sensitive to the variation in stress. Characteristics of the material's energy band are described in further detail in the following sections.

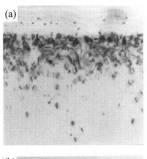

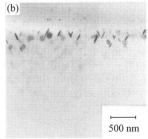

FIGURE 6.4 Bright-field cross-section TEM micrograph of Si(100) implanted with Fe to a dose of 2×10^{15} cm^{-2} at 300°C and annealed at 800°C for 23 h. Panel (a) is taken in two-beam configuration. Panel (b) is analysed far from the Bragg condition (from Reference 28).

6.3.1 Source of stress

In general, there are three sources of stress in a crystal, namely lattice mismatch, volume reduction and thermal expansion [29], and these are described in the following sections.

6.3.1.1 Lattice mismatch

As its name suggests, the stress from lattice mismatch occurs when the lattice constant or the lattice structure differs during the process of epitaxial growth. It is found that for β-FeSi$_2$ and Si interface, such stress is likely to occur only in certain crystal directions such as β(101)||Si(111), β(110)||Si(111) and β(100)||Si(001). The stresses generated in these crystal directions all differ from one another. Thus, the relationship between the illuminating characteristics β-FeSi$_2$ and its stress and crystal directions is worthy of further exploration.

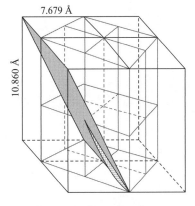

FIGURE 6.5 Correspondence between four tetragonal β-FeSi$_2$ conventional cells before the Jahn–Teller distortion and eight cubic γ-FeSi$_2$ cells. The (110) or (111) plane of the former superimposes to the (111) plane of the latter. β-FeSi$_2$(100) corresponds to γ-FeSi$_2$(100), but for an in-plane rotation by 45°.

6.3.1.2 Volume reduction

Stress can also be induced by volume reduction during the process in which Fe undergoes phase transformation to form β-FeSi$_2$.

Light-emitting iron disilicide

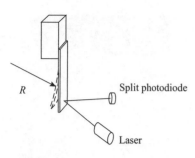

FIGURE 6.6 The curvature of radius R (stress) of the sample is monitored by reflecting the laser beam from the sample surface to a split photodiode (from Reference 30).

Sander et al. [30] conducted an experiment to study this phenomenon. They coated Fe on Si(111) specimens at a temperature of 600 K. Variation of the specimen curvature (measurement shown in FIGURE 6.6) during the coating process was measured to observe the changes of stress. The experimental results are shown in FIGURE 6.7. The temperature of 600 K is sufficient to induce phase transformation of both Fe and Si substrate. Hence, AES and magnetic Kerr effect analysis were used to first ascertain the emergence of β-FeSi$_2$ phase. Curvature was then measured. A considerable level of tensile stress was measured regardless of the coating rates of 0.4 or 0.1 nm/min. The tensile stress values were 11.2 and 18 N/m, respectively. According to Sander's explanation of the cause behind the tensile stress, the number of Si atoms reduces from 0.2007 to 0.187 nm^3 as the formation process of β-FeSi$_2$ proceeds. The reduction of approximately 4.5% causes tensile stress after Fe transforms into β-FeSi$_2$.

6.3.1.3 Thermal expansion

Stress may also be induced by different thermal expansion coefficients of the materials during the growth. When specimens finish high temperature annealing and return to room temperature from 900°C, stress occurs because of the different thermal expansion coefficients of β-FeSi$_2$ and Si. In addition, since the thermal expansion coefficient of β-FeSi$_2$ is roughly twice that of Si, the stress generated is tensile stress. But the thermal expansion coefficient of

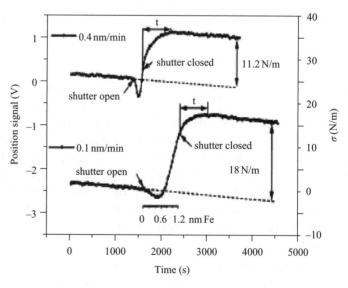

FIGURE 6.7 Stress measurement taken during the silicide formation. Fe was deposited on a Si(111) sample heated to 600 K (from Reference 30).

6.3.2 Effect of stress on β-FeSi$_2$

The energy band simulation results discussed earlier were all obtained under the condition of bulk materials. The results not only confirm that β-FeSi$_2$ is not a direct bandgap material, but also reveal that the energy gap between the direct and indirect bandgaps is very small. Hence, it is worth investigating what the energy band graphs may look like if β-FeSi$_2$ exists in the form of quantum dot or thin film. Generally speaking, when materials exist in the form of thin film or quantum dot, they are usually subject to the stress imposed by the substrate, which distorts the lattice. Therefore, a number of research teams have further simulated the energy band graphs formed by distorted lattice for comparison with the results involving bulk materials. The results of two teams are described in the following sections to explain the relationship between stress and illumination effect. It is worth noting that the stress imposed on the material represents the combined stress from three sources – that is, lattice mismatch, volume reduction and thermal expansion. However, to simplify the explanation, both the teams reported in the following assumed that the stress came only from lattice mismatch.

6.3.2.1 Clark and coworkers' model

Clark et al's simulation [31] assumed that in β-FeSi$_2$(100) epitaxially grown on the Si(001) substrate, given this condition, the a axis is free of the strain. Only b and c axes are subject to two-dimensional stress. It is known that the lattice constants of β-FeSi$_2$ are $a = 9.863$ Å, $b = 0.7884$ nm and $c = 0.7791$ nm, while that of Si is $a = 0.543$ nm. We can further introduce two possible epitaxial growth modes that may be called Type A and Type B. For the growth mode of Type A, β-FeSi$_2$[010] or [001] overlaps in Si[110] direction. Let the crystal axis of Si be a and those of β-FeSi$_2$ be a', b' and c', respectively. Then, its growth mode should be $b' = (1, -1, 0)a$ and $c' = (1, 1, 0)a$ as shown in FIGURE 6.8.

The length of the axis in Si[110] direction is 0.76792 nm, slightly smaller than those of β-FeSi$_2$ at 0.7884 and 0.7791 nm. In this case, β-FeSi$_2$ undergoes a compressive stress around 1.8% of the lattice mismatch.

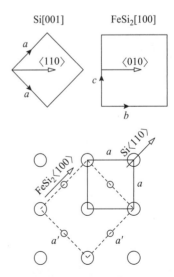

FIGURE 6.8 Type A heteroepitaxial relationship (from Reference 31).

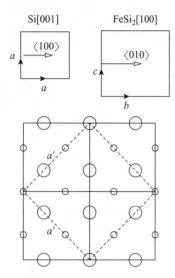

FIGURE 6.9 Type B heteroepitaxial relationship (from Reference 31).

As for the growth mode of Type B, β-FeSi$_2$[010] or [001] overlaps on top of Si[100], with a length ratio of $2b' = 3a$ and $2c' = 3a$ as shown in FIGURE 6.9. The triple of the lattice constant for Si a axis equals 1.629 nm, larger than both the double of b axis at 15.768 Å and that of c axis at 1.5582 nm for β-FeSi$_2$. Thus, β-FeSi$_2$ is subject to tensile stress around 4% of the lattice mismatch. Clark et al. simulated the energy band structures of both Type A and Type B for comparison.

In simulating Type B, Clark et al. assumed that β-FeSi$_2$ is completely epitaxially grown on Si substrate. Hence, parameters b and c axes of β-FeSi$_2$ were set at 1.5 times the axis length of the Si[100] direction, that is, 0.8145 nm. The simulated diagram is shown in FIGURE 6.10(a). Clark et al. found that both the direct and indirect bandgaps in the diagram were smaller than the bandgap of bulk β-FeSi$_2$ (please refer to Clark et al.'s bulk material simulation results discussed earlier). This suggests that as β-FeSi$_2$ undergoes tensile stress, the bandgap shrinks. The simulated direct bandgap is 0.62 eV and the indirect bandgap 0.53 eV. The difference around 0.1 eV between the two is quite significant. Thus, Type B is an indirect bandgap with a limited chance of illumination.

As for Type A simulation, Clark et al. assumed that β-FeSi$_2$ is completely epitaxially grown on Si substrate, setting parameters b and c axes of β-FeSi$_2$ at the exact axis length of the Si[110] direction, that is, 0.76792 nm. The simulated diagram is shown in FIGURE 6.10(b). They found that both the direct and indirect bandgaps in the diagram were larger than the bandgap of bulk β-FeSi$_2$. Thus, the bandgap widens when β-FeSi$_2$ undergoes compression stress. The direct bandgap measures 0.85 eV, while the indirect bandgap measures 0.82 eV, making them almost identical. Besides, the value of direct bandgap is very close to the experimental value. These two pieces of evidence both explain the illumination possibility of Type A.

To sum up, Clark et al. drew two key conclusions based on their simulations of Type A and Type B:

1. The energy band graph of β-FeSi$_2$ is extremely sensitive to the effect of stress. It widens under the effect of compression stress, but shrinks when sustaining tensile stress. This explains why bulk β-FeSi$_2$ is not capable of illumination, whereas thin film or quantum dot β-FeSi$_2$ is.
2. There are two types of epi-growth modes on the β-FeSi$_2$(100)/Si(001) interface, namely, Type A and Type B. Verification through simulation indicates that Type A growth mode has a greater illumination potential because its direct and indirect bandgaps have similar values and the direct bandgap value

Light-emitting iron disilicide

is consistent with the experimental value. Clark et al. named such illumination phenomenon as quasi-direct transition.

Thus, Clark et al. maintains that the illuminating feature of β-FeSi$_2$ is contributed by β-FeSi$_2$ Type A. The presence of Type A β-FeSi$_2$ therefore becomes essential in the manufacture of illumination devices.

6.3.2.2 Miglio and Meregalli's model

Miglio and Meregalli [32] used two different conditions to identify the effect of lattice distortion on the energy band diagram. The two conditions were the assumption that β-FeSi$_2$ epitaxial grown on Si substrate β-FeSi$_2$(100)/Si(001)) and that β-FeSi$_2$ epitaxially grown on SiGe substrate. The Si/Ge ratio was 60 : 40. The assumption of Type A growth mode applied in both conditions. Among them, the lengths of b and c axes on Si substrate are 7.6792 Å, but the ones on the Si$_{0.6}$Ge$_{0.4}$ are 7.81 Å. In comparison, β-FeSi$_2$ sustained compression stress in Si. In Si$_{0.6}$Ge$_{0.4}$, the b axis was subject to a slight compression stress, while the c axis was subject to a slight tensile stress. FIGURE 6.11 illustrates the simulations conducted by Miglio and Meregalli.

Miglio and Meregalli's simulation results indicate that point Y is very sensitive to stress variation. In FIGURE 6.12, the position of Y descends as the lattice constant decreases. More explanations are needed to determine the role point Y plays on the illumination quality of β-FeSi$_2$. As shown in FIGURE 6.12, the other two direct bandgaps, Γ and Λ, experience limited variation relative to changes in the lattice constant. However, the bandgap of point Y increases rapidly as the lattice constant decreases. Thus, the effect

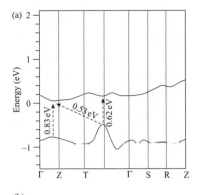

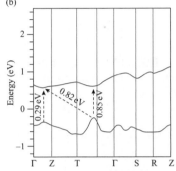

FIGURE 6.10 (a) Band structure of Type B β-FeSi$_2$, the arrows indicate the smallest direct and indirect bandgaps. (b) Band structure of Type A β-FeSi$_2$ constrained to the experimental lattice parameter of Si (from Reference 31).

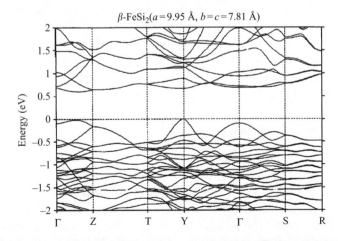

FIGURE 6.11 Energy band diagram of β-FeSi$_2$ from simulation done by Miglio and Meregalli [32].

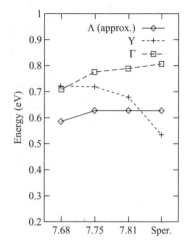

FIGURE 6.12 Trends of the direct gap of β-FeSi$_2$ (from Reference 32).

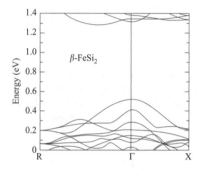

FIGURE 6.13 Band structure (LDA) of β-FeSi$_2$ for energies close to the bandgap (from Reference 14).

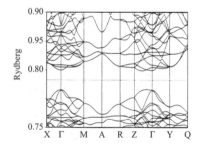

FIGURE 6.14 Band structure of β-FeSi$_2$ near the gap (from Reference 33).

of optical illumination comes primarily from the control of the position of point Y. In other words, by imposing the right level of stress to control the position of point Y, one can satisfy the condition required to achieve illumination. This is the exact discovery by Miglio and Meregalli.

Based on the results of the two teams in question, we can be certain that the energy band structure of β-FeSi$_2$ is very sensitive to stress. Thanks to the limited difference between the direct and indirect bandgaps, it takes only a slight stress to transform β-FeSi$_2$ from an indirect bandgap material to a direct one and achieve illumination. Furthermore, the effect of stress is particularly significant on point Y, thus pointing to a possible correlation between the illumination phenomenon and variation at point Y. The proper control of the energy gap at point Y becomes the decisive factor in determining whether β-FeSi$_2$ is capable of illumination. It is worth noting that the above simulations were all conducted under the assumption that the stress came only from lattice mismatch. But in reality, stress comes from the combination of three sources mentioned earlier. Therefore, in the realistic condition, net stress induces distortion in the lattice, which alters the position of the energy band. The situation of such alteration is more complex and harder to calculate than the simulations.

There are several other simulated band structure models worth mentioning and these are discussed in the following sections.

6.3.2.3 Christensen's model

Christensen [14] used LMTO to calculate the energy band under the assumption of local density approximation (LDA). His simulation results are illustrated in FIGURE 6.13. According to his findings, the bandgap of β-FeSi$_2$ is indirect. But the direct bandgap (Γ) is only slightly larger than the indirect bandgap. The simulated direct bandgap of 0.8 eV is not much different from the experimental value of 0.85 eV. In addition, the bandgap simulated under the assumption of LDA is generally smaller than the actual value. Hence, the simulation result is by and large consistent with the experimental result.

6.3.2.4 Eppenga's model

Eppenga [33] used *ab initio* calculations to simulate the energy band under the assumption of LDA. The results, shown in FIGURE 6.14, indicate β-FeSi$_2$ to be an indirect bandgap between Γ and Z with the value of 0.44 eV, while the direct bandgap is located at Γ with the value of 0.46 eV. The values calculated are significantly different from the experimental results. The assumption of LDA usually yields simulation results about 50% smaller

Light-emitting iron disilicide

than the actual bandgap. In spite of inaccuracy, Eppenga's diagram shows a trend very similar to Christensen's findings.

6.3.2.5 Filonov and coworkers' model

Under the same LDA assumption, Filonov et al. [7] simulated the energy band diagram, shown in FIGURE 6.15, using LMTO. Based on the simulation results, Filonov found that β-FeSi$_2$ possessed two sets of direct bandgaps, namely 0.742 eV at position Λ and 0.825 eV at position Y. Position Λ between Γ and Z has a lower energy level, representing the primary bandgap. The value of 0.742 eV is slightly smaller than the 0.8 eV value derived by Christensen [14] and the experimental value of 0.85 eV. Likewise, the LDA assumption may have contributed to the difference. Indirect bandgaps, shown in FIGURE 6.16, are located in all three directions near Λ, namely Λ–G, Λ–H and Λ–T. The simulated bandgaps are 0.741, 0.738 and 0.740 eV, respectively. All of these simulation values are slightly smaller than the 0.742 eV at position Λ, making β-FeSi$_2$ an indirect bandgap material.

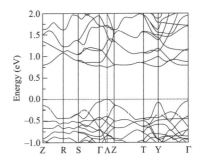

FIGURE 6.15 Band structure of β-FeSi$_2$ (from Reference 7).

6.3.3 Using Raman optical measurement to obtain the stress value

The previous sections discussed the effect of stress on the illuminating properties of β-FeSi$_2$. But how do we measure the stress value? The mentioned curvature approach is helpful in obtaining the stress caused by volume reduction, but not in determining the overall stress variation. To solve this problem, Birdwell and Glosser [29] proposed the use of Raman optical measurement to obtain information related to stress variation in β-FeSi$_2$.

In his experiment, Birdwell implanted Fe$^+$ ions through implantation onto a p-type Si(100) specimen with an energy level of 200 keV. During the implantation process, the specimen's temperature rose to 350°C because of the implanted ion beams. Birdwell's study is divided into two parts. First, the implantation density is set at 2×10^{17} Fe$^+$ cm^{-2}. The annealing temperature is then changed while the annealing time is set at 18 h. The experimental parameters are listed in TABLE 6.2. In the other part, the implantation density is changed to 2–4×10^{17} Fe$^+$ cm^{-2}, while the annealing temperature is set at 920°C for 18 h. The parameters are listed in TABLE 6.3. Such a density distribution generates three forms of β–FeSi$_2$: (1) the density of 2×10^{17} cm^{-2} yields polycrystal precipitates of the size of roughly 100 nm; (2) the density of 3×10^{17} cm^{-2} yields cross-linking precipitates; and (3) the density of 4×10^{17} cm^{-2} yields continuous β-FeSi$_2$ thin film. The Raman

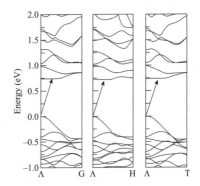

FIGURE 6.16 Band structure along some symmetry lines. The arrows indicate the global minimum position of the lowest conduction band (from Reference 7).

Light-emitting iron disilicide

TABLE 6.2   IBS β-FeSi$_2$ samples of a constant implant dosage that have been annealed at different temperatures (from Reference 29).

Sample	Dosage ($\times 10^{17}$ Fe$^+$ cm^{-2})	Anneal (°C/h)
F12A	2	600/18
F12B	2	800/18
F12	2	920/18
F12C	2	1000/18

TABLE 6.3   IBS β-FeSi$_2$ samples of varying dosages that have been annealed at the same temperature (from Reference 29).

Sample	Dosage ($\times 10^{17}$ Fe$^+$ cm^{-2})	Anneal (°C/h)
F12	2	920/18
F15	3	920/18
F11	4	920/18

measurement is carried out later on these specimens to determine the sustained stress.

While conducting Raman measurement, Birdwell first measured the bulk β-FeSi$_2$ sample to obtain the benchmark specimen. According to the measurement results, the peak value of β-FeSi$_2$ occurred at 247 cm^{-1}, which was consistent with the experimental data reported by other researchers. In the following measurement, position 247 cm^{-1} served as the benchmark for β-FeSi$_2$.

Results of the measurement of the first part, that is, TABLE 6.2, are shown in FIGURE 6.17. The figure shows that as the annealing temperature rises, the intensity of peak value increases and the full width at half maximum (FWHM) narrows. This explains why as the specimen's temperature rises, the highly disordered state gradually transforms to a more regular state. In addition, the peak value occurs at 246.3 cm^{-1}, showing a slightly red-shift phenomenon compared with bulk material. Though the reason for red-shift is yet to be ascertained, it may come from the tensile stress sustained by β-FeSi$_2$ itself or the stress caused by implantation damage (poor crystallinity). However, it is certain that at least part of red-shift is caused by stress.

As for the measurement of the second part (TABLE 6.3), the peak value position and implantation density are shown in

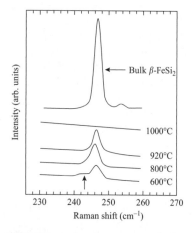

FIGURE 6.17   The 247 cm^{-1} Raman peak of 185 β-FeSi$_2$ as a function of anneal temperature. The Raman spectrum for β-FeSi$_2$ bulk is shown for comparison (from Reference 29).

FIGURE 6.18. Birdwell found that as the implantation density increased from 2×10^{17} to 3×10^{17} cm^{-2}, the level of red-shift also increased. However, as the density further rose from 3×10^{17} to 4×10^{17} cm^{-2}, a slight blue-shift emerged and the peak value position also regressed slightly. We cannot be certain that these phenomena are caused by stress. They could possibly have been the result of β-FeSi$_2$ crystal quality. In other words, the increase of red-shift can be the result of deteriorating quality of crystallinity. On the contrary, the phenomenon of blue-shift may suggest an improving crystalline quality. To distinguish the effect of stress from that of the quality of crystallinity, Birdwell also produced a diagram of FWHM values and implantation density, as shown in FIGURE 6.19. The figure shows that FWHM declines as the implantation density increases from 2×10^{17} to 3×10^{17} cm^{-2}. However, as the density increases from 3×10^{17} to 4×10^{17} cm^{-2}, the FWHM rises. FWHM is an indicator of the quality of crystallinity. Thus, the quality of crystallinity of continuous thin film is usually better than that of precipitate. According to Birdwell's measurement data, the quality of crystallinity deteriorates as the density increases from 3×10^{17} to 4×10^{17} cm^{-2}. One possible reason is that Si atoms on the surface layer are driven out of the surface during implantation, resulting in an insufficient amount of Si on the surface and declining quality of crystallinity. In any case, it is a fact that poor crystalline quality occurs as the density changes from 3×10^{17} to 4×10^{17} cm^{-2}. Thus, the blue-shift phenomenon observed during Raman measurement between 3×10^{17} and 4×10^{17} cm^{-2} cannot be interpreted as an improvement of the quality of crystallinity, but the result of stress relief. Meanwhile, the effect of stress from the lattice damage caused by implantation often generates the blue-shift. Thus, we may attribute the red-shift measured at 246.3 cm^{-1} to the tensile stress sustained by β-FeSi$_2$ itself.

Based on the Raman measurements, Birdwell concluded that overall β-FeSi$_2$ was subject to the effect of tensile stress. We know that volume reduction exerts a tensile stress effect on β-FeSi$_2$ [30]. As for lattice mismatch, the precipitate formed through implantation is buried inside the substrate, rendering the precipitate subject to the effect of three-dimensional stress. Thus, it is impossible to estimate the actual stress value imposed. Nevertheless, it may be concluded that the combined stress from the three sources of lattice mismatch, volume reduction and thermal expansion is a tensile stress. However, we have learned from energy band simulations that only a slight compression stress can achieve the effect of illumination. How to generate the compression stress required becomes a crucial issue.

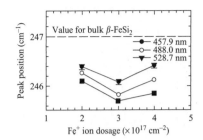

FIGURE 6.18 The 247 cm^{-1} Raman peak of β-FeSi$_2$ as a function of Fe$^+$ ion dosage for the three laser wavelengths used (from Reference 29).

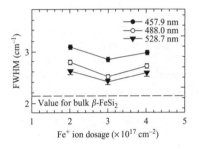

FIGURE 6.19 The FWHM of the 247 cm^{-1} Raman peak of β-FeSi$_2$ as a function of Fe$^+$ ion dosage (from Reference 29).

6.3.4 Effect of stress on the feature of β-FeSi$_2$

It is well known that β-FeSi$_2$ thin film has an indirect bandgap while strain free, and that the energy band graph of β-FeSi$_2$ is extremely sensitive to stress. Thus, it simply takes the appropriate stress to transform the indirect bandgap into a direct one. Suemasu et al. [26] also investigated the relationship between stress and the illumination feature as discussed below.

They started by varying the process conditions. They changed the film deposition temperature while covering p-type Si (0.3 μm) in order to observe the effect of different temperatures on β-FeSi$_2$ dots. They compared four different temperatures – 400, 500, 630 and 750°C – and found that the lower the temperature the stronger the photoluminescence (PL) signals (shown in FIGURE 6.20).

Cross-sectional TEM images are used here to clarify and explain the reasons. FIGURE 6.21 shows the TEM images at 500°C, on the left-hand side, and 750°C, on the right. The findings indicate that the specimen heated to 500°C must undergo 14 h of annealing at 900°C to completely form dots (FIGURE 6.21(e)). Island-like structures are formed before annealing (FIGURE 6.21(d)). But the specimen heated to 750°C does not require annealing to form dots, which approximate spheres even more than the dots formed on the 500°C specimen after the annealing duration of 14 h. Besides, the dots retain their shape when the 750°C specimen undergoes annealing. Thus, it is concluded that coating p-type Si at low temperatures allows β-FeSi$_2$ dots to acquire sufficient strain that transforms the indirect bandgap into a direct one. Coating p-type Si at high temperatures produces strain-free dots, which reduce the illumination effect.

To further understand the effect of strain on β-FeSi$_2$ particles, the researchers observed the specimens through HRTEM (FIGURE 6.22). FIGURE 6.22(a) shows the 750°C specimen and FIGURE 6.22(b) the 500°C specimen. They calculated the lattice constants of both specimens in the [100] direction and derived the constant of 0.99 nm for the 750°C specimen and 1.08 nm for the 500°C specimen. In other words, strain-free particles are found at 750°C, whereas the particles sustain around 9% tensile strain in the [100] direction at 500°C. Owing to such differences in strain, the consequent illuminating qualities differ between the two specimens. Therefore, it is widely believed that as long as β-FeSi$_2$ nanodots are given the appropriate stress during the process of growing dots, the illumination phenomenon can be achieved.

The above method applies to the preparation of ordinary specimens. To an LED device [25], p$^+$ Si (10^{18} cm^{-3}) 0.7 μm and n$^+$ Si (10^{18} cm^{-3}) 1 μm must be deposited as ohmic contacts.

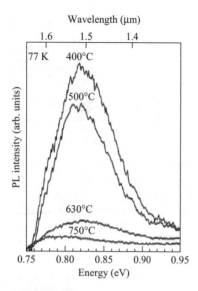

FIGURE 6.20 Dependence of the PL spectrum measured at 77 K on MBE-Si growth temperature for embedding β-FeSi$_2$ in Si is shown (from Reference 26).

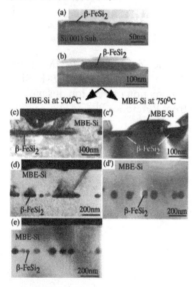

FIGURE 6.21 TEM images observed after (a) the 10 nm β-FeSi$_2$ growth and (b) the 850°C UHV annealing are shown. (c) and (c′) are images after 0.1 μm, and (d) and (d′) are those after 0.3 μm Si over growth, for the 500 and 750°C samples; (e) is after the 900°C anneal of the 500°C sample (from Reference 26).

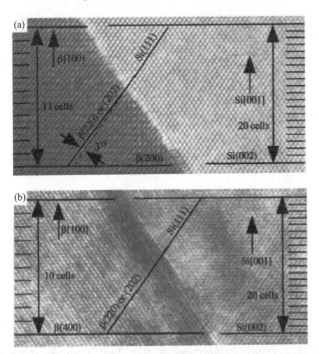

FIGURE 6.22 High-resolution TEM images of the β-FeSi$_2$/Si interface for (a) the 750°C and (b) the 500°C samples are shown (from Reference 26).

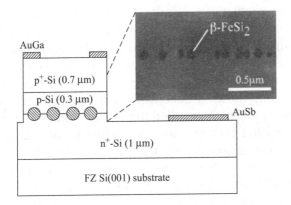

FIGURE 6.23 Schematic cross-section and TEM photograph of the fabricated diode (from Reference 25).

Then AuGa and AuSb contacts must also be deposited, as shown in FIGURE 6.23.

6.3.5 Optical properties

The most direct method to obtain the basic optical qualities is using optical transmission (OT) measurement [19] to measure the absorption coefficient. OT measurement allows us to conclude that the β-FeSi$_2$ continuous layer deposited by IBS is a direct bandgap

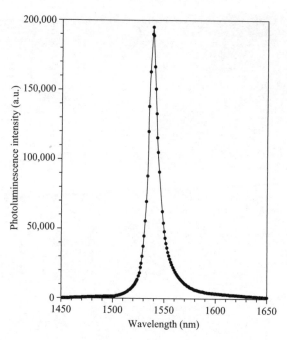

FIGURE 6.24 Typical PL spectrum, at 10 K, from a sample containing precipitates excited only by tungsten lamp (from Reference 19).

from low to room temperature. The bandgap measures around 0.87 eV at room temperature and 0.9 eV at 80 K. In addition to bandgap information, OT measurement also allows us to obtain information related to material qualities by producing a diagram of the absorption coefficient and energy. The pattern of absorption coefficient shows a decreasing tail beneath the location of bandgap energy, which is called the Urbach tail [19]. It is primarily caused by the energy states in the middle of the gap for two reasons: phonon disorder and structure disorder. The influence of phonon disorder is very distinct at room temperature, but negligible at 80 K. Thus, the Urbach tail reflects only structure disorder at low temperature. This difference provides valuable information on material properties.

At present, PL measurement has the most critical demand on optical properties. FIGURE 6.24 illustrates a typical example of β-FeSi$_2$ illumination using IBS [19]. The specimen undergoes implantation at a density of 3×10^{16} cm^{-2} under 1.5 MeV.

6.3.5.1 PL measurement

The PL peak (FIGURE 6.25) can be obtained by measuring the PL of a specimen at 77 K, regardless of whether it is an LED structure or a specimen directly grown on a floating zone (FZ) substrate [25]. The illuminating wavelength measures around 1.53 μm. As shown

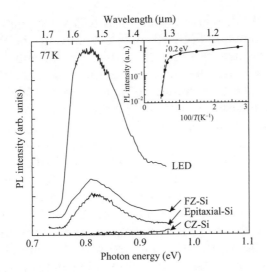

FIGURE 6.25 Si substrate dependence of PL spectrum measured at 77 K. The inset shows the temperature dependence of integrated PL intensity obtained for the epitaxial Si sample (from Reference 25).

in the figure, the PL intensity of the LED is greater than that of its FZ substrate. The reason may be that the LED structure grows on top of n^+ Si. The amount of electrons provided is greater than that on the specimen directly grown on the FZ substrate. The higher quantity of electron–hole pair recombination results in stronger PL signals.

Suemasu et al. further compared the influence of different substrates on the variation of PL signal intensity. They utilised three different substrates, namely n^+-type Czochralski (CZ), n^+-type FZ and n-type epitaxial Si(20 μm)/CZ n^+-type Si(001). Comparison of the findings shown in FIGURE 6.25 indicates that the specimen with FZ substrate shows the best illuminating quality. Suemasu et al. believe that the result can be linked to the oxygen content in the substrate surface. They maintain that FZ-Si has the lowest oxygen content and, thus, gives the best illuminating quality.

FIGURE 6.25 also shows the relationship between $100/T$ and PL intensity. The activation energy is calculated to be around 0.2 eV. Thus, the average defect level of β-$FeSi_2$ dots can be estimated at around 0.2 eV from the conduction band. As a result, the illumination effectiveness drops rapidly after the temperature reaches 100 K.

6.3.5.2 EL measurement

The electroluminescence (EL) measurement was conducted on the specimen in question by measuring the EL illumination at room

Light-emitting iron disilicide

temperature [25], which corresponds to a wavelength of around 1.6 μm (as shown in FIGURE 6.25). Relative to the PL signal of 1.53 μm measured at 77 K, the former wavelength suggests the red-shift. The reason behind this phenomenon is that the bandgap decreased as the temperature rose. As shown in the figure, signals can be measured once the electric current density exceeds 10 A/cm^2. After that, the intensity of the EL peak climbs rapidly as the electric current density increases. Suemasu et al. offered an explanation for this phenomenon. According to them, a number of defects still exist even after β-FeSi$_2$ dots have been treated by high temperature annealing. All of these defects act as non-radiative recombination centres. As electric currents are introduced, most of them pass through the defects first, preventing illumination. Only when the defects are saturated with the currents, will they begin to pass the radiative recombination centres (β-FeSi$_2$ dots), thus generating EL illumination. We may speculate the reason why illumination at room temperature cannot be achieved. Most of the excited electrons choose the non-radiative recombination centre as the top priority for recombination. Therefore, the highest barrier to achieve illumination at room temperature remains the defects inside the β-FeSi$_2$ dots. How to obtain higher quality β-FeSi$_2$ dots will be the key issue in the future.

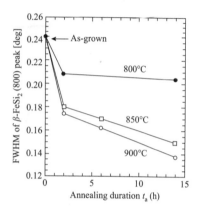

FIGURE 6.26 Dependence on annealing duration t_a on the FWHM of an XRD β-FeSi$_2$ (800) peak intensity for all the samples (from Reference 34).

6.3.5.3 Effect of high temperature annealing on PL illumination

Further treatment of high temperature annealing is necessary after specimen preparation is completed. High temperature annealing improves the crystallinity of β-FeSi$_2$ and eliminates non-radiative defects, leading to better PL measurements. Suemasu et al. compared different annealing temperatures and duration in search of the annealing conditions required to obtain the sharpest PL signals [34]. Their experimental results are described below.

FIGURE 6.26 shows the correlation between FWHM and duration of annealing. It is reported that FWHM would reach close to saturation at the annealing temperature of 800°C and a time period of over 2 h. However, at 850 and 900°C, FWHM continued to drop as the annealing duration increased. In addition, as the annealing temperature rises, it is expected to take longer time to reach saturation. The FWHM also narrows. Thus, it is widely believed that the higher the temperature, the sharper the PL peak obtained. FIGURE 6.27 shows the PL measurement after an annealing duration of 14 h at different temperatures. The sharpest PL peak occurs at 900°C, indicating that the temperature of 900°C and the duration of 14 h constitute the optimal high temperature annealing conditions.

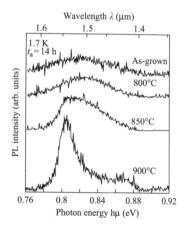

FIGURE 6.27 Dependence on annealing temperature of the PL spectra for the samples annealed at 800, 850 and 900°C for 14 h (from Reference 34).

REFERENCES

[1] S.P. Murarka, M.H. Read, C.J. Doherty, D.B. Fraser [*J. Electrochem. Soc. (USA)* vol.129 (1982) p.293]
[2] S. Wolf, R.N. Tauber [*Silicon Processing for the VLSI Era* (Lattice Press)]
[3] R. Iscoff [Photomask and Reticle Materials Review, *Semicond. Int.* (1986) p.82]
[4] D.A. McGillis [Lithography in *VLSI Technology* Ed. S.M. Sze (McGraw-Hill Book Co, New York, 1983) Ch. 7, p.267–301]
[5] A.B. Filonov, I.E. Tralle, N.N. Dorozhkin, D.B. Migas, V.L. Shaposhinkov [*Phys. Stat. Sol. (b) (Germany)* vol.186 no.1 (1999) p.209–15]
[6] M.C. Bost, J.E. Mahan [*J. Electron. Mater. (USA)* vol.16 no.6 (1987) p.389–95]
[7] A.B. Filonov, D.B. Migas, V.L. Shapohnikov, N.N. Dorozhkin, G.V. Petrov [*J. Appl. Phys. (USA)* vol.79 no.10 (1996) p.7708–12]
[8] H.C. Cheng, T.R. Yew, L.J. Chen [*J. Appl. Phys. (USA)* vol.57 (1985) p.5246]
[9] M.E. Schlesinger [*Chem Rev. (UK)* vol.90 (1990) p.607]
[10] W.B. Person [*Handbook of Lattice Spacing and Structure of Metals and Alloys* (Pergamon, New York, 1967)]
[11] H. Von Kanel, N. Onda, H. Sirringhaus, E. Mullergubler, S. Goncalvesconto, C. Schwarz [*Appl. Surf. Sci. (USA)* vol.70/71 (1993) p.559–63]
[12] J. Desimoni, H. Bernas, M. Behar, X.W. Lin, S. Washburn, Z. Lilientalweber [Ion-Beam Synthesis of Cubic $FeSi_2$, *Appl. Phys. Lett.* Vol.62 (1993) p.306–8]
[13] J. Alvarez, J.J. Hinarejos, E.G. Michel, R. Miranda [*Surf. Sci. (USA)* vol.287/288 (1993) p.490–4]
[14] N.E. Christensen [Electronic Structure of Beta-$FeSi_2$, *Phys. Rev. B (USA)* vol.42 (1990) p.7148–53]
[15] Y. Tomm, L. Ivaneko, K. Irmscher, et al. [*Mater. Sci. Engng B (Netherlands)* vol.37 (1996) p.215–18]
[16] J.H. Oh, S.K. Lee, K.P. Han, K.S. An, C.Y. Park [*Thin Solid Films (Netherlands)* vol.341 (1999) p.160–4]
[17] H. Sirrighaus, N. Onda, E. Muller-Gubler, P. Muller, R. Stalder, H. Von Kanel [*Phys. Rev. B (USA)* vol.47 (1993) p.10567]
[18] H.U. Nissen, E. Muller, H.R. Deller, H. Von Kallel [*Phys. Stat. Sol. A (Germany)* vol.150 (1995) p.395]
[19] K.P. Homewood, et al. [*Thin Solid Films (Netherlands)* vol.381 (2001) p.188–93]
[20] T. Yoshilake, T. Nagamoto, K. Nagayama [*Thin Solid Films (Netherlands)* vol.381 (2001) p.236–43]
[21] T. Suemasu, N. Hiroi, T. Fujii, K. Takakura, F. Hasegawa [*Jpn. J. Appl. Phys. (Japan)* vol.38 (1999) p.878–81]
[22] K. Takaura, T. Suemasu, Y. Ikura, F. Hasegawa [*Jpn. J. Appl. Phys. (Japan)* vol.39 (2000) p.789–91]
[23] J.Y. Natoli, I. Berbezier, A. Ronda, J. Derrien [*J. Crystal Growth (Netherlands)* vol.146 (1995) p.444–8]
[24] C.E. Zybill, W. Huang [*Inorg. Chim. Acta (Netherlands)* vol.291 (1999) p.380–7]
[25] T. Suemasu, Y. Negishi, K. Takaura, F. Hasegawa [*Jpn. J. Appl. Phys. (Japan)* vol.39 (2000) p.1013–15]
[26] T. Suemasu, Y. Nigishi, K. Takakura, F. Hasegawa [*Appl. Phys. Lett. (USA)* vol.79 (2001) p.17]

[27] M. Tanaka, Y. Kumagai, T. Suemasu, F. Hasegawa [*Jpn. J. Appl. Phys. (Japan)* vol.36 (1997) p.3620–4]
[28] M.G. Grimaldi, S. Coffa, C. Spinella, et al. [*J. Lumin. (Netherlands)* vol.80 (1999) p.467]
[29] A.G. Birdwell, R. Glosser [*J. Appl. Phys. (USA)* vol.89 (2000) p.965]
[30] D. Sander, A. Enders, J. Kirschner [*Appl. Phys. Lett. (USA)* vol.67 (1995) p.1834]
[31] S.J. Clark, H.M. Al-Allak, S. Brand, R.A. Abram [*Phys. Rev. B (USA)* vol.58 (1998) p.10389]
[32] L. Miglio, V. Meregalli [*J. Vac. Sci. Technol. B (USA)* vol.16 (1998) p.1064]
[33] R. Eppenga [*J. Appl. Phys. (USA)* vol.68 (1990) p.3027]
[34] T. Suemasu, Y. Iikura, T. Fujii, K. Takaura, N. Hiroi, F. Hasegawa [*Jpn. J. Appl. Phys. (Japan)* vol.38 (1999) p.620–2]

Chapter 7

Silicide contacts for Si/Ge devices

J.E. Burnette, M. Himmerlich and R.J. Nemanich

7.1 SCOPE OF THE CHAPTER

Electronic devices based on Si/Ge alloys are already important for high frequency applications [1] and infrared detectors [2]. Heterojunction bipolar transistors (HBTs) which were first reported in 1988 [3], have demonstrated operating frequencies significantly greater than 100 GHz. Moreover, the electronic properties of Si/Ge with respect to and in combination with silicon offer significant advantages for integrated circuit technology [4, 5]. Some examples of the potential of Si/Ge in CMOS technology are displayed in FIGURE 7.1. In fact, the combination of HBT and CMOS will provide the potential for monolithic silicon-based RF systems [1].

The development of ohmic or rectifying contacts to these semiconducting Si/Ge layers is a critical step in the development of the device technologies. Understanding the formation and electronic states of contacts to Si/Ge semiconducting layers begins with the clean surfaces. The formation and stability of the contacts will depend on the interface chemistry, and for Si/Ge, the system shows significant complexity in comparison to silicide contacts to silicon surfaces. Finally, the device operation will depend on the electrical properties of the interface. This chapter presents summaries of the current understanding of each of these areas.

Bulk crystalline silicon and germanium in the diamond crystal structure form a continuous alloy solid solution with no other phases in the binary phase diagram. While the cubic lattice constant varies continuously and nearly linearly from that of Si to that of Ge, the alloys show three different average bond lengths representative of Si/Si, Si/Ge, and Ge/Ge bonds [6, 7]. There is only a slight variation of these bond lengths over the alloy concentration. Evidence suggests that the bulk alloys are indeed random alloys without preferential ordering or bonding. Using the common chemical notation we would write $Si_{1-x}Ge_x$ to represent the alloy.

Scope of the chapter p.175

Surface properties p.176
  $Si_{1-x}Ge_x$/Si(001) surfaces p.176
  $Si_{1-x}Ge_x$-Si(111) surfaces p.178
  Stoichiometry of $Si_{1-x}Ge_x$ surfaces p.179

Formation and stability p.181
  Interface thermodynamics p.181
  Ti(Si/Ge) on Si/Ge p.184
  $CoSi_2$ on Si/Ge p.187
  NiSi on Si/Ge p.190
  Other metals p.192

Electrical properties p.194

Summary p.196

Acknowledgment p.197

References p.197

Silicide contacts for Si/Ge devices

However, the notation usually employed for semiconducting alloy films of silicon and germanium has been written as $Si_{1−x}Ge_x$. We will adopt this tradition in this report. We will also write Si/Ge to mean $Si_{1−x}Ge_x$ alloys without a specific x value.

The solid-state reaction of a metal with silicon can lead to the formation of a silicide compound. This process has formed the basis of silicide contacts to silicon. Similarly, the solid-state reaction of a metal with Si/Ge, can lead to a compound with the metal, and Si and Ge. These compounds have sometimes been termed germanosilicides. We will use this terminology also.

7.2 SURFACE PROPERTIES

This section presents a description of Si/Ge surfaces from the perspective of growth on silicon surfaces. Here, we present the structure and growth modes of Si/Ge on the two common surfaces for silicon device technology, namely the (001) and (111) faces. The stoichiometry of the surfaces is then considered.

7.2.1 $Si_{1−x}Ge_x$/Si(001) surfaces

Silicon atoms on the Si(001) surface form buckled dimers to minimise the surface energy by reducing the number of dangling bonds (DBs) to one per atom. The dimers align in rows resulting in an LEED pattern characteristic of a 2×1 unit cell mesh. As a result of the diamond structure, at every mono-atomic step of the substrate the dimerisation axis is rotated by 90° leading to alternating (2×1) and (1×2) surface diffraction patterns. The two different step structures have been labelled as S_A when the dimer axis on the upper terrace is perpendicular to the step edge, and S_B when the axis is aligned parallel to the step edge [8].

The growth of $Si_{1−x}Ge_x$ and Ge on silicon is strongly influenced by the lattice mismatch between the layer and the substrate (4.2% between Ge and Si) resulting in successive strain relief mechanisms. Depending on the mismatch, after the formation of a thin wetting layer, a transition occurs to three-dimensional growth, which results in strain relaxation. This process is typical of the Stranski–Krastanov growth mode. The details of the growth of Si/Ge heteroepitaxial layers has been recently reviewed [9].

In the first stages of the deposition of Ge on Si(001) in the sub-monolayer region (0.1–0.2 ML) (ML = monolayer), mixed asymmetric (buckled) Si/Ge dimer rows form with Ge on the upper positions [10, 11]. Chen et al. [12] determined the Si/Ge bond length to be 2.43 ± 0.10 Å with a tilt angle of $31 \pm 2°$. The compressive stress that develops during further deposition

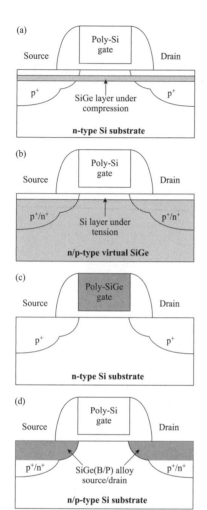

FIGURE 7.1 Schematic MOSFETs with the grey regions showing where the Si/Ge alloys can be present to enhance the device performance: (a) compressively strained channel, (b) relaxed virtual substrate for tensile-strained Si atop, (c) tuneable workfunction gate, and (d) low-resistivity source/drain extension. Reprinted with permission from Reference 5, S.-L. Zhang, *Microelectron. Engng. (UK)* vol.70 (2003) p. 174–85.

Silicide contacts for Si/Ge devices

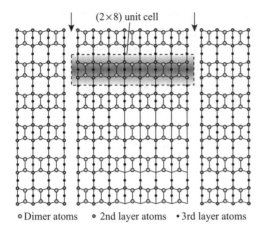

FIGURE 7.2 Schematic model of the atomic configuration of a Ge/Si(001)-(2 × 8) reconstructed surface. The missing dimer rows are indicated by the arrows.

leads to the formation of defects at a Ge thickness of less than 1 ML. The defects appear as periodic arrays of missing dimer rows. This results in a $2 \times n$ surface reconstruction, as confirmed by STM and LEED measurements [13, 14], where every nth dimer is missing. The periodicity, n, strongly depends on the layer thickness, decreasing with increasing thickness to a lower limit of $n = 8$ for 3 ML of Ge, as well as on the Ge content of the $Si_{1-x}Ge_x$ alloy [15, 16]. The alignment of these trenches follows the structure of the Si substrate, which is rotated by 90° at every single layer step edge. A schematic example of a (2 × 8) reconstructed surface and an STM image of Si(001) covered with Ge are shown in FIGURES 7.2 and 7.3, respectively. The absence of the dimer rows leads to two competing mechanisms: stress relaxation and increased surface energy. The structure allows for partial relief of compressive strain, but at the cost of an increase in the number of DB, which increases the surface energy. Theoretical calculations taking these two aspects into account find values for the periodicity in good agreement with the experimental results [17–19].

Since the electronic structure of semiconductor surfaces and interfaces plays a crucial role in the performance of semiconductor devices, it is surprising that there have been only a few studies of the electronic states of the Ge/Si(001) system. In a photoemission study of strained and relaxed $Si_{1-x}Ge_x$ alloys two surface states were observed with energies near the valence band maximum [20]. Through hydrogen adsorption, it was deduced that these states are due to the DB of the surface dimer atoms. A second surface state associated with the dimer back-bond state was observed. This study showed that the dispersion of the surface states was relatively independent of the composition of $Si_{1-x}Ge_x$ as well as the strain of

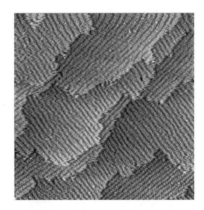

FIGURE 7.3 An STM image showing the Ge wetting layer on a Si(001) surface with a (2 × n) surface reconstruction (image area: 1600 Å × 1600 Å, $T = 575$ K, Ge coverage: 1.26 Å). Adapted from Reference 19.

the layer. A recent photoemission spectroscopy study used polarised synchrotron light and single domain Ge:Si (2 × 1) surfaces [21]. Additional surface states were observed which appeared to be similar to those of the clean Si(100) 2 × 1 surface. This study detected splitting of the DB surface state, which was attributed to the buckled dimer.

7.2.2 $Si_{1-x}Ge_x$-Si(111) surfaces

The deposition of Ge on Si(111) is also initiated with the growth of a two-dimensional pseudomorphic wetting layer. Initially, triangular islands form which then merge into a complete layer [22]. For Ge on Si(111), two types of reconstructions have been observed: the (7 × 7)-pattern analogous to the surface of clean Si(111), and a reconstruction with a (5 × 5) unit mesh [23, 24]. The amount of Ge on the surface apparently determines which Ge stabilised structure is observed. The Ge:Si (7 × 7) is observed if there is an excess of Si atoms on the Ge covered surface. This can occur for low Ge deposition [25, 26], for $Si_{1-x}Ge_x$ alloys with low x [23], or when the surface is annealed at high temperatures. In the latter case, the annealing may lead to intermixing of Si and Ge atoms on the surface caused by Si diffusion from the substrate [27, 28]. The Ge:Si (5 × 5) surface may be characterised by the presence of both Si and Ge in random adatom positions but with a predominance of Ge adatoms [25, 29]. The 7 × 7 and 5 × 5 structures are both very similar and can be described by related dimer-adatom-stacking (DAS) fault models [30, 31]. The surface reconstruction consists of a double layer with two triangular subunits, which are faulted or unfaulted with respect to the stacking sequence of the Si substrate. Most of the atoms in the first layer have their DB saturated by bonds with the adatoms while the so-called "restatoms" retain their threefold coordination with one DB. The atoms in the second layer are dimerized along the boundaries of the subunits. At the corner of each unit cell exists a vacancy in the first layer, which leads to the "corner hole" with a free DB. A schematic of the (5 × 5) reconstruction is shown in FIGURE 7.4, and an STM image of a surface with 3 ML of Ge on Si(111) is shown in FIGURE 7.5.

The surface electronic structure of Si(111) (7 × 7)-Ge and of Si(111) (5 × 5)-Ge has been investigated by electron energy loss spectroscopy [27]. A peak observed at a loss energy of 1.4 eV was attributed to the Ge DB, and peaks at loss energies of 9 and 14 eV were attributed to back-bond surface states. Miller et al. [32] found evidence of surface states on the Ge covered surface using angle-integrated photoemission measurements. In a detailed photoemission study, Mårtensson et al. [28] observed the

Silicide contacts for Si/Ge devices

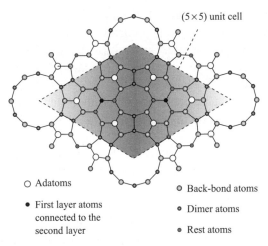

○ Adatoms
● First layer atoms connected to the second layer
◉ Back-bond atoms
◌ Dimer atoms
● Rest atoms

FIGURE 7.4 Plan view schematic of the DAS model of the Ge-stabilised 5×5 structure. Adatoms and atoms in the stacking-fault layer are occupied by Ge. Adapted with permission from Reference 31.

adatom DB surface state at ~ 0.3 eV below the Fermi level, another DB state arising from the restatom at 1.05 eV, and the back-bond state at ~ 1.7 eV. A comparison of the electronic states of the 5×5 and 7×7 surfaces indicated that the spectra are similar with only slight shifts in binding energy. Inverse photoemission spectroscopy was employed and an empty surface state located at ~ 0.7–0.8 eV above E_F was observed. These results were confirmed by later experiments [26, 33]. Theoretical studies [34, 35] indicated that the DAS models can explain the density of states of the clean Si(111) surface and that Ge addition or permutation on the Si surface is rather favourable compared to substitution of Si atoms.

7.2.3 Stoichiometry of $Si_{1-x}Ge_x$ surfaces

To employ $Si_{1-x}Ge_x$ structures in semiconductor technology, the realisation of homogeneous films and atomically sharp interfaces plays an important role in determining the quality of the devices. Various studies have shown that these two basic demands are not always fulfilled during the growth of $Si_{1-x}Ge_x$-Si and Si-Ge-Si structures.

The indications of Si/Ge intermixing during the growth of a Si capping layer on Ge were observed by Raman spectroscopy and medium energy ion scattering (MEIS) [36]. Using secondary-ion mass spectroscopy (SIMS) Zalm et al. [37] found segregation of Ge during the growth of a Si layer on $Si_{1-x}Ge_x$. It was suggested that the minimisation of the surface free energy causes the enrichment of Ge at the surface layer during the initial growth of the $Si_{1-x}Ge_x$ layer which results in a depletion of the germanium in the first layers causing a degradation of the interface ("leading edge")

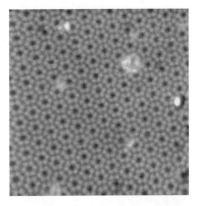

FIGURE 7.5 STM image of a Ge/Si(111) surface after deposition of 3 ML of Ge at 500°C. $I = 1$ nA, $V = +3$ V, (230×230) Å^2. Image reproduced with permission from Reference 22.

until the Ge concentration achieves equilibrium with the deposition flux. Similarly, in the growth of a Si capping layer, further Ge segregation takes places in combination with incorporation of Ge into the Si layer causing a smeared Si-$Si_{1-x}Ge_x$ interface ("trailing edge"). Detailed experiments showed that the slope of the trailing edge depends on the substrate as well as on the growth temperature and the deposition rate [38] and can be optimised by changing these parameters. The segregation effect was stronger for Si(001) as compared to Si(111). Fukatsu and coworkers investigated Ge segregation at the Si-Ge interface, observing a non-exponential incorporation of Ge at high concentrations in a Si capping layer which they attributed to a self limited process in a two-state exchange model [39, 40]. Their model appears successful in describing the temperature dependence of the Ge segregation at the trailing edge. An extension to the Si-$Si_{1-x}Ge_x$ interface was performed by Godbey and Ancona [41]. They found a strong dependence of the width of the leading edge on the composition of the $Si_{1-x}Ge_x$ alloy while the self limited process resulted in a similar profile for the trailing edge.

Additional experiments indicated that the surface segregation is not limited to the topmost monolayer as assumed in the initial model. Without the Si capping layer it was found that the surface of the $Si_{1-x}Ge_x$ alloy consisted of several layers that were highly enriched in Ge [42]. Jernigan et al. [43] found that the top layer consisted of only Ge atoms while the second surface layer also had a higher concentration of Ge relative to the $Si_{1-x}Ge_x$ alloy. The observation of more than one layer of enriched Ge on the surface does not match the two-state model. They proposed that the segregation behavior is strongly influenced by the film stress leading to stronger Ge segregation in the case of higher Ge concentration and that the temperature dependence of the leading edge is influenced by intermixing of Ge with the Si buffer layer. Further indication of strain driven Ge surface segregation was deduced from measurements of the local Ge composition of undulating $Ge_{0.2}Si_{0.8}$ layers. Both vertical and lateral Ge segregation was found [44]. The Ge concentration on top of the ripples was found to be higher than at the troughs indicating the diffusion of Ge to the top of the ripples allowing a reduction of the strain at the surface.

Recently, X-ray photoelectron spectroscopy (XPS) combined with low-energy ion scattering spectrometry (LEISS) was used to examine the composition of the top layers of $Si_{1-x}Ge_x$ [45]. The result is a top surface layer consisting of nearly 100% Ge while the amount of Ge incorporated in the second layer is still strongly enriched depending on the composition of the desired $Si_{1-x}Ge_x$ alloy. The layers below exhibit the equilibrium composition until the region of the $Si_{1-x}Ge_x$-Si interface is reached where a depletion of Ge atoms is present in the leading edge. To describe the observed results, a new model was proposed [45, 46] assuming a continuum of Ge states with binding energies that decreased from the surface to the bulk.

The use of column III or V atoms as surfactants during Si/Ge growth can result in suppression of Ge segregation. Pre-deposition of a monolayer of materials like Ga, Sb and As [37, 47, 48] supersedes Ge as the segregating material by lowering the surface free energy during growth. This approach can lead to sharper interfaces and a more uniform material distribution. It seems that the incorporation of the extrinsic atoms is almost negligible due to the strong tendency to segregate to the surface. The presence of hydrogen during GSMBE and chemical vapour deposition (CVD) has also been found to decrease the Ge segregation and, therefore, lead to an improvement of the precision of the interfaces [49–51].

7.3 FORMATION AND STABILITY

7.3.1 Interface thermodynamics

One of the key issues in the formation of high-quality ohmic and rectifying contacts is the thermodynamic stability at the contact/semiconductor interface. Stability at the interface is governed by the solid phase reactions between constituent components at the phase separation boundary. The solid phase reactions which occur are influenced by both reaction kinetics and thermodynamic factors that determine the final outcome, and hence, the stability in a particular system.

In current VLSI technology, silicide forming metals are often used to form contacts in the gate and source/drain regions on doped Si. The most commonly used metals are Ti and Co, due in part to the formation of low resistivity silicides through solid phase reaction with Si. Similar solid phase reactions of metals with Ge can be employed for contacts to germanium devices. These germanide compounds are in some cases structurally isomorphic to their silicide counterparts.

Since Si and Ge are structurally isomorphic, have similar electronic properties, and are miscible in one another, the concept of forming germanosilicide contacts through a solid phase reaction between a metal and Si/Ge may be realised when both the silicide and germanide phases share a common crystal structure. TABLE 7.1 lists the composition and structure of some terminal metal silicide and germanide phases.

Interface thermodynamics play a crucial role in the formation of a stable contact interface. The thermal stability between two

TABLE 7.1 Composition and structure of terminal refractory silicide and germanide phases. The silicide and germanide phases of Ni, Co, Pt, and Pd do not exist in the C54, C11$_b$, C40, or C49 structures, but their respective structures are listed.

Refractory metal	Final silicide		Final germanide	
	Phase	Structure	Phase	Structure
Titanium	TiSi$_2$	C54	TiGe$_2$	C49
Zirconium	ZrSi$_2$	C49	ZrGe$_2$	C49
Hafnium	HfSi$_2$	C49	HfGe$_2$	C49
Vanadium	VSi$_2$	C40	VGe$_2$	C40
Niobium	NbSi$_2$	C40	NbGe$_2$	C40
Tantalum	TaSi$_2$	C40	TaGe$_2$	C40
Chromium	CrSi$_2$	C40	–	–
Molybdenum	MoSi$_2$	C11$_b$	MoGe$_2$	C23
Tungsten	WSi$_2$	C11$_b$	WGe$_2$	C11$_b$
Cobalt	CoSi$_2$	C1	CoGe$_2$	C$_e$
Nickel	NiSi	B31	NiGe	B31
	NiSi$_2$	C1	–	–
Platinum	Pt$_2$Si	C22	Pt$_2$Ge	C22
	PtSi	B31	Pt$_2$Ge	B31
	PtSi$_2$	B31	PtGe$_2$	C35
Palladium	Pd$_2$Si	C22	Pd$_2$Ge	C22
	PdSi	B31	PdGe	B31

coexisting phases can be determined by considering changes in the Gibbs free energy of the two-phase system. The change in the Gibbs free energy for each phase can be expressed as:

$$\Delta G = \Delta H - T \Delta S \quad (7.1)$$

where ΔG is the change in the Gibbs free energy, usually given in units of kJ/mol, ΔH is the enthalpy of formation for a compound phase, and ΔS is the entropy of mixing. Equilibrium in the two-phase system can be established by determining the minimum of the total Gibbs free energy of the system, given by EQN (7.2),

$$\Delta G_{\text{TOT}} = \Delta G_{\text{RMGS}} + \Delta G_{\text{Si/Ge}} \quad (7.2)$$

where ΔG_{RMGS} is the free energy change in the metal germanosilicide compound and $\Delta G_{\text{Si/Ge}}$ is the free energy change in the Si/Ge solid solution. Experimentally, the achievement of equilibrium may be marked by the formation of precipitates, as the system liberates Ge or a Si/Ge compound to minimise the total free energy.

The concept of thermodynamic equilibrium at the interface can be studied by assuming that the phases of interest can be modelled such that the expressions for the free energies are functions of the Ge concentration in both the Si/Ge solid solution and the metal germanosilicide phase. In systems where the refractory metal silicides and germanides form a continuous solid solution, a one-phase domain connects the silicide and germanide solid phases. In general, it can be assumed that the Si/Ge solid solution behaves as an ideal solid solution, and its contribution to the total free energy of the system is due to the mixing of Si and Ge atoms, and is given by

$$\Delta G_{Si/Ge} = RT[x_{Ge} \ln x_{Ge} + (1 - x_{Ge}) \ln(1 - x_{Ge})] \quad (7.3)$$

where x_{Ge} is the concentration of Ge by atomic percentage of the Si/Ge solid solution. The free energy of the metal germanosilicide compound solid solution can be expressed as

$$\Delta G_{RMGS} = \Delta H_{RMGS} + bRT[y_{Ge} \ln y_{Ge} + (1 - y_{Ge}) \ln(1 - y_{Ge})] \quad (7.4)$$

where y_{Ge} is the concentration of Ge in the metal germanosilicide layer, $\Delta H_{RMGS} \equiv \Delta H[y_{Ge}]$ is the enthalpy of formation of the metal germanosilicide compound, and is, in general, a function of y_{Ge}, and b is the fraction of Si/Ge atoms available to participate in mixing. The equilibrium solution of these relations leads to a situation where the relative Si and Ge concentrations are different in the Si/Ge solid solution and the metal germanosilicide. The formation of the Si/Ge precipitates may then occur in these systems in order to bring the interfaces into equilibrium.

In other systems, a metal germanosilicide compound may be initially formed in a metal-Si-Ge solid phase reaction, however, the metal germanosilicide phase created may not be stable in contact with the Si/Ge solid solution. In this case, the metal germanosilicide compound might phase segregate, sometimes in stages, leading to the formation of silicide phases that are thermodynamically stable with the solid solution, and also to the expulsion of Ge atoms or a Si/Ge precipitate. The reduction in the total Gibbs free energy for such an event can be expressed as

$$\Delta G_A \Rightarrow \Delta G_B \Rightarrow \cdots \Rightarrow \Delta G_Z \quad (7.5)$$

where A, B, ..., Z represent the successive free energy gains associated with the formation of the various silicides leading to

Silicide contacts for Si/Ge devices

the formation of the most stable silicide on Si/Ge, and $\Delta G_Z < \cdots < \Delta G_B < \Delta G_A$. Other materials systems may adopt these behaviours or a combination of these behaviours in order to decrease the total free energy of the system.

7.3.2 Ti(Si/Ge) on Si/Ge

The Ti/Si/Ge solid phase reaction has been reported to result in the formation of a Ti(Si$_{1-y}$Ge$_y$)$_2$ solid solution adjacent to the Si$_{1-x}$Ge$_x$ substrate [52–55]. Experimental results have shown that initially, the titanium germanosilicide compound contains the same ratio of Si and Ge as the substrate

$$\frac{Si}{Ge} : \frac{1-y}{y} : \frac{1-x}{x} \qquad (7.6)$$

After further annealing at a constant temperature, the formation of Ge-rich Si/Ge precipitates was observed at the titanium germanosilicide grain boundaries, resulting in a decrease in the overall Ge concentration in the Ti(Si$_{1-y}$Ge$_y$)$_2$ layer, that is, $y < x$. This occurrence, which is shown in FIGURE 7.6, is thought to be driven by differences in the enthalpy of formation between TiSi$_2$ and TiGe$_2$ compounds in the expression for the total Gibbs free energy of the system.

A Gibbs free energy model was developed by Aldrich et al. [56, 57], in which the total free energy of the system was given as a sum of the Gibbs free energies of 1 mol of the formed Ti(Si$_{1-y}$Ge$_y$)$_2$ equilibrium compound and 1 mol of the Si$_{1-x}$Ge$_x$ solid solution. The thermal stability was explored at a temperature of 700°C, where the formation of TiSi$_2$ and TiGe$_2$ has been observed. The Gibbs free energy of the Ti(Si$_{1-y}$Ge$_y$)$_2$ equilibrium compound is expressed as

$$\Delta G_1 = m(y-z) + \frac{2}{3}RT[(y-z)\ln(y-z) + (1-(y-z))\ln(1-(y-z))] \qquad (7.7)$$

where m is equal to the difference in the enthalpy of formation between TiSi$_2$ and TiGe$_2$ compounds, while that of the Si$_{1-x}$Ge$_x$ solid solution is expressed as,

$$\Delta G_2 = RT\left[\left(x+\frac{2}{3}z\right)\ln\left(x+\frac{2}{3}z\right) + \left(1-\left(x+\frac{2}{3}z\right)\right) \times \ln\left(1-\left(x+\frac{2}{3}z\right)\right)\right] \qquad (7.8)$$

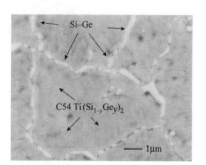

FIGURE 7.6 Micrograph of the surface morphology of C54 Ti(Si$_{1-y}$Ge$_y$)$_2$ on Si$_{1-x}$Ge$_x$. The C54 TiM$_2$ film was formed by reacting 40 nm of Ti with ~220 nm of Si$_{0.67}$Ge$_{0.33}$ at 700°C. Upon heating, the formation of Ti(Si$_{0.67}$Ge$_{0.33}$)$_2$ is observed initially. During further annealing the composition of the compound changes and germanium-rich Si/Ge alloy precipitates form along the grain boundaries of the compound. Reproduced with permission from Reference 56.

and the total free energy of the system is given by $\Delta G_{TOT} = \Delta G_1 + \Delta G_2$. As the reaction proceeds, the Ge concentration of the Ti(Si$_{1-y}$Ge$_y$)$_2$ layer decreases by an amount z. Since two-thirds of the germanosilicide is composed of Si and Ge atoms, the Ge concentration of the solid solution is expected to increase by an amount $\frac{2}{3}z$ as atoms are exchanged across the interface. Given the initial condition $y = x$, the minimisation of the total free energy allows the determination of the equilibrium value of the Ge concentration in both the Ti(Si$_{1-y}$Ge$_y$)$_2$ equilibrium compound and the Si$_{1-x}$Ge$_x$ solid solution.

This model was then used to determine stability in the Ti-Si/Ge system at different temperatures, and to describe the Ti-Si/Ge thin film reaction in relation to the formation of the Ge-rich Si/Ge precipitates. A ternary phase diagram indicating stability between compounds in the two-phase field for different substrate compositions is shown in FIGURE 7.7. The precipitation process is related to differences in the mobilities of the Si and Ge atoms in both the Ti germanosilicide compound and the Si$_{1-x}$Ge$_x$ substrate. The mobilities of Si and Ge in the substrate are assumed to be negligible relative to the mobilities in the compound.

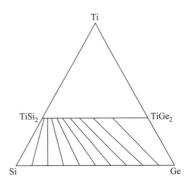

FIGURE 7.7 The Ti/Si/Ge ternary phase diagram at 700°C, indicating phase stability between a Ti(Si$_{1-y}$Ge$_y$)$_2$ film and Si$_{1-x}$Ge$_x$ substrate for compositions $x = 0.10, 0.20, \ldots, 0.90$.

If the atomic mobilities of Si and Ge in both the Ti(Si$_{1-y}$Ge$_y$)$_2$ compound and Si$_{1-x}$Ge$_x$ substrate were similarly high, the system would be directly driven to equilibrium, with a stable interface and the absence of precipitation. However, this is not the case for the temperatures commonly employed for the reaction. Diffusion in the Si/Ge layer is limited, and upon forming a Ti(Si$_{1-y}$Ge$_y$)$_2$ equilibrium compound, the diffusion of Si and Ge from the substrate into the Ti(Si$_{1-y}$Ge$_y$)$_2$ layer is due to consumption of the substrate. The replacement of Ge for Si on the Ti(Si$_{1-y}$Ge$_y$)$_2$ lattice due to enthalpy of formation differences is then accompanied by the formation of the Ge-rich Si/Ge precipitates. The initial concentration of Ge in the Si/Ge precipitate will correspond to the composition that is in equilibrium with the initial Ti(Si$_{1-y}$Ge$_y$)$_2$ compound. As the system evolves, and precipitation continues, the replacement of Ge for Si on the lattice will result in a shift in the Ge concentration of the Ti(Si$_{1-y}$Ge$_y$)$_2$ layer towards that value, which is in equilibrium with the substrate, and the concentration of the Ge-rich Si/Ge precipitates will evolve towards that of the substrate. The precipitation phenomenon occurs until the Ti(Si$_{1-y}$Ge$_y$)$_2$ layer, the Si$_{1-x}$Ge$_x$ substrate, and the precipitates are in thermodynamic equilibrium.

Similar results were obtained in a study by Lai and Chen [58], where the effects of composition on the formation temperature and electrical resistivity in the Ti/Si/Ge system were examined using AES, TEM, and glancing-angle X-ray diffraction. Both the C54 Ti(Si$_{1-y}$Ge$_y$)$_2$ formation temperature and the temperature at which Si/Ge precipitation occurred were shown to decrease with increasing Ge content. Precipitates were formed in all samples annealed above 800°C.

The thermal stability of $Ti(Si_{1-y}Ge_y)_2$ thin films on strained $Si_{1-x}Ge_x$ was explored at a temperature of 700°C; a temperature high enough to ensure the formation of the $Ti(Si_{1-y}Ge_y)_2$ compound while minimising morphological degradation [59]. Based on previous work [56, 57], the incorporation of an a-Si interlayer before the final Ti metal deposition was investigated as a means of eliminating the formation of Si/Ge precipitates. It was found that the presence of the amorphous layer did indeed affect the formation of precipitates in the solid phase reaction of $Ti(Si_{1-y}Ge_y)_2$ on $Si_{0.8}Ge_{0.2}$ and $Si_{0.7}Ge_{0.3}$ substrates.

Depending on the amorphous layer thickness, the initial value for the concentration of Ge in the $Ti(Si_{1-y}Ge_y)_2$ layer could be altered. The Gibbs free energy model [56, 57] was invoked to determine the concentration of the $Ti(Si_{1-y}Ge_y)_2$ film that would be in equilibrium with $Si_{0.8}Ge_{0.2}$ and $Si_{0.7}Ge_{0.3}$ substrates. Once the equilibrium values had been established, the change in the Ge concentration due to the a-Si layer incorporation was determined using the following relation

$$y' = \left[\left(1 - \frac{t_{Si}}{2.27\, t_{Ti}}\right) x\right] \qquad (7.9)$$

where y' is the Ge concentration of the $Ti(Si_{1-y}Ge_y)_2$ layer, t_{Si} is the thickness in angstroms of the a-Si layer, t_{Ti} is the thickness of the Ti layer, and the factor of 2.27 accounts for differences in density and atomic weight between Ti and Si. It was determined that the precipitation phenomena occurred in samples when y' was greater than the equilibrium value for the concentration in the $Ti(Si_{1-y}Ge_y)_2$ layer, and precipitation did not occur when y' was less than the equilibrium value, which was confirmed by XRD and SEM measurements.

It was found that the precipitation phenomenon was directly controlled by the availability of Ge in the $Ti(Si_{1-y}Ge_y)_2$ layer. When y' was greater than the equilibrium value, thermodynamic stability was achieved by the replacement of Ge with Si in the $Ti(Si_{1-y}Ge_y)_2$ lattice, as both Si and Ge diffused from the substrate into the equilibrium metal germanosilicide compound, which led to the formation of the Si/Ge precipitates as in the first case [53, 54]. With y' less than the equilibrium value, thermodynamic stability would be achieved by replacing Si with Ge in the lattice, which would result in the formation of Si-rich Si/Ge precipitates. However, Ge is not available in sufficient quantities in the Ti(Si$_{1-y}$Ge$_y$)$_2$ layer to accommodate the formation of the Si-rich precipitate. Moreover, the incorporation of Si and Ge from the substrate would lead to the phase segregation of the Ti(Si$_{1-y}$Ge$_y$)$_2$ compound into TiSi$_2$ and Ge, both of which are unstable with the Si$_{1-x}$Ge$_x$ substrate. As a result, the Ge concentration of the Ti(Si$_{1-y}$Ge$_y$)$_2$ compound retains its value of y' in order to maintain stability with the substrate.

An investigation on the formation of amorphous interlayers during Si and Ge diffusion in Ti was performed by Lai and coworkers [60]. Thin films of Ti were deposited by UHV e-beam evaporation on both Sb-doped(001) and (111) wafers and on UHV CVD grown 1.5 µm thick Si$_{1-x}$Ge$_x$ ($x = 0.3, 0.4$ and 0.7) and 1.5 µm thick Si$_{1-y}$Ge$_y$ (y varies from 1 to $1-x$) layers on phosphorus-doped Si(001) wafers.

The samples were annealed in a diffusion furnace under N$_2$ ambient at temperatures between 300 and 500°C, and examined using XRD, AES, STEM, HRTEM, and XTEM, and the results were compared to those observed during the amorphisation of Ti layers on Si [61]. Amorphous interlayers less than 2 nm thick formed in all of the as-deposited samples, and a non-monotonic variance in amorphous layer thickness as a function of concentration was seen at temperatures up to 430°C. In addition, the crystalline phase formation temperature decreased with increasing Ge content, and a maximum interlayer thickness between 11 and 14 nm was observed.

A qualitative analysis correlated interlayer thickness with the chemical driving force for amorphisation. The formation of a thicker interlayer in the Ti-Ge system as compared with the Ti-Si system was attributed to differences in the mobility of Si and Ge atoms in Ti.

7.3.3 CoSi$_2$ on Si/Ge

In the solid phase reaction of Co on a Si$_{1-x}$Ge$_x$ substrate, the formation of CoSi$_2$ has been observed [62]. The formation of the CoSi$_2$ phase was preceded by the formation of CoSi, and was found to be affected by the initial thickness of the Co layer and the Ge concentration of the substrate. Both thin film reaction processes were accompanied by segregation of Ge, which is thought to be driven by changes in the Gibbs free energy that lead to the formation of the most stable phase on Si/Ge.

Initially, the reaction of Co with the substrate leads to the formation of a Co(Si$_{1-y}$Ge$_y$) compound, which has been observed at a temperature of 400°C [62]. The Ge concentration in the Co(Si$_{1-y}$Ge$_y$) layer was also observed to have a value that was less than that of the substrate ($y < x$), and it was found to decrease with increasing annealing temperature. This change was associated with the expulsion of Ge from the $Co(Si_{1-y}Ge_y)$ compound, which leads to the formation of a CoSi compound adjacent to the substrate. The reaction can be expressed as

$$Co(Si_{1-y}Ge_y) + Si_{1-x}Ge_x \to CoSi + Ge \qquad (7.10)$$

The expulsion of Ge is driven by a Gibbs free energy difference between CoGe and CoSi, which is obtained assuming that the heat of formation for the ternary phase is a linear interpolation between the heats of formation for both binary compounds.

Further annealing at higher temperatures, in the neighbourhood of 700°C, resulted in the formation of a polycrystalline $CoSi_2$ layer. The formation of $CoSi_2$ occurs during the reaction of the newly formed CoSi layer with the substrate. The reaction can be expressed as

$$CoSi + mSi_{1-x}Ge_x \to CoSi_2 + (m-1)Si_{1-y}Ge_y \qquad (7.11)$$

where m is the number of moles of $Si_{1-x}Ge_x$ that react with CoSi, and $y = (m/(m-1))x$.

The driving force for the formation of $CoSi_2$ is given by ΔG, the total change in the Gibbs free energy:

$$\Delta G = \Delta G_{CoSi \to CoSi_2} + RT[(m-1)(y \ln y + (1-y)\ln(1-y)) \\ - m(x \ln x + (1-x)\ln(1-x))] \qquad (7.12)$$

where $\Delta G_{CoSi \to CoSi_2}$ is the driving force for the formation of $CoSi_2$ from CoSi in the Co-Si solid phase reaction, and the remainder is the entropy change which describes the expulsion of Ge and the formation of Ge-rich $Si_{1-y}Ge_y$. A ternary phase diagram, which indicates stability between alloys below 50 at.% Co at 760°C and above 50 at.% Co at 950°C is shown in FIGURE 7.8 [63].

It was established that an amorphous Si interlayer could again be employed to limit phase segregation. Wu and coworkers [64] compared the properties of 15 nm Co layers deposited directly on $Si_{0.7}Ge_{0.3}$ and with an amorphous Si interlayer. Here, the interlayer was chosen such that the formation of $CoSi_2$ would lead to essentially complete consumption of the interlayer. In this case the amorphous Si layer was 50 nm thick. The results showed that Ge segregation could be avoided and that low resistivity $CoSi_2$ films could be prepared with this approach.

The $CoSi_2$ layers formed in these studies were found to be polycrystalline. Other approaches have been employed for the formation of epitaxial $CoSi_2$ on Si/Ge. The growth of epitaxial

Silicide contacts for Si/Ge devices

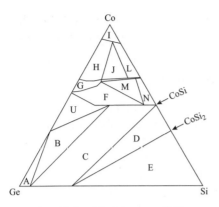

A: $CoGe_2$ + Si–Ge solid solutions up to 6% Si
B: CoGe + Co(Si, Ge) + Si–Ge solid solution of 6% Si
C: Co(Si, Ge) + Si–Ge solid solutions of 6 to 29% Si
D: CoSi + $CoSi_2$ + Si–Ge solid solution of 29% Si
E: $CoSi_2$ + Si–Ge solid solutions of 29 to 100% Si

FIGURE 7.8 Co/Si/Ge ternary phase diagram at 760°C for alloys below 50 at.% Co and at 950°C for alloys above 50 at.% Co. Adapted with permission from Reference 63.

$CoSi_2$ on Si/Ge(001) has been achieved [65] by the use of a modified templating method which had been previously developed as a means of forming epitaxial $CoSi_2$ on Si(100) [66]. The deposition of the modified template at room temperature, with a layered template structure of 2 ML Si-1 ML Co-2 ML Si, followed by a codeposited 0.2 nm/0.73 nm Co/Si layer was shown to result in the formation of an epitaxial $CoSi_2$ film.

The use of a Ti interlayer has also been shown to result in the formation of epitaxial $CoSi_2$ on Si/Ge [67]. The Ti interlayer method has been previously shown to promote the formation of $CoSi_2$ on Si(100) [67–70]. Ti layers of varying thickness between 1 and 10 nm were deposited onto a c-Si-$Si_{1-x}Ge_x$ layer pseudomorphically strained to Si(100), before the final Co metallisation step. The samples were annealed in UHV at 700°C for 15 min, and were characterised by AES, SEM, XRD and XAFS measurements. It was found that the presence of the Ti interlayer promoted the formation of epitaxial $CoSi_2$, and the ideal interlayer thickness was estimated to be between 5 and 6 nm. Previous studies of the Co-Ti-Si(100) system lend credence to the possibility that the primary role of the interlayer is to control the diffusion of Co atoms towards the substrate, as well as influence the nucleation of $CoSi_2$ by slowing down CoSi grain boundary and interface diffusion. Since it has been estimated that the solid solubility of Ti in CoSi and $CoSi_2$ is less than 1% [69, 71], its presence is not expected to influence the formation of CoSi$_2$. It may be expected that the same holds true in the Co-Ti-c-Si-Si$_{1-x}$Ge$_x$ solid phase reaction.

7.3.4 NiSi on Si/Ge

NiSi has been shown to be a promising candidate in the formation of deep submicron integrated circuits, due to its low resistivity and its stability during thermal processing necessary for the fabrication of device structures. In investigations of the Ni-Si$_{1-x}$Ge$_x$ solid phase reaction by Seger and Zhang [72], Zhang [73], and Jarmar et al. [74], the morphological and phase stability of Ni/Si/Ge compounds on strained and polycrystalline Si$_{1-x}$Ge$_x$ was examined. In the study from Reference 72, a 54 nm thick Si$_{0.8}$Ge$_{0.2}$ layer pseudomorphically strained to Si(100) was metallised with 20 nm thick Ni layers, and annealed under an RTP process at temperatures between 400–850°C. The samples were characterised by four-point probe sheet resistance measurements and XRD, where the formation of a Ni(Si$_{1-x}$Ge$_x$) phase whose composition matched that of the substrate was observed. In addition, the films were found to be highly textured along the [010] direction, suggesting an epitaxial relationship to the [100] orientation of the substrate. It was determined that the morphological stability of the Ni(Si$_{1-x}$Ge$_x$) films under the compressive strain provided by the Si$_{0.8}$Ge$_{0.2}$ substrate was improved as compared to NiSi grown on a tensile strained Si layer on a relaxed Si$_{0.8}$Ge$_{0.2}$ substrate [75].

The complete formation of the Ni(Si$_{0.8}$Ge$_{0.2}$) layer was achieved at 450°C, and the Ni(Si$_{0.8}$Ge$_{0.2}$) layer remained stable at temperatures up to 850°C. This was attributed to entropy of mixing effects in the Ni/Si/Ge alloy, which suppressed the phase separation of the Ni(Si$_{0.8}$Ge$_{0.2}$) layer.

The phase stability and morphology of Ni(Si$_{1-u}$Ge$_u$) films on polycrystalline Si$_{1-x}$Ge$_x$ was investigated by Jarmar et al. [74], in which 20 nm thick Ni films were deposited by e-beam evaporation onto 42.5–57.5 nm thick polycrystalline Si$_{1-x}$Ge$_x$ layers, with $x = 0.00$, 0.29 and 0.58, which were grown on thermally oxidised Si(100). The samples were annealed in N$_2$ within a rapid thermal processing chamber at temperatures of 400–850°C, and they were characterised by cross-sectional transmission electron microscopy (XTEM), X-ray diffraction (XRD), energy dispersive spectroscopy (EDS), and sheet resistance measurements. It was determined that for Si$_{0.42}$Ge$_{0.58}$ substrates, the Ge concentration of both the germanosilicide and the Si$_{1-x}$Ge$_x$ layer are equal, or $u = x$, at temperatures of 500–600°C. However, the Ge concentration u of the Ni(Si$_{1-u}$Ge$_u$) layer shifted to a value of 0.12–0.15 in the temperature range of 650–850°C. This shift in concentration was correlated with the agglomeration of the Ni(Si$_{1-u}$Ge$_u$) layer and the formation of Ge rich Si/Ge precipitates in the vicinity of the germanosilicide grains. A separation of the Ni(Si$_{1-u}$Ge$_u$) grains, in addition to an increase in grain size and the migration of the grains towards the poly-Si$_{1-x}$Ge$_x$ layer was also observed. This grain migration towards the substrate was deemed responsible for the further expulsion of Ge, which proceeds until the germanosilicide layer is in thermodynamic equilibrium with the Si$_{1-x}$Ge$_x$ substrate.

The experimental results were described qualitatively by a Gibbs free energy model, similar to the model used by Aldrich et al. [56, 57], which analysed stability in the Ti-Si/Ge system. According to the calculated ternary phase diagram in the Ni/Si/Ge system [74], stability exists between NiSi$_2$ and Ni(Si$_{1-u}$Ge$_u$), NiSi$_2$ and Si$_{1-x}$Ge$_x$, and Ni(Si$_{1-u}$Ge$_u$) and Si$_{1-x}$Ge$_x$. However, the absence of NiSi$_2$ within the range of formation temperatures suggests that a pseudophase ternary diagram, which excludes NiSi$_2$, would be more appropriate. The pseudophase diagrams for the ternary Ni/Si/Ge system at 600 and 700°C are shown in FIGURES 7.9(a) and (b).

To avoid the effects of phase segregation, Wu and coworkers [76] employed an amorphous Si interlayer between the Ni layer and a Si$_{0.7}$Ge$_{0.3}$ epitaxial layer on Si(001) substrates. The results established that phase segregation was avoided when an appropriate thickness amorphous Si layer was used. If the layer was too thin, Ge was observed to segregate at the interface and an uneven interface was measured.

In a related approach, the formation of NiSi on strained Si has been examined [75]. In the Ni-Si solid phase reaction, a Ni-rich Ni$_2$Si phase forms at 200°C, followed by the formation of NiSi and NiSi$_2$ at 500 and 800°C, respectively. An increase in sheet resistance occurs as the monosilicide–disilicide transformation takes place, and it is affected by the morphology of the NiSi surface, which is influenced by the NiSi-Si interfacial energy.

The effect of strain on the thermal stability of NiSi on Si was examined by growing pseudomorphically strained 40 nm Si layers on a relaxed Si$_{0.8}$Ge$_{0.2}$ substrate, which was 300 nm thick, and grown on a Si(100) substrate at 550°C. This was followed by the deposition of a 10 nm Ni film under UHV conditions, with anneals between 300 and 1000°C. Control samples, consisting of a 10 nm Ni film on Si(100) were also grown.

The increased in-plane lattice constant of the Si$_{1-x}$Ge$_x$ layer relative to that of bulk Si, due to the presence of Ge, ensured that the pseudomorphically grown Si layer was under tensile strain. The bi-axial strain in the Si layer affects its electronic properties as well, since the strain decreases the crystal symmetry of the

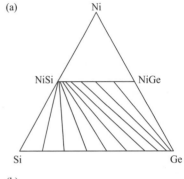

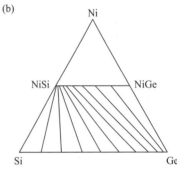

FIGURE 7.9 (a) Ternary phase diagram of the Ni/Si/Ge system at 600°C, indicating stability between a Ni(Si$_{1-u}$Ge$_u$) compound and a Si$_{1-x}$Ge$_x$ solid solution. (b) Ternary phase diagram of the Ni/Si/Ge system at 700°C, indicating stability between a Ni(Si$_{1-u}$Ge$_u$) compound and a Si$_{1-x}$Ge$_x$ solid solution. Reproduced with permission from information obtained from Reference 72.

Si unit cell, thereby causing a splitting of the conduction band degeneracy, resulting in higher electron mobilities.

The films were characterised by SEM, Raman spectroscopy, LEED, AES, XPS, and PEEM, to observe morphology, confirm the absence of oxygen contamination, determine phase transition temperatures, and to measure the amount of strain present in the pseudomorphically strained layer. The sheet resistivity of the samples was also measured using a four-point probe apparatus.

Experimental results indicate that the NiSi phase was found to be more stable on the strained Si(100) substrate than on the bulk Si(100) substrate. A measurement of the sheet resistivity as a function of temperature showed a difference of 100°C in the onset of increased resistivity in strained NiSi as compared to the NiSi film grown on bulk Si. It was proposed that islanding of the NiSi film was responsible for the increase in resistivity, since the formation of $NiSi_2$ was delayed up to 900°C. The delay in the resistivity increase was related to the fact that the interface stress of the strained film is less than that of the film grown on bulk Si. In this case, higher temperatures would be required to drive the islanding process.

7.3.5 Other metals

The formation of other types of refractory metal silicide or germanosilicide thin films on Si/Ge may be predicted, based on similarities in crystal structures between the silicide and germanide phases, as given in TABLE 7.1. Both zirconium and hafnium have terminal C49 silicide and germanide phases, and the vanadium, niobium, and tantalum silicide and germanide phases exhibit a terminal C40 structure.

The formation of $Zr(Si_{1-x}Ge_x)_2$ thin films on $Si_{1-x}Ge_x$ has been reported [77]. In contrast to the Ti-Si/Ge system, the formation of the zirconium germanosilicide film was found to occur with the absence of the Ge-rich Si/Ge precipitates, and independent of the Ge concentration of the $Si_{1-x}Ge_x$ layer. It was also observed that the Ge concentration of the C49 $Zr(Si_{1-x}Ge_x)_2$ film matched that of the substrate ($y = x$) at temperatures up to 700°C. The absence of phase segregation was related to the larger average enthalpy of formation between $ZrGe_2$ and $ZrSi_2$, the lower mobility of Si and Ge in Zr, and a higher barrier for Ge segregation due to a smaller driving force for phase segregation, all of which are related to the nature of Zr-Si/Ge chemical bonding. A ternary phase diagram of the Zr/Si/Ge system, indicating stability between the $Zr(Si_{1-x}Ge_x)_2$ and $Si_{1-x}Ge_x$ phases for different substrate compositions at 700°C, is shown in FIGURE 7.10.

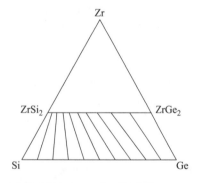

FIGURE 7.10 The Zr/Si/Ge ternary phase diagram calculated at 700°C, indicating phase stability between the C49 $Zr(Si_{1-y}Ge_y)_2$ film and $Si_{1-x}Ge_x$ substrate for compositions $x = 0.10, 0.20, \ldots, 0.90$.

There have been limited results concerning the formation of other types of refractory metal silicides and germanosilicides through solid phase reaction with $Si_{1-x}Ge_x$. However, a promising candidate for study includes Cr, as the ternary phase diagram in the Cr/Si/Ge system predicts stability between $CrSi_2$, a Cr/Si/Ge ternary compound, and a Ge-rich $Si_{1-x}Ge_x$ alloy phase [78]. However, a lack of miscibility between Si-rich and Ge-rich Cr compounds excludes the existence of a continuous Cr/Si/Ge solid phase.

An examination of the Cr-Si/Ge solid state reaction [79] details the formation of a Si-rich Cr/Si/Ge compound, which was located between the $Si_{1-x}Ge_x$ layer and a Ge-rich CrGe phase at temperatures as low as 350°C. The layered phase separation in this system was attributed to the high mobility of Ge in the Cr/Si/Ge layer. Perhaps a viable scheme to employ in the formation of thermodynamically stable $CrSi_2$ contacts to $Si_{1-x}Ge_x$ would be the use of an amorphous-Si layer which would be deposited before the final metallisation step, in the stoichiometric amounts required for the formation of $CrSi_2$.

Other refractory metals, such as vanadium, niobium, and tantalum have been shown to exhibit superconducting properties in their various silicide and germanide phases [80–86], and the formation of stable compounds to $Si_{1-x}Ge_x$ could be important in future technological developments. Of these, vanadium has been shown to exhibit partial miscibility between its silicide and germanide phases, so it is probable that the formation of a stable V/Si/Ge compound on $Si_{1-x}Ge_x$ can be achieved.

In addition, the metallisation of $Si_{1-x}Ge_x$ by other transition metals, such as platinum and palladium, has been explored [72, 74, 79, 87–92]. The thermal stability of PtSi contacts to $Si_{1-x}Ge_x$ was investigated by Hong et al. [79], where both Pt and Si were codeposited to form PtSi films to a final thickness of 10 nm under Pt-rich, Si-rich, and stoichiometric conditions on relaxed $Si_{0.5}Ge_{0.5}$ substrates of 260 nm thickness. The samples were annealed at temperatures which ranged from 300 to 800°C and were characterised by RBS, glancing-angle XRD, SEM and TEM measurements. The measurements indicated stability of both the stoichiometric and Si-rich PtSi layers with the substrate at temperatures up to 650°C, while the interface became unstable above 700°C. This transition was marked by a change in surface morphology and the eventual penetration of the PtSi layer into the substrate with further increases in temperature, all of which were associated with the low eutectic temperature between PtSi and Si. However, the intermixing of the Pt-rich PtSi layer and substrate was found to occur at temperatures up to 650°C.

The $Pd-Si_{1-x}Ge_x$ solid phase reaction has been investigated by Buxbaum et al. [89], in order to determine the reaction products. The $Si_{0.84}Ge_{0.16}$ layers were pseudomorphically strained to Si(100) and grown to a thickness of 230 nm, and were metallised by 130 nm thick Pd layers. The samples were annealed at temperatures ranging from 200 to 650°C between 2 and 4 h, and were characterised by XRD, TEM, AES, RBS and EDS, where the formation of $Pd_2(Si_{1-y}Ge_y)$ and Pd_2Ge phases adjacent to the substrate was confirmed. The germanosilicide films were composed mostly of highly textured $Pd_2(Si_{1-y}Ge_y)$ with the c-axis oriented along the [100] direction of the substrate. An AES depth profile of samples reacted at 250 and 550°C for 4 h indicated that the Ge concentration in the $Pd_2(Si_{1-y}Ge_y)$ layer decreased with an increase in annealing temperature, and indicated the formation of a Ge-rich layer between the germanosilicide film and substrate. The phase separation was attributed to the low solubility of Ge in the $Pd_2(Si_{1-y}Ge_y)$ matrix and to a lower enthalpy of formation for Pd-silicides as opposed to Pd-germanides.

7.4 ELECTRICAL PROPERTIES

The silicide or germanosilicide contacts to Si/Ge layers in device structures will be used for both ohmic and rectifying contacts. The electrical properties of a metal–semiconductor interface are typically described in terms of the Schottky barrier that develops at the interface. The Schottky barrier will determine the rectifying characteristics of the junction and will also contribute to the resistance of an ohmic contact. An ohmic contact is typically obtained through the use of a highly doped region of the semiconductor. This highly doped region leads to a thin depletion region in the semiconductor, and electrons from either the metal or semiconductor can readily tunnel through the barrier. A rectifying contact is obtained by choosing a metal layer with an appropriate Schottky barrier on a lightly doped semiconductor. For rectifying and ohmic contacts it is necessary to have a metal–semiconductor interface that is both chemically stable and electrically uniform. In both cases an intimate contact is usually preferred over a diffuse interface or one with substantial contamination. The prior sections have presented the considerations for controlling the surface structure and chemical stability of the silicide-Si/Ge or germanosilicide-Si/Ge interface.

While there have been many studies of the electrical properties of metal contacts on Si/Ge, it is disconcerting to note that there have been very few studies of the Schottky barrier of stable silicide-Si/Ge or germanosilicide-Si/Ge contacts. We are left then with reviewing the data of the metal layers on Si/Ge alloys and attempting to glean some understanding of the Schottky barrier that might develop from the chemically stable and uniform contacts.

Silicide contacts for Si/Ge devices

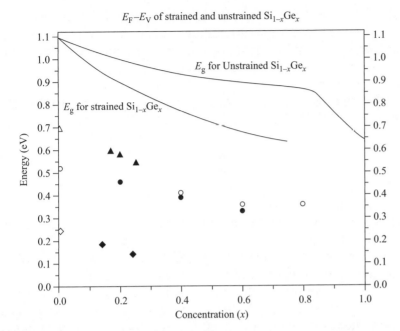

FIGURE 7.11 The p-type Schottky barrier heights of Al, Co, and Pt on strained and relaxed $Si_{1-x}Ge_x$ alloys, and the bandgaps as reported by Schwartzentruber et al. [8]. The diamonds represent Pt, the triangles represent Al, and the circles represent Co [93–95]. The open symbols represent contacts to relaxed layers, and the filled symbols represent strained layers.

For $Si_{1-x}Ge_x$ layers the bandgap depends on both the Ge concentration (x) and on the strain in the film. FIGURE 7.11 displays the variation in bandgap for two limits: bulk $Si_{1-x}Ge_x$ (i.e. unstrained), and $Si_{1-x}Ge_x$ that is pseudomorphically strained such that the in-plane lattice constant matches that of a Si(001) substrate.

We consider a simple picture of the Schottky barrier, where the barrier is the energy difference between the metal Fermi level and the conduction band or valence band of the semiconductor. Within this picture we would expect that the Schottky barriers of a specific metal to p- and n-Si/Ge would add up to the bandgap. The "ideal" Schottky barrier is described by the Schottky–Mott model where the n-type barrier is equal to the difference of the metal work function and the electron affinity of the semiconductor. However, for silicon based materials, this model does not describe the measurements, and a more complex model that can account for the development of an interface dipole is typically adopted. The metal induced gap states (MIGS) model has been considered to describe the general trends observed for metal Schottky barriers on Si.

The Schottky barrier can be deduced from several different techniques. Some of the most common are current–voltage, capacitance–voltage, photo-response, and electron photoemission.

All of the measurements have advantages and disadvantages and they measure somewhat different quantities. Still, through careful measurements and analysis of uniform structures with intimate interfaces, similar results are often deduced from each of the different approaches.

The reported p-type Schottky barrier results for Al [93], Co [94], and Pt [95] on Si/Ge are presented in FIGURE 7.11. In this figure the symbols each represent a measured value with respect to the Si/Ge valence band. The position of the points is expected to be the metal Fermi level position at the interface for both n- and p-type substrates. The n-type Schottky barrier would be expected to be the difference of the bandgap and the p-type Schottky barrier. Note that the strain dependence of the bandgap will be significant.

Several trends are evident in the data. For low values of x and for strained layers, the p-type Schottky barrier decreases as x increases. It appears that the n-type Schottky barrier remains relatively constant in this range. There have been a few studies where the results have compared the Schottky barrier on strained and relaxed Si/Ge of the same x. In some cases there was no measured difference, and in others there was a tendency for the p-type Schottky barrier to be smaller on the strained substrates, but not to the degree expected for the predicted change in bandgap. This dependence on strain evidently requires more detailed study. We note that this strain dependence is predicted to be a very significant effect for devices with $x > 0.2$.

7.5 SUMMARY

With the significant potential of Si/Ge in a range of semiconducting devices, the need for stable ohmic and rectifying contacts will be of increasing importance. While the bulk crystal properties of Si/Ge alloys are now largely established, the properties of surfaces and interfaces are only now being studied. Certainly, the role of strain is an important factor for device structures based on epitaxial layers. This study has established that at every level the properties are significantly more complex than those for contacts on silicon or germanium. The surfaces show additional reconstructions and morphology that develop due to the strain in the films. The interface stability and reactions involve complex thermodynamics and kinetics, and establishing chemical equilibrium is more complex than seeking adjoining phases in a binary phase diagram. Moreover, the electrical properties of these stable interfaces have not been established. While we can draw upon our knowledge of silicide and germanide contacts to Si or Ge, the properties of a specific interface may vary substantially from these estimates. However, there have been significant advances in our understanding. Through thermodynamic calculations and analysis based on the ternary phase diagram, it appears that we can determine regions of stability and instability, and we can predict the products of a solid-state reaction between a metal layer and Si/Ge. More than this, the route to establish thermodynamically stable interfaces on Si/Ge is truly a model system, which will lead to the development of the scientific basis for stability with binary alloys and also the relation to the electronic properties.

ACKNOWLEDGMENT

We gratefully acknowledge the support of the National Science Foundation.

REFERENCES

[1] J.D. Cressler [*IEEE Trans. Microwave Theory Technol. (USA)* vol.46 (1998) p.572–89]
[2] J.R. Jimenez, X. Xiao, J.C. Sturm, P.W. Pellegrini [*Appl. Phys. Lett. (USA)* vol.67 (1995) p.506–8]
[3] G.L. Patton, S.S. Iyer, S.L. Delage, S. Tiwari, J.M.C. Stork [*IEEE Electron Device Lett. (USA)* vol.9 (1988) p.165–7]
[4] P.M. Mooney, J.O. Chu [*Annu. Rev. Mater. Sci. (USA)* vol.30 (2000) p.335–62]
[5] S.-L. Zhang [*Microelectron. Engng. (UK)* vol.70 (2003) p.174–85]
[6] D.B. Aldrich, R.J. Nemanich, D.E. Sayers [*Phys. Rev. B (USA)* vol.50 (1994) p.15026–33]
[7] C. Tzoumanekas, P.C. Kelires [*Phys. Rev. B (USA)* vol.66 (2002) p.1952009-1–1952009-11]
[8] B.S. Schwartzentruber, N. Kitamura, M.G. Lagally, M.B. Webb [*Phys. Rev. B (USA)* vol.47 (1993) p.13432]
[9] C. Teichert [*Phys. Rep. (USA)* vol.365 (2002) p.335]
[10] L. Patthey, E.L. Bullock, T. Abukawa, S. Kono, L.S.O. Johannsson [*Phys. Rev. Lett. (USA)* vol.75 (1995) p.2538]
[11] X.R. Qin, B.S. Swartzentruber, M.G. Lagally [*Phys. Rev. Lett. (USA)* vol.84 (2000) p.4645]
[12] X. Chen, D.K. Saldin, E.L. Bullock, et al. [*Phys. Rev. B (USA)* vol.55 p.R7319]
[13] U. Köhler, O. Jusko, B. Müller, M. Horn-von Hoegen, M. Pook [*Ultramicroscopy (Netherlands)* vol.42–44 (1992) p.832]
[14] F. Iwawaki, M. Tomitori, O. Nishikawa [*Ultramicroscopy (Netherlands)* vol.42–44 (1992) p.902]
[15] R. Butz, S. Kampers [*Appl. Phys. Lett. (USA)* vol.61 (1992) p.1307]
[16] L.W. Guo, Q. Huang, Y.K. Li, S.L. Ma, C.S. Peng, J.M. Zhou [*Surf. Sci. (USA)* vol.406 (1998) p.L592–L596]
[17] J. Tersoff [*Phys. Rev. B (USA)* vol.45 (1992) p.8833]

[18] F. Liu, M.G. Lagally [*Phys. Rev. Lett. (USA)* vol.76 (1996) p.3156]
[19] B. Voigtländer, M. Kästner [*Phys. Rev. B (USA)* vol.60 (1999) p.R5121]
[20] J.-H. Ku, R.J. Nemanich [*Phys. Rev. B (USA)* vol.54 (1996) p.14102]
[21] H.W. Yeom, J.W. Kim, K. Tono, I. Matsuda, T. Ohta [*Phys. Rev. B (USA)* vol.67 (2003) p.085310]
[22] N. Motta, A. Sgarlata, R. Calarco, et al. [*J. Vac. Sci. Technol. B (USA)* vol.16 (1998) p.1555]
[23] H.-J. Gossmann, J.C. Bean, L.C. Feldman, W.M. Gibson [*Surf. Sci. (USA)* vol.138 (1984) p.L175]
[24] R.S. Becker, J.A. Golovchenko, B.S. Swartzentruber [*Phys. Rev. B (USA)* vol.32 (1985) p.8455]
[25] R. Cao, F. Bozso, Ph. Avouris [*J. Vac. Sci. Technol. A (USA)* vol.10 (1992) p.2322]
[26] R. Cao, F. Bozso, Ph. Avouris [*J. Vac. Sci. Technol. A (USA)* vol.10 (1992) p.2322]
[27] S. Hasegawa, H. Iwasaki, S.-T. Li, S. Nakamura [*Phys. Rev. B (USA)* vol.32 (1985) p.6949]
[28] P. Mårtensson, W.-X. Ni, G.V. Hansson, J.M. Nicholls, B. Reihl [*Phys. Rev. B (USA)* vol.36 (1987) p.5974]
[29] T. Fukuda [*Surf. Sci. (USA)* vol.351 (1996) p.103]
[30] K. Takayanagi, Y. Tanishiro, M. Takahashi, S. Takahashi [*J. Vac. Sci. Technol. A (USA)* vol.3 (1985) p.1502]
[31] K. Kajiyama, T. Tanishiro, K. Takayanagi [*Surf. Sci. (USA)* vol.222 (1989) p.47]
[32] T. Miller, T.C. Hsieh, T.-C. Chiang [*Phys. Rev. B (USA)* vol.33 (1986) p.6983]
[33] S. Van, D. Steinmetz, D. Bolmont, J.J. Koulmann [*Phys. Rev. B (USA)* vol.50 (1994) p.4424]
[34] L. Stauffer, S. Van, D. Bolmont, J.J. Koulmann, C. Minot [*Surf. Sci. (USA)* vol. 307–308 (1994) p.274]
[35] L. Stauffer, P. Sonnet, C. Minot [*Surf. Sci. (USA)* vol.371 (1997) p.63]
[36] S.S. Iyer, J.C. Tsang, M.W. Copel, P.R. Pukite, R.M. Tromp [*Appl. Phys. Lett. (USA)* vol.54 (1989) p.219]
[37] P.C. Zalm, G.F.A. van de Walle, D.J. Gravesteijn, A.A. van Gorkum [*Appl. Phys. Lett. (USA)* vol.55 (1989) p.2520]
[38] K. Nakagawa, M. Miyao [*J. Appl. Phys. (USA)* vol.69 (1991) p.3058]
[39] S. Fukatsu, K. Fujita, H. Yaguchi, Y. Shiraki, R. Ito [*Appl. Phys. Lett. (USA)* vol.59 (1991) p.2103]
[40] S. Fukatsu, K. Fujita, H. Yaguchi, Y. Shiraki, R. Ito [*Surf. Sci. (USA)* vol.267 (1992) p.79]
[41] D.J. Godbey, M.G. Ancona [*J. Vac. Sci. Technol. B (USA)* vol.11 (1993) p.1392]
[42] D. Aubel, M. Diani, M. Stoehr, et al. [*J. Phys. III (France)* vol.4 (1994) p.733]
[43] G.G. Jernigan, P.E. Thompson, C.L. Silvestre [*Surf. Sci. (USA)* vol.380 (1997) p.417]
[44] T. Walther, C.J. Humphreys, A.G. Cullis [*Appl. Phys. Lett. (USA)* vol.71 (1997) p.809]
[45] A.M. Lam, Y.-J. Zheng, J.R. Engstrom [*Appl. Phys. Lett. (USA)* vol.73 (1998) p.2027]
[46] Y.-J. Zheng, A.M. Lam, J.R. Engstrom [*Appl. Phys. Lett. (USA)* vol.75 (1999) p.817]
[47] K. Fujita, S. Fukatsu, H. Yaguchi, T. Igarashi, Y. Shiraki, R. Ito [*Jpn. J. Appl. Phys. (Japan)* vol.29 (1990) p.L1981]

[48] M. Copel, M.C. Reuter, M. Horn von Hoegen, R.M. Tromp [*Phys. Rev. B (USA)* vol.42 (1990) p.11682]
[49] M. Copel, R.M. Tromp [*Appl. Phys. Lett. (USA)* vol.58 (1991) p.2648]
[50] N. Ohtani, S.M. Mokler, M.H. Xie, J. Zhang, B.A. Joyce [*Surf. Sci. (USA)* vol.284 (1993) p.305]
[51] H. Kim, N. Taylor, J.R. Abelson, J.E. Greene [*J. Appl. Phys. (USA)* vol.82 (1997) p.6062]
[52] Z. Wang, D.B. Aldrich, Y.L. Chen, D.E. Sayers, R.J. Nemanich [*Thin Solid Films (Netherlands)* vol.270 (1995) p.555–60]
[53] D.B. Aldrich, Y.L. Chen, D.E. Sayers, R.J. Nemanich, S.P. Ashburn, M.C. Öztürk [*J. Mater. Res. (USA)* vol.10 (1995) p.2849–63]
[54] D.B. Aldrich, H.L. Heck, Y.L. Chen, D.E. Sayers, R.J. Nemanich [*J. Appl. Phys. (USA)* vol.78 (1995) p.4958–65]
[55] W. Freiman, A. Eyal, Y.L. Khait, R. Beserman [*Appl. Phys. Lett. (USA)* vol.69 (1996) p.3821–3]
[56] D.B. Aldrich, F.M. d'Heurle, D.E. Sayers, R.J. Nemanich [*Phys. Rev. B (USA)* vol.53 (1996) p.16279–82]
[57] D.B. Aldrich, F.M. d'Heurle, D.E. Sayers, R.J. Nemanich [*Mater. Res. Soc. Symp. Proc. (USA)* vol.402 (1996) p.21–6]
[58] J.B. Lai, L.J. Chen [*J. Appl. Phys. (USA)* vol.86 (1999) p.1340–5]
[59] J.E. Burnette, R.J. Nemanich, D.E. Sayers [Unpublished results]
[60] J.B. Lai, C.S. Liu, L.J. Chen, J.Y. Cheng [*J. Appl. Phys. (USA)* vol.78 (1995) p.6539–42]
[61] W. Lur, L.J. Chen [*Appl. Phys. Lett. (USA)* vol.54 (1989) p.1217–19]
[62] B.I. Boyanov, P.T. Goeller, D.E. Sayers, R.J. Nemanich [*J. Appl. Phys. (USA)* vol.84 (1998) p.4285–91]
[63] F. Wald, J. Michalik [J. Less-Common Metals (Netherlands) vol. 24, p.227–289]
[64] W.W. Wu, T.F. Cjiang, S.L. Cheng, et al. [*Appl. Phys. Lett. (USA)* vol.81 (2002) p.820–2]
[65] B.I. Boyanov, P.T. Goeller, D.E. Sayers, R.J. Nemanich [*J. Appl. Phys. (USA)* vol.86 (1999) p.1355–62]
[66] S.M. Yalisove, R.T. Tung, J.L. Batstone [*Mater. Res. Soc. Symp. Proc. (USA)* vol.116 (1988) p.439–45]
[67] J.E. Burnette, D.E. Sayers, R.J. Nemanich [Unpublished results]
[68] A. Vantomme, M.-A. Nicolet, N.D. Thoedore [*J. Appl. Phys. (USA)* vol.75 (1994) p.3882–91]
[69] C. Detavernier, X.P. Qu, R.L. VanMeirhaeghe, B.Z. Li, K. Maex [*J. Mater. Res. (USA)* vol.18 (2003) p.1668–78]
[70] C. Detavernier, R.L. VanMeirhaeghe, F. Cardon, K. Maex [*Phys. Rev. B (USA)* vol.62 (2000) p.12045–51]
[71] C. Detavernier, R.L. VanMeirhaeghe, F. Cardon, K. Maex, W. Vandervorst, B. Brijs [*Appl. Phys. Lett. (USA)* vol.77 (2000) p.3170–2]
[72] J. Seger, S.-L. Zhang [*Thin Solid Films (Netherlands)* vol.429 (2003) p.216–19]
[73] S.-L. Zhang [*Microelectron. Engng. (UK)* vol.70 (2003) p.174–85]
[74] T. Jarmar, J. Seger, F. Ericson, D. Mangelinck, U. Smith, S.-L. Zhang [*J. Appl. Phys. (USA)* vol.92 (2002) p.7193–9]
[75] E. Maillard-Schaller, B.I. Boyanov, S. English, R.J. Nemanich [*J. Appl. Phys. (USA)* vol.85 (1999) p.3614–18]
[76] W.W. Wu, S.L. Cheng, S.W. Lee, L.J. Chen [*J. Vac. Sci. Technol. B (USA)* vol.21 (2003) p.2147–50]

[77] Z. Wang, D.B. Aldrich, R.J. Nemanich, D.E. Sayers [*J. Appl. Phys. (USA)* vol.82 (1997) p.2342–8]
[78] *Properties of Metal Silicides* Ed. K. Maex, M. van Rossum, INSPEC (1995) p.124–5
[79] Q.Z. Hong, J.G. Zhu, C.B. Carter, J.W. Mayer [*Appl. Phys. Lett. (USA)* vol.58 (1991) p.905–7]
[80] C.J. Chien, H.C. Cheng, C.W. Neih, J.L. Chen [*J. Appl. Phys. (USA)* vol.57 (1985) p.1887–9]
[81] A. Noya, M. Takeyama, K. Sasaki, T. Nakanishi [*J. Appl. Phys. (USA)* vol.76 (1994) p.3893–5]
[82] J.C. Ousset, V. Pupuis, H. Rakoto, et al. [*Phys. Rev. B (USA)* vol.39 (1989) p.4484–8]
[83] S. Paidassi [*Appl. Phys. Lett. (USA)* vol.35 (1979) p.886–8]
[84] M. del Giudice, R.A. Butera, M.W. Ruckman, J.J. Joyce, J.H. Weaver [*J. Vac. Sci. Technol. A (USA)* vol.4 (1986) p.879–81]
[85] J.Y. Cheng, J.L. Chen [*J. Appl. Phys. (USA)* vol.69 (1991) p.2161–8]
[86] H.J. Trodahl, H.L. Johnson, A.B. Kaiser, C.K. Subramaniam, B.J. Ruck, P. Lynam [*Phys. Rev. B (USA)* vol.53 (1996) p.15226–30]
[87] Q.Z. Hong, J.W. Mayer [*J. Appl. Phys. (USA)* vol.66 (1989) p.611–15]
[88] H.K. Liou, X. Wu, U. Gennser, et al. [*Appl. Phys. Lett. (USA)* vol.60 (1992) p.577–9]
[89] A. Buxbaum, M. Eizenberg, A. Raizman, F. Schaffler [*Appl. Phys. Lett. (USA)* vol.59 (1991) p.665–7]
[90] D. Chen, J. Luo, W. Lin, C.Y. Chang, P.S. Shih [*Appl. Phys. Lett. (USA)* vol.73 (1998) p.1355–7]
[91] J. Seger, S.-L. Zhang, D. Mangelinck, H.H. Radamson [*Appl. Phys. Lett. (USA)* vol.81 (2002) p.1978–80]
[92] J.-S. Luo, W.-T. Lin, C.Y. Chang, W.C. Tsai [*J. Appl. Phys. (USA)* vol.82 (1997) p.3621–3]
[93] R.L. Jiang, J.L. Liu, J. Li, Y. Shi, Y.D. Zheng [*Appl. Phys. Lett. (USA)* vol.68 (1996) p.1123–5]
[94] J.-H. Ku, R.J. Nemanich [*Appl. Surf. Sci. (USA)* vol.104/105 (1996) p.262–6]
[95] O. Nur, M. Willander, R. Turan, M.R. Sardela Jr., G.V. Hansson [*Appl. Phys. Lett. (USA)* vol.68 (1996) p.1084–6]

Chapter 8
Silicide technology for SOI devices

L.P. Ren and K.N. Tu

8.1 OVERVIEW

The swift and continuous progress of silicon-on-insulator (SOI) technology in recent years has made it competitive to bulk silicon technology for ULSI applications. Some of the inherent properties in favour of SOI technology over bulk include the reduction of junction capacitance, the ease to make shallow junctions, radiation hardness, and latch-up immunity [1]. The ultra-thin silicon film also results in reduction of the short-channel effects, which cause the gate to lose channel control with decreasing gate length and prevent scaling into the deep sub-micron regime. These advantages boosted SOI from merely a substitute for silicon-on-sapphire (SOS) to achieving acceptance in military applications and beyond.

However promising SOI technology may have become, there are always challenges that deter the widespread acceptance of SOI semiconductors. When scaling guidelines of SOI suggest that thinner films enjoy more improved sub-threshold slopes and reduced drain-induced-barrier-lowering (DIBL) than thicker fully depleted films [2], one major obstacle is the ability to obtain low source/drain series resistance. A high series resistance can easily offset the advantages to be gained from using the ultra-thin SOI material and it seriously degrades SOI current drivability. This is especially severe for short channel devices.

To overcome this problem, silicide technology on SOI devices has been developed and investigated. In this chapter, the SOI silicide technology is described. Starting from the review of silicide design analysis on SOI CMOS, challenges to implement existing bulk silicide technology to SOI devices are introduced. Then various silicide technologies on SOI are discussed.

Overview p.201

Source/drain engineering for SOI CMOS p.202
  Introduction of SOI device structure p.202
  Impact of series resistance p.203
  Silicide design analysis p.206

Challenges to implement existing bulk silicide technology to SOI devices p.208
  Titanium silicide process and voids p.208
  Silicide thickness control p.209
  Thin silicide thermal stability p.209
  Alternative silicide choice p.210

Advanced silicide technology for SOI devices p.211
  Ti silicide formation on pre-amorphised silicon p.211
  Co silicide using Ti/Co or Co/Ti as source materials p.217
  Ni silicide as a suitable candidate p.220
  Low-barrier silicides of $ErSi_2$ and PtSi for ultra-thin devices p.221
  Selective deposition of silicide on SOI p.222

Summary p.225

References p.225

8.2 SOURCE/DRAIN ENGINEERING FOR SOI CMOS

8.2.1 Introduction of SOI device structure

FIGURE 8.1 shows the cross-section of SOI CMOS devices. As shown in the figure, on the bulk silicon substrate, a buried oxide layer is formed. On the top of the buried oxide layer there is a silicon thin film, where active MOS devices and circuits are located. In the early stage of SOI technology, this insulator is made from silicon nitride or sapphire. Nowadays, the insulator in the SOI is based on the oxide layer – the buried oxide layer is used to isolate the active device thin film from substrate. Owing to the excellent isolation provided by the buried oxide, immunity of the SOI device against high-energy particle illumination is excellent and the parasitic capacitance of SOI MOS is also smaller than those of bulk ones. For bulk CMOS devices, between devices and between device and substrate, depletion regions of the reverse-biased pn-junction have been used for isolation. Therefore, device density of the bulk CMOS technology cannot be high. In addition, the parasitic npn and pnp bipolar junction transistors (BJTs) in the well and the substrate may be accidentally turned on to cause latch-up. In contrast, for SOI CMOS devices, buried oxide has been used for isolation, no latch-up exists and device density can be greatly increased as shown in FIGURES 8.1 and 8.2.

The SOI devices can be divided into thick film and thin film, depending on the thickness of the silicon thin film. As shown in the FIGURE 8.3, if the silicon thin film is thick, only the top portion of the silicon thin film is depleted and the bottom portion is neutral. This type of SOI CMOS devices is called partially depleted SOI CMOS devices. If the silicon thin film is fully depleted, this type of the SOI CMOS is called fully depleted SOI CMOS devices [3].

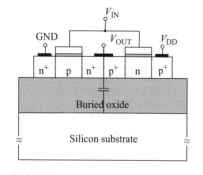

FIGURE 8.1 Cross-section of SOI CMOS devices.

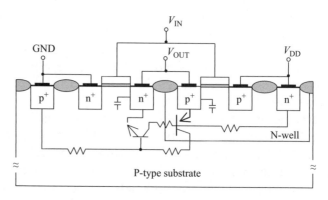

FIGURE 8.2 Cross-section of bulk CMOS devices.

Silicide technology for SOI devices

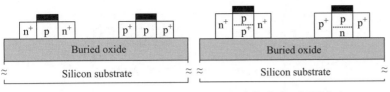

FIGURE 8.3 SOI CMOS devices: fully depleted versus partially depleted.

In a partially depleted SOI CMOS device, the threshold voltage is insensitive to the thin film thickness since the depletion region is independent of the film thickness. However, due to the neutral region in the thin film, the floating body effect is more serious. In addition, the depletion region is more susceptible to the influence from the source and the drain regions. Therefore, the device second-order effects are more serious and kink effect can be identified. In a fully depleted thin film SOI CMOS device, regardless of the biasing condition, the silicon thin film is fully depleted. Due to the large control capability over the silicon thin film, a larger transconductance can be expected. Therefore, the device second-order effects including the short channel effect and the narrow channel effect are smaller. In addition, the subthreshold slope is improved. Thus, thin film fully depleted SOI devices are superior to their partially depleted counterparts and bulk devices in many aspects.

8.2.2 Impact of series resistance

Thin film SOI CMOS has the potential of high current drivability, making it suitable for low power operation. However, the thin film nature of SOI possesses several challenges, one of them is the degradation of drain current due to the effect of source/drain series resistances R_{SD}, which is the series combination of the intrinsic channel resistance and the parasitic series resistance associated with diffusions and contacts. FIGURE 8.4 shows the sheet resistance versus SOI film thickness at the conditions with and without silicidation. Measurement of the sheet resistance of conventional SOI MOSFETs for SOI film thickness varies, from 40 to 170 nm which shows that the source/drain resistances are aggravated when the film thickness is less than 50 nm, where the increase of source/drain parasitic resistances is no longer linearly proportional to the inverse of film thickness as predicted by Ohm's law. This is due to surface scattering and the subsequent reduction of the carrier mean free path. However, when titanium silicide is used for a 40 nm SOI film, the silicide source/drain regions show a sheet resistivity less than 5 Ω/sq, which is three orders of magnitude lower than its non-silicided parts.

Silicide technology for SOI devices

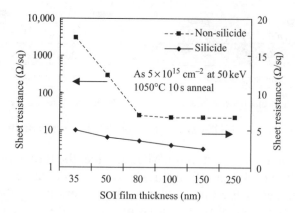

FIGURE 8.4 Sheet resistance versus SOI film thickness.

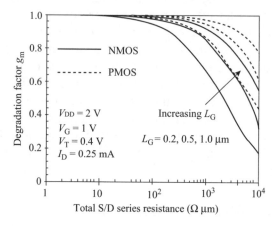

FIGURE 8.5 Source/drain series resistance impact on transconductance [7] (© 2004 IEEE).

To investigate the degradation due to source/drain series resistance, an analytic MOS transistor model (the EKV model [5]), which is applicable to SOI CMOS, was used to simulate the current–voltage (I–V) characteristics of NMOS and PMOS transistors of various channel length [6]. FIGURE 8.5 shows the extracted transconductance as a function of R_{SD} since the transconductance is directly related to the device speed. The device parameters of modelled transistors are listed in TABLE 8.1. The degradation factor, g_m, begins to drop precipitously when R_{SD} is larger than 300 Ω μm for 0.2 μm NMOS and about 1000 Ω μm for 0.2 μm PMOS. The simulation results clearly indicate the strong effect of source/drain series resistance on transconductance for short-channel devices.

Another important parameter related to the silicide process is the gate resistance, which is related to the signal delay time. Liu et al. [7] have calculated the gate electrode RC delay factor in a

TABLE 8.1 Model parameters.

Gate oxide d_{OX} (nm)	Buried oxide d_{BOX} (nm)	SOI thickness d_{Si} (nm)	Channel doping N_a (cm^{-3})	Threshold voltage V_T (V)	NMOS mobility μ_n (cm^2/V/s)	PMOS mobility μ_p (cm^2/V/s)
8	200	50	3×10^{17}	0.4	500	150

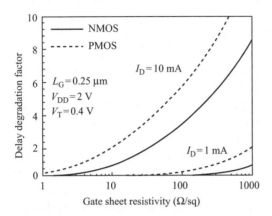

FIGURE 8.6 Gate resistance induced RC delay degradation. The gate width is 1.1 μm for NMOS and 3.8 μm for PMOS [7] (© 2004 IEEE).

basic inverter configuration [8]. The transistor I–V characteristics are again calculated with the EKV model and the same device parameters listed in TABLE 8.1. The propagation delay time t_{pd} was estimated as the time required to discharge or charge a load capacitance connected to the output common drain of the parallel transistors formed by the π-ladder network. The gate resistance R_G induced degradation factor DF is defined as $(t_{pd} - t_{pd0})/t_{pd0}$, where t_{pd} and t_{pd0} are the delay times with and without R_G, respectively. So degradation factor DF is the fractional increase in switching time due to R_G. From FIGURE 8.6 we can see, in the case of the 0.25 μm NMOS with 1 mA current drive, the gate-induced degradation does not begin to increase significantly until the sheet resistivity increases above 100 Ω/sq. So the gate sheet resistivity is not a speed-limiting factor for small current drive. For a similar gate length PMOS, the degradation onset is slightly lower (about 80 Ω/sq) due to the wider gate required to generate the same current for PMOS. However, for wide transistors like those used in output driver circuits, the effects of gate sheet resistivity become much more serious. The curves for 10 mA I_D indicate that to stay below a delay degradation factor of 2, the sheet resistivity needs to be maintained below 10 Ω/sq for PMOS and 30 Ω/sq for NMOS. Although these resistivities are relatively easy to achieve on bulk

CMOS, they are non-trivial in thin film CMOS because of the very thin active Si layer on source/drain region.

8.2.3 Silicide design analysis

For a lightly-doped drain (LDD) structure MOSFET, the total resistance includes the channel resistance (R_{ch}), the accumulation and spreading resistance (R_{ac} and R_{sp}), the resistance of the drain extension tab (R_{ext}) and the source/drain junction resistance (R_{sd}), as well as the contact resistance (R_{co}) plus the silicide resistance ($R_{silicide}$) as shown in FIGURE 8.7. The accumulation resistance is generally small and neglected. The spreading resistance depends on the doping profile of the junction. The drain extension resistance is given by the sheet resistance of the drain extension and the number of square. The contact resistance depends on the contact resistivity, the sheet resistance of the source/drain, and the area of the contact hole [9]. When the effective length and width are scaled in each new generation of technology, the contact resistance will ultimately dominate the total device resistance.

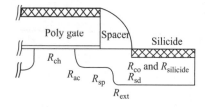

FIGURE 8.7 Cross-section of a typical MOSFET showing series resistance components.

To estimate a set of silicide parameters that will satisfy the design constraints, analytical expressions based on the transmission line model for calculating the source/drain contact resistance of silicided thin film SOI MOSFETs have been used [10]. They allow one to evaluate the effects of design geometry of silicide quality on the overall source/drain series resistance. As discussed earlier, to minimise the impact of series resistance on device size, one would like to have the total source/drain series resistance below 300 Ω μm (normalised to a 1 μm gate width), especially for NMOS. With this series resistance and a 0.2 μm gate length, the NMOS transconductance is already reduced as a matter of fact by more than 15% compared to the ideal case. As the gate length continues to be scaled down, the requirement becomes even more stringent. FIGURE 8.8 shows the configuration of silicide thickness design. The filled area within the curve represents possible operating parameters to achieve source/drain series resistance below 300 Ω μm and sheet resistivity below 20 Ω/sq.

During silicidation, the silicon consumption increases the sheet resistivity of the underlying Si layer and decreases the effective contact area, which results in a constrained design space. For a given specific contact resistance between the silicide and Si, there is a required minimum remaining Si thickness underneath the silicide to achieve a tolerable series resistance. Moreover, there is a constraint on the maximum allowed Si thickness, because of the minimum silicide thickness required to achieve a given sheet resistivity in order to minimise the gate series resistance induced propagation delay effect. With the 50 nm thick active

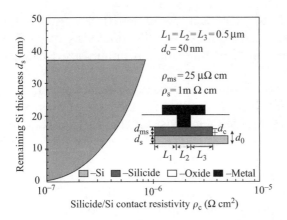

FIGURE 8.8 Silicide thickness design space. Inset is the contact structure used for the calculation along with the device geometrical parameters [7] (© 2004 IEEE).

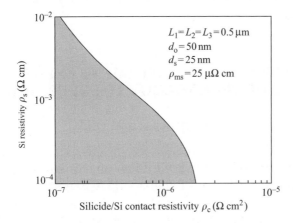

FIGURE 8.9 Silicide resistivity design space. The contact structure's geometrical parameters are same as FIGURE 8.8 [7] (© 2004 IEEE).

layer, a minimum of 20 nm remaining Si is required for a specific contact resistivity of $5 \times 10^{-7}\,\Omega\,\text{cm}^2$.

Another important parameter is the resistivity of the silicon layer underneath the silicide, ρ_s. Again, the required Si resistivity ρ_s of the remaining Si layer underneath the silicide along with the specific contact resistivity ρ_c at the silicide/Si interface is calculated. The filled area within the curve in FIGURE 8.9 represents possible operating parameters for achieving source/drain series resistance below $300\,\Omega\,\mu\text{m}$. For a specific contact resistance of $5 \times 10^{-7}\,\Omega\,\text{cm}^2$, the required resistivity of the Si layer is below $1\,\text{m}\Omega\,\text{cm}$. This condition implies that the active dopant concentration needs to exceed $10^{20}\,\text{cm}^{-3}$.

Both FIGURES 8.8 and 8.9 also indicate the importance of minimising the specific contact resistivity ρ_c at the silicide/Si interface. If ρ_c exceeds a few times $10^{-6}\,\Omega\,\text{cm}^2$ [11], there is no silicide thickness or reasonable Si thickness that will satisfy the design criterion of having source/drain series resistance less than $300\,\Omega\,\mu\text{m}$.

Finally, the effects of the silicide sheet resistivity itself need to be evaluated. It turns out that for the case of 50 nm thick silicide process, the silicide sheet resistivity has little impact on the overall source/drain series resistance. However, for ultra-thin SOI films, the silicide with low resistivity is still desired, especially for long polysilicon gates. Therefore, in terms of the silicidation on SOI devices, the major parameters are: (1) the silicide thickness; (2) the silicide/Si specific contact resistivity; and (3) the silicon resistivity. Both the interface-specific contact resistivity and the silicon resistivity are directly related to the dopant concentration in the silicon layer. To minimise these resistivities, we need to maximise the degenerate doping in the source/drain regions to the order of $10^{20}\,\text{cm}^{-3}$.

8.3 CHALLENGES TO IMPLEMENT EXISTING BULK SILICIDE TECHNOLOGY TO SOI DEVICES

8.3.1 Titanium silicide process and voids

The standard silicide process for metal–oxide–semiconductor field-effect transistors (MOSFETs) on Si substrates is well developed and fairly well understood. A two-step rapid thermal annealing (RTA) method is used to minimise lateral overgrowth of silicide onto the oxide sidewall. Among the choice of metal silicides, titanium silicide is widely used in industry due to its low resistivity, high thermal stability, ability to remove the native oxide on Si, and being easy to form and etch for pattern generation. However, for thinner SOI films, the design window for Ti salicide process is narrowed. One problem with titanium silicide results from silicon being the dominant diffuser during the formation of $TiSi_2$ [12, 13]. If the titanium layer is too thin, it will form a non-uniform $TiSi_2$ layer where thermal agglomeration and discontinuous islanding of titanium silicide will occur [14–16], resulting in poor electrical contact. If the titanium layer is too thick, bridging between source/drain and gate electrodes may occur, shorting the device. Moreover, additional lateral migration of the silicide into the source/drain junctions rapidly consumes the finite volume of silicon and forms voids at the interface of silicided and non-silicided regions [14, 17]. A schematic representation of this phenomenon is shown in FIGURE 8.10. During first anneal step, the C49-$TiSi_2$ advances downward toward the buried SiO_2

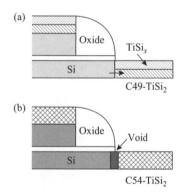

FIGURE 8.10 Silicide contact formation for a thin film SOI transistor: (a) formation of a layered structure during the first annealing step. Si diffuses from channel region; (b) void formation as a result of net volume shrinkage.

TABLE 8.2 Basic properties of some popular silicides.

Properties	TiSi$_2$	CoSi$_2$	NiSi	PtSi
nm of Si consumed per nm of metal	2.27	3.64	1.84	1.3
Silicide thickness formed per nm of metal	2.51	3.52	2.22	1.95

layer and may actually reach it. At this point, a TiSi$_x$ with $x \sim 1.0$ and C-49 TiSi$_2$ has been formed. At the second annealing step ($T > 800°C$), replenishment of Si to the precursor layer for TiSi$_2$ formation and to promote re-crystallisation to the C54 structure will occur. The mean difference between the bulk and thin film SOI devices is that Si in bulk devices may diffuse a very short distance from the substrate into the silicide layer. In thin film SOI devices, this process can occur only so long as underlying Si remains. If this underlying layer has been completely consumed, then Si must diffuse from the channel region, which will lead to voiding.

8.3.2 Silicide thickness control

From previous discussions we can see that a controllable silicide depth is very important for thin film SOI devices. The most popular methods to control the silicide depth in bulk devices are either by depositing the limited silicide metal layer, or by controlling the first anneal time and temperature. TABLE 8.2 gives the formed silicide amount for per unit deposited metal. For a required silicide thickness, the required metal film can be calculated.

However, for Ti silicide process, part of the deposited Ti film reacts with the N$_2$ ambient in the Ti silicide process. It makes it difficult to precisely determine how much Ti film reacts with the Si layer. For cobalt and nickel silicide, even though all the metal films can react with Si layer, to precisely control the initial metal thickness is challenging for the thin SOI film when the typical thickness of Si in SOI is less than 50 nm. In addition, it is reported that the silicide thickness should be less than 80% of the SOI thickness when the contact resistivity between silicide and Si is greater than $1 \times 10^{-7}\,\Omega\,cm^2$ [18]. From this requirement a highly stable and controlled formation of ultra-thin silicide under a limited Si film thickness must be established.

8.3.3 Thin silicide thermal stability

An important aspect of silicide process on bulk Si substrate is that the supply of Si for silicide formation is effectively infinite. The silicide thickness at this condition can be much thicker than that on SOI devices and the uniformity is easy to control. Experimental data show that the thermal stability for TiSi$_2$, CoSi$_2$ and NiSi silicide on bulk devices can be up to 950, 900 and 650°C, respectively [19], which is compatible to CMOS process and satisfies the application requirements. However, for thin film SOI devices, this condition does not hold anymore, where only limited Si film is available. In the case of titanium silicidation, if the titanium layer is too thin, it will form a non-uniform $TiSi_2$ layer where thermal agglomeration and discontinuous islanding of titanium disilicide will occur [14–16], resulting in poor electrical contacts. If the titanium layer is too thick, bridging between source/drain and gate electrodes may occur, shorting the device. Moreover, additional lateral migration of the silicide onto the source/drain regions rapidly consumes the finite volume of silicon and results in either voids at the interface of silicide and non-silicided regions or uncompleted replenishment, leading to high contact resistance.

8.3.4 Alternative silicide choice

Since Ti silicide has the limitation of void formation and segregation on narrow poly gate lines, alternative silicide choices like cobalt silicide and nickel silicide have been investigated [20–25]. They have the advantages of having the lowest resistivities ($\sim 14\,\mu\Omega$ cm) [26], good thermal stability (up to 700–900°C), low forming temperature ($\sim$400–600°C) and little or no resistivity degradation on narrow gate lines. Moreover, for $CoSi_2$, it can have low film stress (lattice mismatch with silicon is only 1.2%), less lateral gate-source/drain silicide overgrowth, good resistance to HF and plasma etching, and it does not react with oxide below 900°C. Su et al. [27] had successfully fabricated low parasitic resistance thin film SOI MOSFETs by using Co/Ti laminate as source materials to form cobalt disilicide. However, the silicide resistance increases dramatically when the amount of cobalt used is in excess of that which is needed to fully consume the film layer, causing the formation of high resistivity CoSi by a phase reversal process [28]. Using Ti/CoSi bimetallic source materials does not guarantee a void free film. Tan et al. showed that voids can form at the lateral interface of $CoSi_2/Si$, suggesting that the source material (CoSi) decomposition rate becomes too slow to maintain a sufficiently large cobalt flux to reach the $CoSi_2$–Si interface [29, 30]. This allows the silicon atom out-diffusion to dominate the $CoSi_2$ formation. Thus, the optimal process of thin $CoSi_2$ films on SOI has yet to be developed. Nickel silicide is a low temperature processing candidate for the silicide process [31]. NiSi can be formed at as low as 450°C, and has a large processing temperature window. It is stable up to 650°C, where the phase transition to high resistive $NiSi_2$ takes place. In addition to achieving low silicide resistivity, nickel consumes less silicon to form mono-silicide than

Ti or Co does in forming disilicide, also there is no reaction with N_2 and no resistivity degradation on gate lines.

8.4 ADVANCED SILICIDE TECHNOLOGY FOR SOI DEVICES

8.4.1 Ti silicide formation on pre-amorphised silicon

For silicide on ultra-thin-film SOI MOSFETs, the key to avoiding the formation of voids and minimising dopant segregation is to lower the silicide formation thermal cycle, and limit the depth of silicide so that phase stability with silicon is maintained. Thus, a novel salicide technology has been proposed and continually developed by forming silicide on damaged layers with a low thermal cycle [32–36]. This is accomplished through the amorphisation of the gate and source/drain regions by a low-dose Ge^+ implantation. Ge^+ was chosen for several reasons: (1) Ge^+ is heavy, and requires a small dose to amorphise the films, compared to Si, and SiGe is also known to have lower contact resistance with metal; (2) both Ge and Si have a diamond crystal structure with small lattice mismatch between Si and SiGe. SiGe has a larger lattice constant than Si so the Si regions below SiGe will be in tensile stress. As a result, it will be difficult for the silicon interstitials (produced by the implantation) from the interstitial–vacancy pairs that have not been annihilated during the thermal cycles to be injected into the SiGe/SiGe-cide regions, leaving the vacancies to form voids. (3) SiGe allows for bandgap engineering [37] to alleviate the floating body effects of SOI devices; and (4) Ge is a neutral species suitable for both p- and n-MOSFETs.

Ge^+ pre-amorphisation has been applied to control the silicide depth by forming a very sharp amorphous/crystalline interface after ion implantation. The damaged source/drain regions lower the silicide formation energy substantially, which allows for a low temperature process. In addition, dopant segregation from the unsilicided region underneath the source/drain sidewall spacer into the silicide is greatly reduced due to the low temperature cycle.

8.4.1.1 Kinetics studies

To prove the effectiveness of the proposed silicide process, Ge^+ implantation was applied to SOI wafers with various background doping levels. FIGURE 8.11 shows the kinetics studies of Ti silicide on n^+ and p^+ diffusion layers with Ge^+ pre-amorphisation. n^+ diffusion was formed by an arsenic implantation of 5×10^{15} cm^{-2} at 110 keV, followed by a furnace anneal at

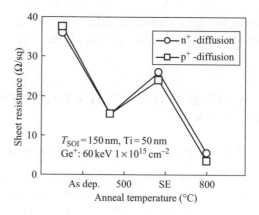

FIGURE 8.11 Low thermal cycle silicide formation by Ge^+ pre-amorphisation. The Ge^+ dose is 1×10^{15} cm^{-2} and SE stands for selective etch [32] (© 2004 IEEE).

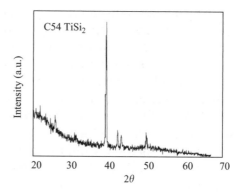

FIGURE 8.12 X-ray diffraction spectrum of $TiSi_2$ after two step RTA. Ge^+ dose is 1×10^{15} cm^{-2}, and the anneal time is 60 s [36] (© 2004 IEEE).

800°C for 30 min. p$^+$ diffusion was formed by a boron difluoride (BF$_2$) implantation of 5×10^{15} cm^{-2} at 90 keV, followed by a furnace anneal at 800°C for 30 min. Then 60 keV Ge$^+$ implantation was applied to 150 nm SOI wafers (the thickness of the amorphous Si layer will be reported later) and 50 nm of Ti was deposited on SOI film. First and second RTA temperatures were 450 and 800°C, respectively. A very low sheet resistance is produced, indicating that this salicide technology can be applicable to both p- and n-MOSFETs.

X-ray diffraction pattern shown in FIGURE 8.12 supports the enhanced phase transformation of TiSi$_2$ from C49 to low resistive C54 at low temperature, suggesting that Ti reacts with the pre-amorphised Si films at 450°C and C-49 transforms to C-54 at 800°C [36]. The damaged film thus broadens the silicide process window.

Silicide technology for SOI devices

FIGURE 8.13 A cross-section TEM micrograph of TiSi$_2$ prepared by Ge$^+$ pre-amorphisation process. The thickness of the silicide is 20 nm [33, 35].

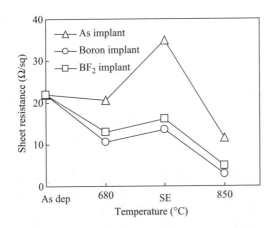

FIGURE 8.14 Comparison of sheet resistance for B$^+$, BF$_2^+$ and As$^+$ doped wafers using conventional high temperatures for first rapid thermal annealing.

The transmission electron microscopy (TEM) micrograph (FIGURE 8.13) further confirms that the silicide depth is controlled by the depth of the amorphous layer (even though a very thick layer of titanium was deposited), not by the thickness of Ti [33, 35]. A silicide film (~20 nm) with excellent interface topology can be obtained, and Ge$^+$ pre-amorphisation makes source/drain resistance less susceptible to the variation of deposited metal thickness. Thus, the demand for silicide layer control can be achieved.

To study the effects of doping species on silicide formation, both conventional high temperature silicidation (680°C plus 850°C) and low temperature silicidation (500°C plus 800°C) were performed on heavily doped wafers. n$^+$ substrate was formed by an arsenic implantation of 5×10^{15} cm^{-2} at 60 keV, followed by RTA at 1050°C for 10 s. p$^+$ substrate was formed by both BF$_2$ and boron implantation. The dose was 5×10^{15} cm^{-2} and the energies were 40 keV for BF$_2$, 20 keV for boron with RTA at 950°C for 15 s. Conventional silicide process was performed at 680°C, 30 s for RTA1 and 850°C, 30 s for RTA2.

FIGURE 8.14 shows that for the conventional high temperature silicidation the sheet resistance of the BF$_2^+$, boron and As-doped wafers are quite different. The sheet resistance for p$^+$-type doped wafers is below 3.5 Ω/sq but for As-doped wafers is about 11 Ω/sq.

Silicide technology for SOI devices

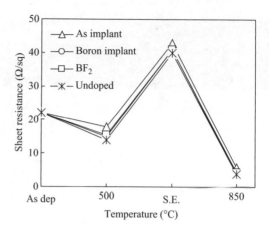

FIGURE 8.15 Comparison of sheet resistance for B^+, BF_2^+ and As^+ doped wafers using low temperatures for first rapid thermal annealing.

FIGURE 8.15 shows the results of the low temperature silicidation at stages of first anneal (500°C for 3 min), selective etch and second anneal (800°C for 1 min). It can be seen that the sheet resistance of different doped wafers after RTA2 is the same as the undoped wafers. Therefore the experiment indicates that low temperature technology suppresses the doping effect on silicide formation. In other words, the difference between As and B implantation as shown in FIGURE 8.14 is gone.

8.4.1.2 Silicide depth control

To understand how different thermal cycles affect sheet resistance and silicide depth, 50 nm Ti was deposited on Ge-damaged silicon substrates. Ge^+ implant energies used in this study were 80 keV with a dose of 1×10^{15} cm^{-2}. The first RTA process was done from 450 to 540°C for various times, and second anneal was done at 800°C for 60 s. FIGURE 8.16 shows the sheet resistance and silicide depth versus first annealing condition. It indicated that at each temperature for the first RTA step, a minimum sheet resistance can be obtained when anneal time reaches a threshold value, for example, at 500°C for 5 min, 520°C for 3 min and 540°C for 2 min. The corresponding $TiSi_2$ thickness is shown in FIGURE 8.16 with light-grey lines.

We can see that when the thermal cycle is chosen so that the completion of the Ti reaction with substrate silicon is reached, prolonging the first RTA time will not change either the final sheet resistance or the depth of $TiSi_2$. This means that the $TiSi_2$ thickness is controlled by the thickness of the amorphous layer generated by Ge^+ implantation. To ensure that Ti only reacts with the amorphous layer, the first RTA temperature has to be chosen carefully.

Silicide technology for SOI devices

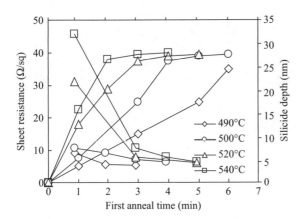

FIGURE 8.16　The sheet resistance and silicide depth versus first RTA time [34] (© 2004 IEEE).

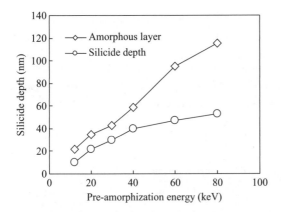

FIGURE 8.17　$TiSi_2$ silicide depth control versus amorphous layer depth for different Ge^+ implant energies. First anneal temperature was 500°C [34] (© 2004 IEEE).

If the first RTA temperature is too high, Ti will gain enough energy to fully react with the amorphous silicon and will continue to react with the crystalline silicon underneath. Ge^+ pre-amorphisation thus will lose its control over the silicide depth.

FIGURE 8.17 shows the final $TiSi_2$ depth versus amorphous layer depth for different Ge^+ implant energies [30, 32]. The deposited titanium was 50 nm, and the thermal cycle was 500°C for the first RTA and 800°C for the second RTA, both for 60 s. It is clear that the silicide depth is always within the amorphous layer for a low temperature first RTA. Compared to low temperature silicidation, Ti silicide for conventional high temperature silicidation is formed with both amorphous layers and crystalline layers. Therefore, the sheet resistance values and silicide depth are almost the same when using the different implant energies as shown in FIGURE 8.18.

Silicide technology for SOI devices

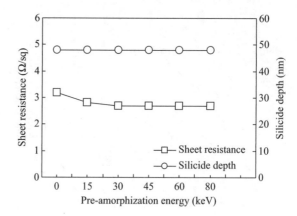

FIGURE 8.18 TiSi$_2$ sheet resistance versus Ge$^+$ pre-amorphisation energies using conventional high temperature silicidation [34] (© 2004 IEEE).

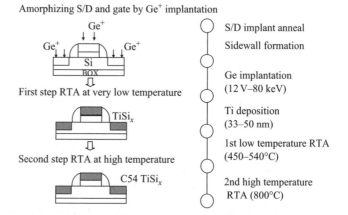

FIGURE 8.19 Silicide process flow for FD SOI CMOS.

8.4.1.3 Process integration

Fully depleted SOI CMOS devices with the developed silicide process were fabricated based on a conventional bulk CMOS process [32, 34]. The silicide process flow is outlined in FIGURE 8.19 and the modules were summarised in TABLE 8.3. The Ge$^+$ pre-amorphisation energy was chosen so that the thickness of the amorphisation layer is about 80% of SOI thickness, which will produce a silicide thickness of about 65% of the amorphous layers. In fabrication, Ge$^+$ implant energies used for 35, 60 and 150 nm films of SOI were 20, 40 and 80 keV, respectively. The Ge$^+$ dose was 1×10^{15} cm^{-2}.

FIGURE 8.20 shows the output characteristics of the CMOS devices. The transistor I_{DS}–V_{DS} results exhibit a very good Ohmic contact for both p- and n-MOSFETs on 35 nm SOI film. Compared to devices using the conventional two-step RTA silicide process

Silicide technology for SOI devices

TABLE 8.3   The summary of silicide modules.

T_{SOI} (nm)	Ti (nm)	Ge$^+$ pre-amorphisation energy (keV)	RTA
35	20	12	500°C 4 min, 800°C 1 min
60	20	40	500°C 4 min, 800°C 1 min
150	50	80	540°C 3 min, 800°C 1 min

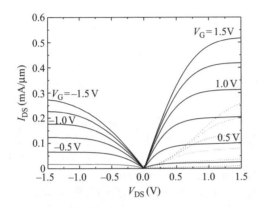

FIGURE 8.20   Output characteristics of the CMOS devices using Ge$^+$ pre-amorphisation silicide process on 35 nm thin film SOI [34] (© 2004 IEEE).

performed with RTA1 at 650°C for 1 min and RTA2 at 800°C for 1 min, it can be seen that since the thin SOI film is fully silicided, the serious Schottky diode behavior is caused by both the void formation [14–16] and dopant segregation at the interface between the silicide and substrate source/drain regions.

One challenge for titanium silicide is the ability to form low sheet resistance on thin gate length [38, 39]. To study the dependence of TiSi$_2$ sheet resistance on thin polysilicon linewidth, the grain size of TiSi$_2$ was checked by SEM and the sheet resistance on gate length from 1.0 to 0.2 μm was measured by using van der Pauw structures. It is clear from FIGURES 8.21 and 8.22 that low temperature silicidation with Ge$^+$ pre-amorphisation has a small grain size averaging 80 nm and a low sheet resistance of 5.8 Ω/sq on 0.2 μm gate lines.

8.4.2   Co silicide using Ti/Co or Co/Ti as source materials

Pre-amorphisation facilitates the titanium silicide phase transition by lowering the silicide formation energy. Similar mechanism can be applied to cobalt silicide. FIGURE 8.23 shows the sheet resistance versus first RTA temperature for 32 nm Co deposited on silicon substrates. Cobalt silicide has more intermediate phases

Silicide technology for SOI devices

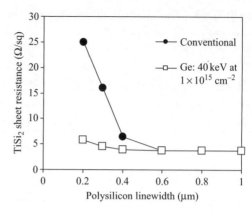

FIGURE 8.21   TiSi$_2$ sheet resistance versus N$^+$ polysilicon linewidth. n$^+$ poly: As$^+$ 50 keV at 1×10^{16} cm^{-2}, Ti: 50 nm [32, 34] (© 2004 IEEE).

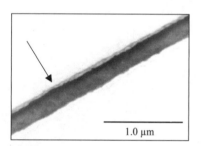

FIGURE 8.22   The grain size of TiSi$_2$ on 0.24 μm gate lines after RTA1 for low temperature silicidation [34] (© 2004 IEEE).

than titanium silicide. The sequence of silicide phases formed by RTA is shown in FIGURE 8.23. Enhanced phase transition with the aid of pre-amorphisation can be observed. However, the first anneal temperature for conventional cobalt silicide process is already at a temperature as low as 450°C. Pre-amorphisation might not have as significant effect as it does to the Ti–Si reaction.

Plain Co processing is prone to thermal agglomeration when the silicide depth is thin [40]. Using Ti/Co as the source material is an alternative way to form cobalt silicide. FIGURE 8.24 shows the sheet resistance at different stages of the silicide process with the first RTA temperature at 600°C. With Ge$^+$ implantation, a low sheet resistance of CoSi$_2$ can be achieved.

A study of fully silicide source/drain structure with Co salicide technology has been presented [41]. In this work, the final body Si thickness, d_{Si}, ranges from 20 to 55 nm. After LOCOS isolation, channel doping, gate oxide growth (4.5 nm), poly gate formation ($L_g = 0.15$–10 μm), spacer formation and source/drain implantation, the fully Co silicidation was performed. The silicidation was carried out with a TiN capping layer for the first RTA. The I–V characterisation of $L_g = 0.2$ μm MOSFETs for various SOI thickness indicated neither serious Schottky diode behavior due to void formation nor effect of high parasitic resistance was evident. The measured total source/drain parasitic resistance for various source/drain SOI thicknesses shows no dependence on the silicide structure and is approximately 500 Ω μm for n-MOSFETs.

Based on the design criteria discussed in Section 8.2.2, thin film silicidation process flows and extracted silicide properties for both Co and Ti using conventional silicide temperature are summarised in TABLES 8.4 and 8.5 [7], respectively.

The extracted data confirmed the design analysis in Section 8.2.3 and show that the major parameters in determining

Silicide technology for SOI devices

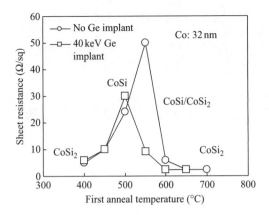

FIGURE 8.23 Cobalt silicide formation on pre-amorphised Si film.

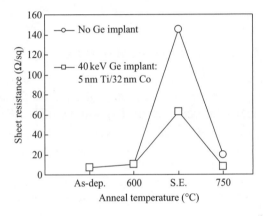

FIGURE 8.24 Cobalt silicide formation using Ti/Co as the source with Ge$^+$ pre-amorphisation.

TABLE 8.4 Process flow for thin cobalt and titanium silicide on 50 nm SOI devices [7].

	$CoSi_2$	$TiSi_2$
Starting metal/thickness	Co/8 nm	Ti/12.5 nm
Capping layer/thickness	Ti/20 nm	TiN/20 nm
Metal deposition tool	CVC AST-601	Electrotech Sigma
Initial RTA	475°C/3 min	675°C/80 s
Wet etch	$NH_4OH : H_2O_2 : H_2O$	$NH_4OH : H_2O_2 : H_2O$
	(1 : 1 : 1) 65°C/1 min	(1 : 1 : 5) 50°C/3.25 min
Final RTA	700°C/1 min	850°C/1 min

the overall source/drain series resistance are the silicide thickness, the interface specific contact resistivity, and the underlying silicon resistivity. Maintaining high silicon dopant concentration is critical in obtaining low series resistance.

Silicide technology for SOI devices

TABLE 8.5 Extracted silicide property [7].

	Type	CoSi$_2$	TiSi$_2$	No silicide
R_{sheet} (Ω/sq)	n$^+$	14	8	118
	p$^+$	14	11	174
ρ_s (mΩ cm)	n$^+$	0.70	0.80	0.59
	p$^+$	1.11	1.28	0.87
ρ_c (μΩ cm^2)	n$^+$	0.31	0.38	0.74
	p$^+$	0.49	0.42	3.40
d_{ms} (nm)	n$^+$	27	34	0
	p$^+$	27	22	0
Calculated R_{SD} (Ω μm)	n$^+$	190	230	450
	p$^+$	290	260	1580

8.4.3 Ni silicide as a suitable candidate

NiSi was chosen as the promising silicide for MOS technologies because it consumes the least amount of Si in its formation compared with other silicides such as TiSi$_2$ and CoSi$_2$. NiSi has a low resistivity of about 14 μΩ cm. It also has no resistivity degradation down to 0.1 μm gate line, low formation temperature, and is thermally stable up to 650°C. In addition, the Si consumption can be well controlled by the amount of the deposited Ni film because the entire Ni film reacts with Si. This is not the case in the Ti salicide process in which part of the deposited Ti film reacts with the N$_2$ ambient, which makes it difficult to precisely determine how much Ti film reacts with the Si layer.

Deng et al. [42] investigated NiSi process for 40 nm fully depleted SOI-MOSFETs. The devices used in the experiment were fully depleted CMOS transistors on SIMOX substrates (d_{si} = 40 nm, d_{box} = 360 nm) with gate oxide d_{ox} = 8 nm, L_g = 0.4 μm, W_g = 20 μm, 200 nm wide oxide spacers and n$^+$-type poly-Si gate for NMOS and PMOS. The n-channel was implanted with boron (20 keV, 4.5 × 10^{12} cm^{-2}), and the source/drain regions were implanted with arsenic (40 keV, 3 × 10^{15} cm^{-2}) without LDD structure. The p-channel was not implanted and source/drain were implanted with BF$_2$ (25 keV, 4 × 10^{15} cm^{-2}). A 16 nm thick Ni film was used to consume 30 nm of the Si layer, to form a 36 nm thick NiSi layer. After one-step annealing at 600°C for 40 s in a N$_2$ ambient, the unreacted Ni film on oxide was selectively etched away using H$_2$SO$_4$: H$_2$O$_2$ = 4 : 1 solution. The device characteristics were measured after Al contact metallisation. Source/drain resistances of NMOS and PMOS were extracted to be 500 and 800 Ω μm with low leakage as shown in FIGURE 8.25. The Ti

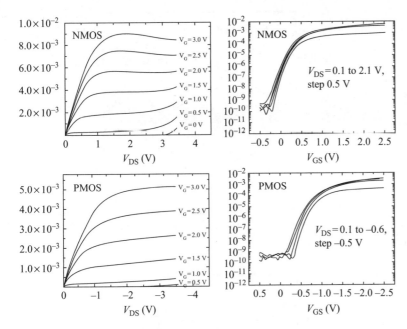

FIGURE 8.25 I–V characteristics and subthreshold plots of Ni-salicided MOS/ SIMOX devices ($L_g = 0.4\,\mu m$, $W_g = 20\,\mu m$ [42]) (© 2004 IEEE).

silicide from the same batch had larger source/drain resistances of 800 $\Omega\,\mu m$ for NMOS and 1400 $\Omega\,\mu m$ for PMOS.

8.4.4 Low-barrier silicides of ErSi$_2$ and PtSi for ultra-thin devices

When the scaling of the traditional bulk design becomes difficult, thin-body gate and double gate devices are regarded as possible extensions of silicon device technology into the sub-20 nm gate length regime. One of the major challenges of the basic thin-body design is the control of the series resistance because the junction depths are limited by the body depth, which is only 5–10 nm. Silicidation on the thin-body is one of the solutions. However, a silicide can introduce a large contact resistance, R_c. FIGURE 8.26 shows the R_c contour plot as a function of silicide barrier height, ϕ_b, and extension doping level [43]. The R_c requirement of $< 200\,\Omega\,\mu m$ cannot be met with a traditional mid-gap silicide such as TiSi$_2$ or CoSi$_2$, which both have a barrier height of 0.6 eV. A silicide barrier of $\phi_b < 0.4$ eV is required to meet the 200 $\Omega\,\mu m$ specification for the doped structure. Unfortunately for a single silicide, lowering ϕ_{bn} increases ϕ_{bp}. By using two different silicides, the barriers for both carriers can be decreased.

For the undoped complementary silicide (CS) structure, the role of the depletion field for a doped contact is played by the e-field,

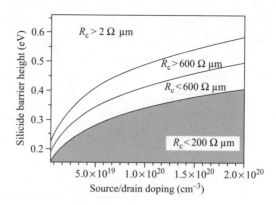

FIGURE 8.26 Design space (in grey) for a doped source/drain thin-body structure. Silicon thin-body is 5 nm. R_c in 5 nm thin-body source [44] (© 2004 IEEE).

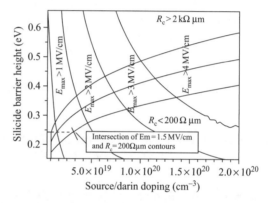

FIGURE 8.27 Design space for undoped CS structure, considering a typical gate induced E-field at source is 1.5 MV/cm. R_c and E_{max} in 5 nm thin-body [44] (© 2004 IEEE).

which is induced by the bias from gate to source. The undoped CS structure needs even a lower silicide barrier than a doped structure because the gate-to-source bias induces a lower electric field than does a highly doped depletion region. FIGURE 8.27 shows the design space for undoped CS structure [44]. Complementary silicide devices have been demonstrated without extension doping, using $ErSi_2$ ($\phi_{bn} = 0.4\,eV$) [45] as the NMOS silicide, and PtSi ($\phi_{bp} = 0.23\,eV$) as the PMOS silicide [46–49].

8.4.5 Selective deposition of silicide on SOI

It is known that titanium salicide process faces more challenges with the shrinkage of device dimensions. First, the process window becomes very narrow, careful process control is needed in all the steps of the salicide process, including titanium sputtering, RTA, and selective wet etch steps [50, 51]. The process becomes more complex to induce phase transformation to low resistance C54 phase for silicide formed on very narrow polysilicon line [51, 52]. The control of silicide thickness becomes more difficult when the reaction rate on n^+ and p^+ silicon is different [53]. There are also problems associated with the process of SOI wafers. In particular silicide thickness is now limited by the thickness of SOI film. A thicker titanium silicide film cannot be formed on a wafer, which has an ultra-thin SOI structure. In this case the silicide film tends to have a higher sheet resistance and poor thermal stability. In view of these, searching for an alternative silicidation technique becomes important.

Selective deposition of $TiSi_2$ on bulk Si wafers by low-pressure chemical vapour deposition (LPCVD) has been reported [54–56]. It is a one-step process, which is favourable in view of process cost. But more importantly, in CVD process the silicide thickness is no longer limited by the silicon thickness of the SOI film. Also, the substrate doping does not have a strong effect on silicide growth; formation of C54 phase occurs during deposition; it is not affected by polysilicon linewidth.

Maa et al. [57] showed the detailed studies of CVD of $TiSi_2$ film on silicon film as thin as 9.7 nm. SOI wafers with 220 nm Si film and 180–200 nm buried oxide layer were used. Then the Si film was thinned down to a final thickness of 20–50 nm. CVD $TiSi_2$ deposition was carried out in an AG Integral One rapid thermal chemical vapour deposition (RTCVD) system. The background pressure of the deposition chamber was around 5×10^{-8} Torr, and the deposition temperature was maintained around 780°C. Process gases consisted of $TiCl_4$, SiH_4, SiH_2Cl_2, and H_2. The flow rates of SiH_4, SiH_2Cl_2, and H_2 were maintained at 61, 5, and 1500 sccm, respectively. It was maintained either at 7 or 8% of its maximum flow range, corresponding to about 0.34 or 0.54 sccm of $TiCl_4$ after subtracting the zero drift which varied slightly from time to time. The pressure was around 300 mTorr.

FIGURE 8.28 shows the thickness of silicide and the amount of Si consumed from or accumulated onto the substrate. An increase of silicide thickness with $TiCl_4$ flow rate and a positive build-up of silicon at the interface were observed. The Si thickness was found to remain unchanged with deposition time at 7% $TiCl_4$ flow (0.34 sccm) and decrease slightly with 8% of maximum flow (0.54 sccm).

The study of CVD silicide film deposition on SOI and on polysilicon structure is shown in FIGURES 8.29 and 8.30. Solid line is as-deposited value. Dashed lines are values after RTA at 850°C for 10, 20 and 30 min. The figures show a decrease of sheet resistance with deposition time in both SOI and polysilicon cases. After 26–28 s the slope of the resistance curves become constant, most

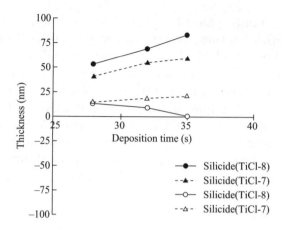

FIGURE 8.28 Silicide thickness and Si consumption (−) or accumulation (+) at 7% (− − −) and 8% (———) of maximum TiCl$_4$ flow [57].

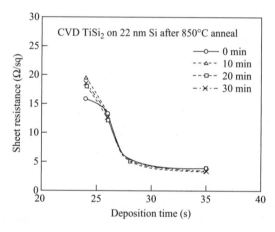

FIGURE 8.29 Sheet resistance of CVD titanium silicide films deposited on 220 Å SOI film [57].

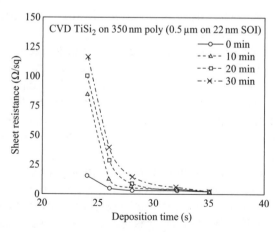

FIGURE 8.30 Sheet resistance of CVD titanium silicide films deposited on 350 nm thick poly-silicon lines [57].

likely due to the completed transition into C54 phase. A low sheet resistance less than 5 Ω/sq can be achieved in both 22 nm SOI film and 0.5 µm polysilicon lines.

The thermal stability of CVD titanium silicide was studied by second anneals at 850°C, as shown in FIGURES 8.29 and 8.30 with dashed lines. It is found that Ti silicide films deposited on SOI wafers were quite stable after annealing at 850°C for 30 min. For 0.5 µm polysilicon structures, increase of sheet resistance depending on silicide deposition time and annealing time is observed. If the deposition time is longer than 35 s, the silicide film is found to be quite stable at 850°C. One remaining issue of this technology is the silicide/Si specific contact resistivity.

8.5 SUMMARY

In this chapter, the SOI silicide technologies are described. Starting from the review of silicide design analysis on SOI, including the effects of source/drain series resistance and gate resistance on SOI device performance, the silicide thickness design analysis and silicon resistivity design space for a specific design constraint. The dominant parameters for silicidation on SOI devices are the silicide/Si specific contact resistivity, the silicide thickness, and the silicon resistivity. Both the interface-specific contact resistivity and the silicon resistivity are directly related to the dopant concentration in the silicon layer. To minimise these resistivities, we need to maximise the degenerate doping in the source/drain regions to the order of 10^{20} cm^{-3}. Then, the challenges to implement existing bulk silicide technology to SOI devices are introduced. Finally, various silicide technologies on SOI including pre-amorphisation source/drain regions to control silicide depth, cobalt and nickel silicide technology for thin film SOI devices, low-barrier silicides of ErSi$_2$ and PtSi for sub-20 nm novel devices, as well as selective silicide deposition on SOI, are discussed.

REFERENCES

[1] J.A. Kittl, Q.Z. Hong [*Proceedings of the 1996 23rd International Conference on Metallurgical Coating and Thin Films*, San Diego, CA, USA. *Thin Solid Films (Netherlands)* vol.290–91 (1996) p.473–6]
[2] J.P. Colinge [*IEEE Electron Device Lett. (USA)* vol.EDL-7 (1986) p.244]
[3] J.B. Kuo, K.-W. Su [*CMOS VLSI Engineering Silicon-on-Insulator (SOI)* (Kluwer Academic Publishers, Dordrecht, 1998) p.50–2]
[4] T.C. Hsiao [PhD dissertation, UCLA (1997) p.38]
[5] C.C. Enz, F. Krummenacher, E.A. Vittoz [*Analog Integra. Circuits Signal Process. (USA)* vol.8 (1995) p.83–114]

[6] J.P. Colinge [*IEEE Int. SOI Conf. Short Course Notes* (1996) p.1–36]
[7] H.I. Liu, J.A. Burns, C.L. Keast, P.W. Wyatt [*IEEE Trans. Electron Devices (USA)* vol.45 (1998) p.1099–104]
[8] T. Sakurai, T. Iizuka [*IEEE Trans. Electron Devices (USA)* vol.ED-32 (1985) p.370–4]
[9] C.M. Osburn, K.R. Bellur [*Thin Solid Films (Netherlands)* vol.332 (1998) p.428–36]
[10] K. Suzuki, T. Tanaka, Y. Tosaka, T. Sugii, S. Andoh [*IEEE Trans. Electron Devices (USA)* vol.41 (1994) p.1007–12]
[11] C.C. Lin, W.S. Chen, H.L. Hwang, K.Y.J. Hsu, H.K. Liou, K.N. Tu [*Appl. Surf. Sci. (USA)* vol.92 (1996) p.660–4]
[12] S.P. Murarka [*Silicides for VLSI Applications* (Academic, New York, 1983)]
[13] P. Fornara, A. Poncet [*IEDM Tech. Dig.* (1996) p.73]
[14] T. Nishimura, Y. Yamaguchi, H. Miyatake, Y. Akasaka [*Proc. IEEE SOS/SOI Tech. Conf.* (1989) p.132–3]
[15] Y. Yamaguchi, T. Nishimura, Y. Akasaka, K. Fujibayashi [*IEEE Trans. Electron Devices (USA)* vol.EDL-39 (1992) p.1179]
[16] J. Foerstner, J. Jones, M. Huang, B.Y. Hwang, M. Racanelli, J. Tsao [*IEEE Int. SOI Conf. Proc.* (1993) p.86]
[17] M.A. Mendicino, E.G. Seebauer [*J. Electrochem. Soc. (USA)* vol.142 (1995) p.28–30]
[18] S. Okazaki [*IEDM Tech. Dig.* (1996) p.57]
[19] K.N. Tu, J.W. Mayer [in *Thin Films—Interdiffusion and Reactions* Ed. J.M. Poate, K.N. Tu, J.W. Mayer (The Electrochemical Society, Wiley-Interscience, Princeton, NJ, 1978) Chapter 10]
[20] K.N. Tu [*IBM J. Res. Dev. (USA)* vol.34 (1990) p.868–74]
[21] J.M. Liang, L.J. Chen, I. Markov, et al. [*Mater. Chem. Phys.* vol.38 (1994) p.250–7]
[22] K. Goto, A. Fushida, J. Watanabe, et al. [*IEDM Tech. Dig.* (1995) p.449]
[23] T. Ohguro, S. Nakamura, E. Morifuji, et al. [*IEDM Tech. Dig.* (1995) p.453]
[24] M. Rodder, E.Z. Holag, M. Nandakumar, S. Aur, J.C. Hu, I.C. Chen [*IEDM Tech. Dig.* (1996) p.563]
[25] G.Z. Pan, E.W. Chang, Y. Rahmat-Samii [*Mater. Res. Soc. Symp. Proc. (USA)* vol.716 (2002) p.451–6]
[26] F. Nava, K.N. Tu, O. Thomas, et al. [*Mater. Sci. Rep. (Netherlands)* vol.9 (1993) p.141–200]
[27] L.T. Su, M.J. Sherony, H. Hang, J.E. Chung, D.A. Antoniadis [*IEDM Tech. Dig.* (1993) p.723]
[28] S.L. Hsia, T.Y. Tan, P.L. Smith, G.E. McGuire [*VLSI Symp. Tech. Dig.* (1994) p.123]
[29] S.L. Hsia, T.Y. Tan, P.L. Smith, G.E. McGuire, W.T. Lynch [*Mater. Res. Soc. Symp. Proc. (USA)* (1994) p.373–8]
[30] T.Y. Tan [*SRC Review* (1993)]
[31] T. Morimoto, T. Ohguro, H.S. Momose, et al. [*IEEE Trans. Electron Devices (USA)* vol.42 (1995) p.915]
[32] T.C. Hsiao, P. Liu, J. Woo [*IEEE Trans. Electron Devices (USA)* vol.45 (1998) p.1092–8]
[33] L.P. Ren, P. Liu, G.Z. Pan, J.C.S. Woo [*Mater. Res. Soc. Symp. Proc. (USA)* vol.514 (1998) p.245–9]
[34] L.P. Ren, B. Chen, J.C.S. Woo [*IEEE International SOI Conference* (1999) p.88–9]

[35] T.C. Hsiao, P.Liu, J.C.S. Woo [*IEEE International SOI Conference Proceedings* (1998) p.153–4]

[36] P. Liu, T.C. Hsiao, J.C.S. Woo [*IEEE Trans. Electron Devices (USA)* vol.45 (1998) p.1280–6]

[37] M. Yoshimi, M. Terauchi, A. Murakoshi, et al. [*IEDM Tech. Dig.* (1994) p.429]

[38] H. Kotaki, M. Nakano, S. Hayashida, et al. [*IEDM Tech. Dig.* (1995) p.457]

[39] J.A. Kittl, Q.Z. Hong, M. Redder, et al. [*VLSI Symp. Tech. Dig.* (1996) p.14]

[40] S.L. Hsia, T.Y. Tan, P. Smith, G.E. McGuire [*J. Appl. Phys. (USA)* vol.72 (1992) p.1864]

[41] T. Ichimori, N. Hirashita [*IEEE International SOI Conference* (2000) p.72]

[42] F. Deng, R.A. Johnson, W.B. Dubbelday, G.A. Garcia, P.M. Asbeck, S.S. Lau [*IEEE International SOI Conference* (1997) p.22–3]

[43] C.Y. Chang, Y.K. Fang, S.M. Sze [*Solid State Electron. (USA)* vol.14 (1971) p.541–50]

[44] J. Kedzierski, M.K. Ieong, P. Xuan, J. Bokor, T.-J. King, C. Hu [*IEEE International SOI Conference* (2001) p.21–2]

[45] K.N. Tu, R.D. Thompson, B.Y. Tsaur [*Appl. Phys. Lett. (USA)* vol.38 (1981) p.626–8]

[46] G. Ottaviani, K.N. Tu, J.W. Mayer [*Phys. Rev. Lett. (USA)* vol.44 (1980) p.284–7]

[47] J. Kedzierski, P. Xuan, E.H. Anderson, J. Bokor, T.J. King, C. Hu [*IEDM* (2000) p.57]

[48] W. Saitoh, A. Itoh, S. Yamagami, M. Asada [*Jpn. J. App. Phys. (Japan)* vol.38 (1999) p.6226]

[49] A. Itoh, M. Saitoh, M. Asada [*Jpn. J. App. Phys. (Japan)* vol.39 (2000) p.4757]

[50] L.P. Hobbs, K. Maex [*Appl. Surf. Sci. (USA)* vol.53 (1991) p.321]

[51] R.W. Mann, L.A. Clevenger [*J. Electrochem. Soc. (USA)* vol.141 (1994) p.1347]

[52] J.A. Kittl, Q.Z. Hong [AIP, American Institute of Physics Conference Proceedings (1998), p.439–50]

[53] H.K. Park, J. Sachitano, M. McPherson, T. Yamaguchi, G. Lehman [*J. Vac. Sci. Technol. A (USA)* vol.2 (1984) p.264]

[54] V. Ilderem, R. Reif [*J. Electrochem. Soc. (USA)* vol.135 (1988) p.2590]

[55] J.L. Regolini, D. Bensahel, G. Bomchil, J. Mercier [*Appl. Surf. Sci. (USA)* vol.38 (1989) p.408]

[56] A. Bouteville, A. Royer, J.C. Remy [*J. Electrochem. Soc. (USA)* vol.134 (1987) p.2080]

[57] J. Maa, B. Ulrich, S.T. Hsu, G. Stecker [*Thin Solid Films (Netherlands)* vol.332 (1998) p.412–17]

Chapter 9

Characterisation of metal silicides

Y.F. Hsieh, S.L. Cheng and L.J. Chen

9.1 SCOPE OF THE CHAPTER

This chapter will describe the fundamental principle of the analytical tools of materials characterisation and specially emphasise the joint applications of two or more techniques applied in various studies of silicide formation. It is aimed to illustrate a specific topic from different perspectives, in order to gain an overall picture of the issue, such as macrostructure versus microstructure examination, electrical properties versus physical characteristics, carrier distribution versus impurity involvement, and lattice imaging versus computer simulation.

9.2 TOOLS OF MATERIALS CHARACTERISATION

9.2.1 Introduction

Analytical tools are indispensable in all kinds of research and development activities. Based on the analytical purposes, the tools of materials characterisation can be categorised into three major fields of applications – physical/structural analyses, chemical/elemental analyses and electrical analyses, as shown in FIGURE 9.1. The most popular techniques are listed as follows [1–3].

9.2.1.1 *Physical/structural analysis*

Scanning electron microscopy (SEM)
Transmission electron microscopy (TEM)
Scanning transmission electron microscopy (STEM)
Focused ion beam (FIB) microscopy
Atomic force microscopy (AFM)
Scanning capacitance microscopy (SCM)
Electrostatic force microscopy (EFM)
X-ray diffraction (XRD)

Scope of the chapter p.229

**Tools of materials
characterisation p.229**
  Introduction p.229
  Fundamental principle of
    beam–solid interaction p.230
  Image resolution p.232
  Probe size and detection limits p.234
  Field of applications p.235

Morphology observation p.237

**Crystal structure of metal
silicides p.240**
  Epitaxial silicides p.240
  Amorphous metal/Si alloy films p.242

Initial silicide formation p.243
  Silicide formation in amorphous
    interlayer p.243

**Phase formation and
identification p.245**
  Silicide formation on patterned
    (001) and (111)Si
    wafers p.245
  Effects of interposing layers on the
    formation of metal silicides p.246
  Silicide formation by metal ion
    implantation into Si wafers p.250

Defect analysis p.252
  Cracks (SEM, FIB, PTEM, XTEM,
    BF, DF, DP) p.252
  Voids (XTEM) p.254
  Pinholes (SEM, AFM, PTEM) p.254
  Planar defects (PTEM, XTEM,
    2B-DF, HRTEM) p.254
  Vacancies (DP, simulation) p.255

Thermal stability (sheet resistance, PTEM, XTEM, BF, DF) p.255

References p.258

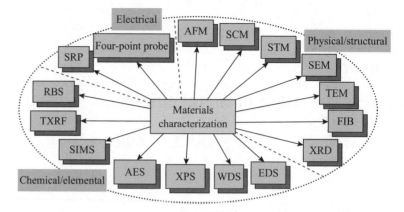

FIGURE 9.1 Popular tools of materials characterisation, which can be categorised as being for physical/structural analyses, chemical/elemental analyses, and electrical analyses.

Rutherford backscattering spectroscopy (RBS)
Scanning tunnelling microscopy (STM)
Reflective high energy electron diffraction (RHEED)
Stress measurement

9.2.1.2 Chemical/elemental analysis

Energy dispersive spectroscopy (EDS) of X-ray
Wavelength dispersive spectroscopy (WDS) of X-ray
Auger electron spectroscopy (AES)
Secondary ion mass spectroscopy (SIMS)
Total reflection X-ray fluorescence (TXRF)
Electron spectroscopy for chemical analysis (ESCA), also called X-ray photoelectron spectroscopy (XPS)
Synchrotron radiation – ultraviolet photoemission spectroscopy (SR-UPS)

9.2.1.3 Electrical properties measurement

Four-point probe (sheet resistance)
I–V measurement (Schottky barrier height, SBH)
Spreading resistance profiling (SRP)

9.2.2 Fundamental principle of beam–solid interaction

Most of the analytical techniques involve a primary beam interaction with a solid material. By detecting the secondary particles excited by high energy bombardment of the primary beam, microscopy and spectroscopy can be obtained by scanning and projection

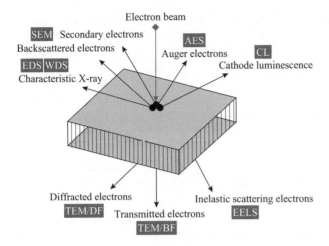

FIGURE 9.2 Interaction of electron beam with solid materials.

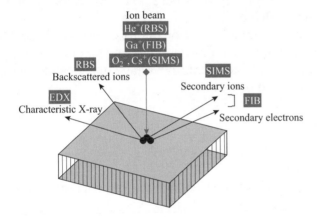

FIGURE 9.3 Interaction of ion beam with solid materials.

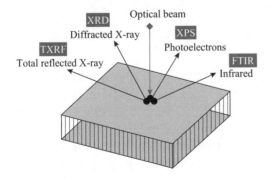

FIGURE 9.4 Interaction of optical beam with solid materials.

coils, or energy/mass analysers. The beam–solid interaction is usually classified into three types, electron beam, ion beam and optical beam. FIGURES 9.2–9.4 show the schematic diagrams of the three types of beam–solid interaction and indicate the corresponding secondary particles associated with the analytical tools.

9.2.3 Image resolution

According to the de Broglie wave theory, the image resolution is dependent on the wavelength of charged particles, which can be derived as a function of accelerating voltage of the primary beam [2]. Thus, high accelerating voltage will enhance the image resolution as a result. Comparing the effective mass of electrons and ions, it is easy to understand the superior resolution of electron microscopy overriding that of ion microscopy. A simple calculation can be used to estimate the electron wavelength, λ_e, as a function of accelerating voltage.

According to de Broglie wave theory, $\lambda_e = h/mv = h/2qmV = 1.22/V$ (nm), where v is the velocity and V the accelerating voltage.

In all kinds of studies, electron microscopes, SEM and TEM, are the most popular tools for morphology and topology examinations. Assuming a certain accelerating voltage, the electron wavelength was calculated to be as follows.

For example,

$$\text{SEM, } 10\,\text{kV}, \quad \lambda_e = 0.012\,\text{nm}$$
$$\text{TEM, } 200\,\text{kV}, \quad \lambda_e = 0.0027\,\text{nm}$$

The wavelengths are much smaller than that of optical microscopy. The wavelength of visible light is typically in the range of 400–700 nm. Thus, it is also justified by Raleigh's criterion, stated as follows:

$$s = 0.61 \frac{\lambda}{n \sin \alpha} = 0.61 \frac{\lambda}{\text{NA}}$$

where s is the resolving power (the minimum distance between two), n the refractive index of the medium, α the half angle between the objective lens and the sample surface and NA the numerical aperture, which is a number indicating the resolution of the lens system and the brightness/quality of the image formation.

The resolving power of optical microscopy was estimated to be about $(0.61 \times 0.5\,\mu\text{m})/0.95 = 0.32\,\mu\text{m}$, assuming $\lambda = 500$ nm for visible light and NA = 0.95 for a typical optical system. However, in order to improve the resolving power, the refractive index, n, can be increased by immersing samples in different media, such as oil ($n = 1.52$) or water ($n = 1.33$). Since the resolving power of human eyes is limited to be about ~ 0.2 mm, the optimum resolution of optical microscopy shall be $\sim 0.2\,\mu$m under normal operating conditions at a magnification of 1000×. A variety of methods have been reported in the literature to enhance the resolution by using laser beam, ultraviolet, X-ray or coherent light sources. Light sources of shorter wavelength are used in order to achieve better resolution. In addition, computer-assisted image processing offers other alternatives to improve the resolution by brightness/contrast enhancement. Thus, low magnification examination, between 1000 and 5000×, can be achieved by optical microscopy, which used to be difficult for SEM, TEM, and/or FIB to make a general morphology survey.

State-of-the-art, commercially available, tools can provide the best resolution of 0.7–1.5 nm for SEM [4] and 0.1–0.2 nm for TEM [5–10], although the electromagnetic lens system might introduce diffraction aberration, spherical aberration, chromatic aberration and stigmatism, which deteriorate the resolution. The most updated design of SEM system usually utilises in-lens or semi-in-lens design to decrease the working distance by immersing the sample in the electromagnetic field of the objective lens. Employing a field emission gun (FEG) is common practice in SEM and TEM, which increases the current density and thus enhances the brightness of the image. The life time of the electron gun is prolonged to a couple of thousand hours as well. For AES, the insertion of a cylindrical multi-channel analyser (CMA) hinders the electron beam from the sample surface and increases the working distance. Commercial tools offer resolution about 10–15 nm. TABLE 9.1 lists the important characteristics of various tools.

The ion beam focusing system utilises the electronic lens control. Reducing the probe size by condenser aperture and multiple evolution of the lens control system has delivered an excellent resolution of 5 nm, which has been successfully demonstrated in FIB. This tool, now available, has emerged as a very powerful tool for IC manufacturing since 1994 [11–14]. It possesses the superior capability of precision cutting and cross-sectioning of any suspected failure site. In addition, the optional attachments can offer software control of the auto navigation system for circuit debugging, auto-FIB cutting/deposition for circuit modification and auto-TEM jobs for precision sample preparation. The image formation mechanisms can be manipulated by detecting signals of secondary ions or secondary electrons of the Faraday cups. In addition, the system configuration can be of a single beam (ion beam) design, or dual beam (ion beam and electron beam) design. However, as previously mentioned, the electron beam image is always better than the ion beam image, due to the effective mass of the particle. The dual-beam system always possesses better image resolution, but the system cost is considerably higher.

TABLE 9.1 Characteristics of the analytical tools for materials characterisation.

Analytical tools	Primary beam	Detected signals	Depth of excitation volume	Image resolution	Detection limits	Detectable elements
SEM	Electrons	Secondary electrons; Backscattered electrons	>1 µm	7–20 Å	–	–
TEM	Electrons	Transmitted electrons; Diffracted electrons	0–1000 Å	1–2 Å	–	–
FIB	Ions	Secondary electrons; Secondary ions	<500 Å	50–100 Å	–	–
EDS	Electrons	Characteristic X-ray	>1 µm	>0.1 µm	1%	>Na
WDS	Electrons	Characteristic X-ray	>1 µm	>0.1 µm	1%	>Li
XRD	X-ray	Diffracted X-ray	>1 µm	>1 µm	5%	Crystalline materials
AES	Electrons	Auger electrons	<50 Å	>0.1 µm	0.1%	>Li
XPS	X-ray or ultraviolet	Photoelectrons	5–10 Å	0.2–1 µm	0.1%	>Li
RBS	He^+ ions	Backscattered ions	<20 µm	1–3 µm	0.1%	>C
SIMS	Cs^+, O_2^- ions	Secondary ions	<500 Å	>0.1 µm	>0.1 ppm	>H
TXRF	X-ray	Total reflected X-ray	5–10 Å	–	>1 ppb	>C

9.2.4 Probe size and detection limits

For chemical analysis/elemental identification, probe size/analytical volume is always a trade-off with detection limits [15–17]. The tools possessing ultimate performance in sensitivity and detection limit require large analytical area/volume [2]. TXRF is excellent in contamination analysis of a couple of atomic layers. The X-ray source will require an area of the order of 100 µm, for collecting sufficient signals. Equipped with an Ar^+ ion sputtering gun, SIMS is a good tool in both surface analysis and depth profiling of any minority species [18, 19]. It can be applied for junction profiling of impurity analysis, identification of diffusion markers, verification of thin interposing layers of a couple of angstroms, etc. It is known that SIMS can reach a detection limit to 10^{16}–10^{18} cm^{-3}, depending on the relative sensitivity factor (RSF) of the targeted species. However, the probing area is still the key factor determining the background noise or/say detection level. A common regime requires an area at least 80–150 μm in diameter to avoid the crater sidewall effect. Similar requirements can be applied to AES as well. A complementary tool of SIMS depth profiling is SRP, which has been widely applied in IC manufacturing for junction profiling of dopant analysis. Since it is operated by a two-point probe, the electrical carrier concentration is basically converted from the resistivity measurement. There is always a discrepancy between these two tools, although it can be presumed that physically present impurities are always more than the electronically active dopants.

For RBS analysis, the most important applications of this tool are in composition and thickness measurement of thin films, such as silicides (WSi$_x$, TiSi$_x$), barrier layers (TiW, TiN), dielectrics (SiO$_x$, SiN$_x$, SiO$_x$N$_y$), and metal layers (Al, AlCu, AlSiCu, Ti, W). It provides an alternative when materials analyses encounter the difficulties in identifying isotopes (SIMS) or emitting X-ray peaks of unresolvable energy channels (EDS). However, due to requirement of relatively large sample size, RBS is seldom used for patterned device analysis. Test pattern design was mandatory, but using control wafer is a more realistic way due to limited chip size of test key and/or test site.

For patterned wafer analyses, EDS, being equipped with SEM or TEM, has become the most intensively used tool nowadays for all industrial needs. Joint applications of FIB/EDS and SEM/EDS [4], serving for in-line and off-line yield enhancement, have been able to analyse any defect down to ∼0.1 μm in size with the aid of FIB single cut or multiple slicing techniques. While, joint applications of TEM/EDS/EELS [10], serving for off-line failure analysis, product debugging and reverse engineering, are able to reveal the image of atomic resolution and to identify foreign elements with a minimum probe size of 5 nm. It is generally recognised as the most useful method in defect analysis among all kinds of elemental characterisation methods, irrespective of the limitation of sample preparation techniques.

9.2.5 Field of applications

For academic study, most of the research work on metal silicides is conducted on either bulk materials or thin films on a substrate. The whole spectrum of analytical tools are mandatory for an in-depth study and the investigation of fundamental mechanisms.

It will certainly need all kinds of techniques with much wider range of capabilities for characterisation. Tools discussed in this chapter, as listed in Section 9.2.1, are the ones frequently used. Those typical examples of joint-tools application onto different research topics will be illustrated in the following sections. Characterisations of XRD, RBS, WDS, ESCA/XPS, SR-UPS, TXRF, STM, RHEED, stress measurement, $I-V$ measurement of Schottky barrier height, and sheet resistance measurement by four-point probe, will generally require sample sizes of the order of 1–2 cm^2, and a relatively large probing area, more than a couple of hundreds of a micrometre. They are seldom used in small area/patterned device analyses. For the investigation of metal silicides, interfacial reaction of metal thin films on Si wafer is known to be highly dependent on temperature, process time, annealing ambient, implanted impurities, pre-amorphisation by ion mixing, pre-cleaning methods, ultra-thin interposing layer, and capping layers. Earlier literature shows that most of the research activities was focused on the investigation of crystal structure, phase identification, defect kinetics, thermal stability, and initial silicide formation. Advanced TEM imaging techniques, including high resolution electron microscopy (HREM) [9], selected area diffraction pattern analysis (SADP), two beam dark field (2BDF), weak beam dark field (WBDF), and computer simulation of reciprocal lattices and high-resolution TEM images [such as fast Fourier transform (FFT) and auto correlation function (ACF) methods] have become major tools of analyses and are often used for the study of material behaviours, especially in the size range of sub-micrometre or nanometre.

For IC manufacturing, analytical tools are employed in many functions of operations, not only in R&D activities, but also in new product development, process monitoring, low yield enhancement, customer return, customer complaint, product debugging, in-coming materials inspection, failure analysis and quality assurance. The most frequently used tools comprise SEM, TEM [20], FIB, AFM, EDS, AES, SIMS, TXRF and SRP. Some of them have been remodelled with clean room body in order to serve as in-line monitoring tools, such as SEM equipped with EDS, FIB equipped with EDS and/or SIMS, stand-alone SIMS, AFM and TXRF. These analytical tools are mostly dedicated for morphological examination, dimension measurement, particle cross-sectioning, surface roughness profiling and contamination characterisations. Specifically for silicide fabrication in IC products, cross-sectional TEM analysis is the most appropriate technique, serving for thickness measurement and uniformity control of the consumed silicon and the degree of silicidation. Plan-view TEM analysis proves to be the most useful for phase identification and grain size measurement.

Meanwhile, FIB is excellent support for precision analysis onto any specific defect sites. However, owing to the small pattern size of device features, applications of the other tools, like AES, SIMS, TXRF, XRD and SRP, will be limited by sample preparation techniques and the probe size of corresponding instrumentation, as has been stated in Section 9.2.4. Chemical analysis, or elements' identification, of a feature size smaller than 0.1 μm in thickness or in size, relies mostly on TEM equipped with EDS or EELS. Verification of phases and crystal structure of miniature regions may use the optional function of microdiffraction for resolving the reacted compounds of various stoichiometry. STEM equipped with EDS could be very useful for line scan and mapping of elements. A recent publication reported that AFM, SCM and EFM were very good at small features analyses, especially for mapping of the dopants of different nature [21].

9.3 MORPHOLOGY OBSERVATION

As film thicknesses lie mostly in the range of 1–100 nm, morphological observation of reacted metal silicides can be done by using direct beam imaging mode (bright field) of OM, SEM,

TABLE 9.2 Imaging modes of structural analyses tool.

Tools	Imaging modes	Purpose of applications
OM	BF, bright field	General examination
	DF, dark field	Surface extrusion, particle examination
SEM	SEI, secondary electron image	Morphological examination
	BEI, backscattering electron image	Composition verification
TEM	2BBF, two-beam bright field	Precipitates, second phase, crystal defects characterisation
	2BDF, two-beam dark field	Precipitates, second phase, crystal defects characterisation
	WBBF, weak-beam bright field	Bubbles, voiding examination
	WBDF, weak-beam dark field	Dislocation characterisation
	MBBF, multi-beam bright field HREM, high-resolution electron microscopy	Lattice imaging, epitaxial crystal orientation, and twinning plane relationships
FIB	SEI, secondary electron image	General examination
	SII, secondary ion image	Ion channeling contrast for grain orientation imaging

Characterisation of metal silicides

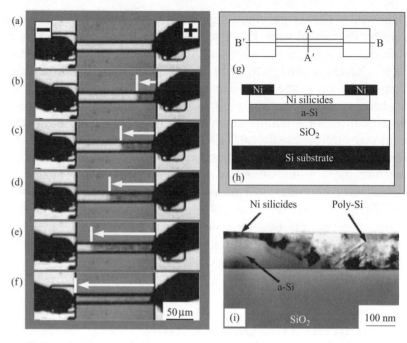

FIGURE 9.5 Plan-view OM images showing the growth sequence of poly-silicide formation with voltage fixed at 60 V taken at (a) 0 s, (b) 0.067 s, (c) 0.1 s, (d) 0.133 s, (e) 0.167 s, (f) 0.2 s with arrows pointing to the growth direction; schematic diagrams of (g) plan-view, and (h) cross-sectional view of the test structure; and (i) XTEM micrographs of the test key structure of current enhanced polysilicon formation (taken along the $B - B'$ direction).

TEM, and FIB. Although alternative imaging modes are available, as listed in TABLE 9.2, they are used for in-depth study for special purposes. Bright field imaging only reveals very limited information as listed below.

OM – as deposited versus reacted phase;
SEM – layer thickness, grain size, uniformity;
TEM – layer thickness, grain morphology, uniformity, phases, crystal structure;
FIB – layer thickness, uniformity.

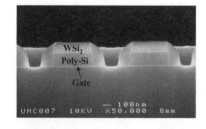

FIGURE 9.6 Cross-sectional SEM micrograph of WSi$_2$ silicide gate structure, taken from an in-processed wafer. The cross-sectional sample was made by mechanical grinding and polishing without chemical staining (courtesy of UMC).

FIGURE 9.5 shows a test key structure of current enhanced poly-Si formation. The reacted poly-Si can be distinguished from the unreacted material by discoloration. The corresponding cross-sectional TEM (XTEM), as shown in FIGURE 9.5(i) has verified the morphology and grain structure of the reacted phase.

It is worth noting that the sample preparation techniques will seriously affect the image quality and, therefore, the interpretation of image must be done very cautiously. Especially, for SEM observation, many pre-treatment procedures of sample preparation, like wafer cleavage/mechanical grinding and polishing/chemical staining, are usually mandatory prior to examination. FIGURES 9.6–9.12 include some typical case studies of WSi_2, $TiSi_2$ and $CoSi_2$ in IC products.

In FIGURE 9.6, a cross-sectional SEM (XSEM) micrograph of WSi_2 gate structure prepared by mechanical grinding and polishing of an in-processed wafer is shown. Contrast of each material only revealed the reflectivity and conductivity of corresponding materials with respect to electron beam interaction. It was difficult to obtain information about the grain size, shape and orientation, owing to flat surface resulting from mechanical cross-sectioning. FIGURE 9.7 is a XSEM micrograph of the WSi_2 gate structure of a finished SRAM product, where the sample was made by cleavage and chemically stained by a dilute HF solution. Interlayer dielectrics, multiple oxide layers, were etched off slightly. The recessed thickness was found to be dependent on the density of different oxide layers as a result. It is seen clearly that the cleaved WSi_2 surface maintained a nice shape of the poly grains. The size and morphology can be recorded for the SEM examination readily. FIGURE 9.8 is a XSEM micrograph of the WSi_2 gate structure of a finished DRAM product. The sample was also made by cleavage, but chemically stained by a dilute HNO_3 and HF solution. Thus, poly-Si of gate and plug, hemispherical Si grains and Si substrate were etched, instead of the oxide layers. It was found that WSi_2 remained unattacked by the staining solution. The grain morphology is still clearly seen.

FIGURE 9.9 shows a direct comparison of SEM and TEM images of a gate structure of flash memory. Fundamental difference in image formation of these two tools can be distinguished. It is apparent that TEM microscopy is more informative and the image resolution is much higher. Certainly, thickness measurement of the stacked ONO structure can be expected to be more accurate.

Regarding the analysis of $TiSi_2$, it is usually difficult to reveal the grain structure from the cross-sectional view. FIGURE 9.10 presents a plan-view OM micrograph and a secondary electron image of a $TiSi_2$ salicide gate structure with a gate length of 0.15 µm obtained by precision FIB image cross-sectioning. The FIB cutting was performed in parallel with the gate line. The secondary electron image of the $TiSi_2$ gate exhibits bright contrast, due to the metallic and conducting natures of material characteristics. Since it is well known that $TiSi_2$ can be easily attacked by corrosive acids, SEM sample preparation using dilute HF solution or dilute HNO_3 and HF solution, will always etch the $TiSi_2$ silicide and result in a dip voiding on top of the poly-Si gate. FIGURE 9.11 shows a series of cross-sectional micrographs

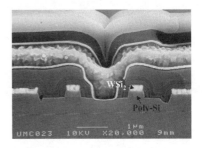

FIGURE 9.7 Cross-sectional SEM micrograph of WSi_2 silicide gate structure of SRAM, taken from a finished product. The cross-section sample was made by cleavage and chemically stained by a dilute HF solution.

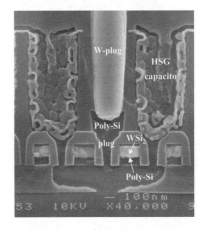

FIGURE 9.8 Cross-sectional SEM micrograph of WSi_2 silicide gate structure of 64 Mb DRAM, taken from a finished product. The cross-section sample was made by cleavage and chemically stained by a dilute HNO_3 and HF solution. The storage capacitor was formed by hemispherical silicon growth (HSG) method. The bit line contact formation employed a stacked plug technology by using an indirect poly-Si plug and a W-plug.

Characterisation of metal silicides

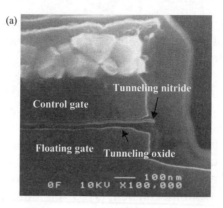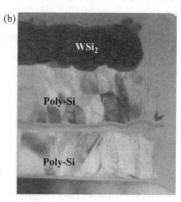

FIGURE 9.9 Cross-sectional micrograph of WSi_2 silicide gate structure of flash memory. (a) The cross-sectional SEM sample was made by cleavage and chemically stained by a dilute HF solution. (b) The TEM micrograph is shown to reveal the image quality and to differentiate the image formation mechanisms.

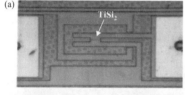
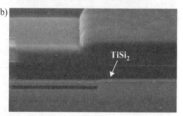

FIGURE 9.10 (a) Plan-view OM micrograph of $TiSi_2$ salicide gate structure with gate length 0.15 μm, (b) secondary electron image of the precision cross-sectioning of the gate structure, where FIB cutting was performed in parallel with the gate line and image was taken from a 45°-tilted view.

of $TiSi_2$ salicide gate structure of a 266 MHz CPU. Obviously, the $TiSi_2$ on gate and source/drain region were etched off. Only morphology of the material configuration was left for examination. The TEM micrograph, as shown in FIGURE 9.11(c), is still perfect for revealing the grain contrast of all material layers and device architecture.

$CoSi_2$ may also be observed using similar sample preparation techniques, although $CoSi_2$ salicide is widely used only in devices 0.18 μm and below. FIGURE 9.12 shows XTEM micrographs of 0.1 μm $TiSi_2$ salicide gate structure, 0.1 μm $CoSi_2$ salicide gate structure and 50 nm WSi_2 gate structure. The integrity of the gate structure is seen to be preserved.

9.4 CRYSTAL STRUCTURE OF METAL SILICIDES

9.4.1 Epitaxial silicides

Epitaxial silicides are defined as those silicides that exhibit similar crystal structures to silicon and comply with special orientation relationships with respect to Si substrate. A silicide is expected to grow epitaxially on silicon if the crystal structures are similar and the lattice mismatches between them are small. Before 1983, PtSi, Pd_2Si, $NiSi_2$ and $CoSi_2$ were the only four generally known epitaxial silicides [22–25]. Since then, almost all transition and rare-earth (RE) metal epitaxial silicides have been successfully grown to a certain extent on silicon substrate [26–40]. In addition, for novel device applications such as metallic base transistors or tunable infrared sensors, it is necessary to grow epitaxial silicon layers on the epitaxial silicide/silicon substrates [41–46]. For RE metal silicides on silicon, the excellent lattice mismatch between many epitaxial RE silicides and silicon substrates offers a good opportunity for growing high-quality double heterostructures.

In this section, the formation of epitaxial silicide thin films as well as epi-Si/epi-silicide/Si double heteroepitaxial structures and their characterisation using different analytical techniques, such as PTEM, XTEM, BF, DF, DP, HREM and XRD will be described. Identification of Pd_2Si epitaxial relationships, verification of twinned versus untwinned $CoSi_2$ and $NiSi_2$ phases, characterisation of interfacial dislocations, and observations of YSi_{2-x} microtwins are also discussed.

For Pd(30 nm)/(111)Si samples annealed at 800°C, from plan-view TEM observation, Pd_2Si was observed to be epitaxially related to the silicon substrate [47]. A high density of interfacial dislocations was found to form. The dislocations were previously determined to possess Burgers vectors of $1/2\langle 1\,1\,2 \rangle$ [48]. From electron diffraction pattern and wide-angle XRD analysis, the orientation relationships of Pd_2Si epitaxy with respect to (111)Si were identified to be $[0001]Pd_2Si//[111]Si$ and $(60\bar{6}0)Pd_2Si//(20\bar{2})Si$. Examples are shown in FIGURES 9.13(a)–(d). In addition, from cross-sectional TEM observation, the surface and interface of epitaxial Pd_2Si thin films on (111)Si are found to be very sharp and smooth. An example is shown in FIGURE 9.14.

For both epitaxial $NiSi_2$ and $CoSi_2$ thin films on silicon substrate, two types of epitaxial films were found to form on (111)Si. The orientations of type A and type B epitaxial films grown on (111)Si are identical to and are rotated by 180° with respect to the substrate normal, respectively [25, 32, 49, 50]. Examples are shown in FIGURE 9.15. From cross-sectional TEM observation, the epitaxial $NiSi_2$ was clearly found to possess heavily faceted structures. The facets are on {111} and {001} planes, which presumably have lower free energy than those of other interfaces. In addition, the interfacial dislocations at epitaxial silicide/Si interface were identified to be of edge or mixed type with $1/2\langle 110\rangle$ or $1/6\langle 112\rangle$ Burgers vectors [25].

For the growth of epitaxial double heterostructures, the a-Si/yttrium/(111)Si samples were annealed at 700°C for 5 min by rapid thermal annealing (RTA). From cross-sectional TEM observation, the epi-Si/epi-YSi_{2-x}/(111)Si double heterostructure was observed to form. Both the epi-Si/epi-YSi_{2-x} and epi-YSi_{2-x}/(111)Si interfaces were found to be very smooth. However, from HRTEM observation, a number of microtwins were found to form in the overgrown epi-Si layer. Examples are shown in FIGURE 9.16. From electron diffraction pattern analysis, the

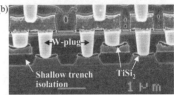

FIGURE 9.11 Cross-sectional micrograph of $TiSi_2$ salicide gate structure of a 266 MHz CPU. (a) and (b) are SEM micrographs, where the cross-sectional SEM sample was made by mechanical grinding and polishing and chemically stained by a dilute HF solution. It is clearly shown in (b) that the $TiSi_2$ on gate and source/drain region were etched off. Only the morphology of the material configuration was revealed as a result. (c) is a TEM micrograph, where the grain contrast of all material layers and device architecture are shown very nicely.

Characterisation of metal silicides

presence of microtwins was also revealed by the appearance of twin diffracted spots in the diffraction pattern. An example is shown in FIGURE 9.17. It is worth noting that the microtwins are difficult to annihilate. The microtwins may affect the electrical properties and thus degrade the reliability of the devices.

9.4.2 Amorphous metal/Si alloy films

Upon co-deposition of metal–silicon thin films at room temperature, the structures of the as-deposited alloy films are mostly amorphous and meta-stable with respect to the crystalline phase. This method of co-depositing amorphous alloy films has enabled certain silicide phases to be formed at significantly lower temperatures [51–55]. The crystallisation processes from the amorphous silicide thin films is, therefore, the most important thing to realise the applications of silicides formed by co-deposition.

In situ TEM was shown to be an extremely powerful technique to investigate the kinetics of nucleation and growth in amorphous silicide films as well as the interfacial reactions of metal thin films on silicon systems. In this section, direct observations of early crystallisation in amorphous $CoSi_2$ thin films using TEM with the aid of a heating stage are discussed [55]. By measuring the nuclei sizes and estimating the growth rate, activation energy of the nucleation and growth process can be extracted from the Arrhenius plot.

It is clearly seen that the crystallisation of $CoSi_2$ spherulites in the amorphous $CoSi_2$ films was found to vary with heat treatment. Examples are shown in FIGURE 9.18. Furthermore, the evolution of sequential nucleation and growth of $CoSi_2$ grains in the amorphous films can be directly investigated using *in situ* TEM. Examples are shown in FIGURE 9.19. The arrows in FIGURE 9.19 indicate that $CoSi_2$ grains nucleated randomly in space. The size and number of grains were found to increase with annealing temperature and time. The growth process continues and leads to the formation of polycrystalline $CoSi_2$.

The nucleation rate can be determined by counting the number of $CoSi_2$ crystallites visible under the *in situ* TEM and by taking the slope of the resulting curves at constant temperature. In addition, the growth rate of the crystallites can also be determined from measurements of the size of selected $CoSi_2$ particles with time. The slopes of the straight lines give the growth rates of the $CoSi_2$ spherulites. By measuring the growth rate at different temperatures, the activation energy for linear growth of the $CoSi_2$ spherulites in amorphous films, obtained from an Arrhenius plot, was found to be 23.9 kcal/mol (1.04 eV).

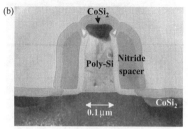
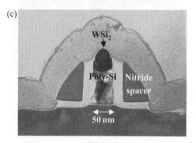

FIGURE 9.12 Cross-sectional TEM micrograph of: (a) 0.1 μm $TiSi_2$ salicide gate structure; (b) 0.1 μm $CoSi_2$ salicide gate structure; and (c) 50 nm WSi_2 gate structure (courtesy of UMC).

9.5 INITIAL SILICIDE FORMATION

9.5.1 Silicide formation in amorphous interlayer

In recent studies, amorphous interlayers have been observed to form by solid-state diffusion in the interfacial reactions of Ti, V, Hf, Y, Gd and a number of other refractory and RE metal thin films on Si substrate [37, 56–63].

The first silicide phase nucleation in the amorphous interlayer has attracted much attention in the past years. However, the determination of the first nucleation phase is still one of the long-standing problems in the interfacial reactions of metal thin films on silicon. In this section, the unambiguous finding of the occurrence of silicide phases in amorphous interlayer has been attributed to the utilisation of HRTEM in conjunction with fast Fourier transform (FFT), auto-correlation function (ACF) analysis and image simulation.

From the direct HRTEM observation and diffraction pattern analysis, Ti_5Si_3 phase was found to be the first nucleated Ti silicide phase in the a-interlayer followed by Ti_5Si_4 and TiSi phases. In addition, these intermediate phases including Ti_5Si_3, Ti_5Si_4, TiSi, and C49-$TiSi_2$, with the exception of the Ti_5Si_3 phase which is located at the a-interlayer/Ti metal thin film interface, were all found to form at the a-interlayer/Si substrate interface. An example is shown in FIGURE 9.20. The observation is consistent with a kinetic model, which predicts that the crystalline phase with composition and structure nearest to that of the a-interlayer will be first nucleated. The combined techniques can also be used for the identification of the nucleated phase in other metal/Si systems [37, 64–66]. The sensitivity of the combined techniques compares favourably with conventional selected-area electron diffraction and microdiffraction, which correlate diffraction patterns with areas about 500 and 100 nm in size in the specimens, respectively [9, 67].

It is worth noting that optical diffraction has been used to detect the presence of microcrystalline crystals in HRTEM micrographs

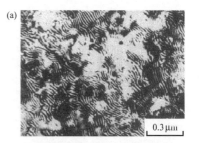

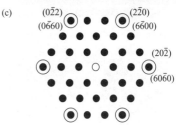
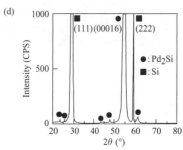

FIGURE 9.13 (a) Plan-view, bright-field TEM image, (b) overlapping [0001]Pd_2Si/[111]Si diffraction pattern, (c) indexed pattern of (b): (●) Pd_2Si, (○) Si, (d) conventional wide-angle XRD spectrum of a Pd/(111)Si sample annealed at 800°C for 1 h.

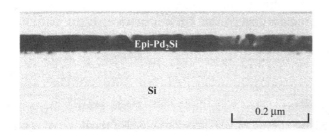

FIGURE 9.14 Cross-sectional TEM image of a Pd/(111)Si sample annealed at 800°C for 1 h.

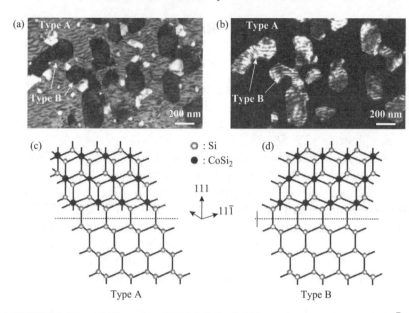

FIGURE 9.15 (a) Plan-view TEM bright-field image (taken along the [11$\bar{2}$] direction) of epitaxial $CoSi_2$ on (111)Si. (b) Corresponding dark-field image of twin regions in epitaxial $CoSi_2$ thin film. (c and d) The ball-and-stick models of $CoSi_2$/(111)Si interface with epitaxially (type A) and twin related (type B), respectively, view is down [$\bar{1}$10].

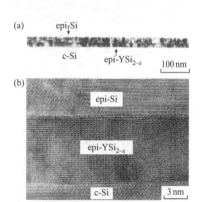

FIGURE 9.16 (a) Cross-sectional TEM and (b) high-resolution TEM images of an a-Si (20 nm)/Y (10 nm)/(111)Si sample annealed by RTA at 700°C for 5 min.

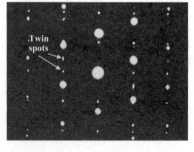

FIGURE 9.17 Cross-sectional electron diffraction pattern of a a-Si/Y/(111)Si sample after annealing at 600°C for 10 min by RTA.

for some time. However, the lattice images and diffraction are often obscured by the noise generated by the amorphous background. Therefore, silicide crystallites as small as 1 nm in size may be embedded in the amorphous interlayer and cannot be detected using the direct imaging technique of HRTEM [68].

For example, for Ti/Si samples after annealing at 400–450 °C for 60 s by RTA, no crystalline silicide phases were seen to be present in the amorphous interlayer by direct HRTEM imaging. However, a high density of periodic structures was observed in the ACF images of outlined regions in the HRTEM micrographs. Examples are shown in FIGURE 9.21. The cross-hatch patterns were found to correspond to Ti_5Si_3, Ti_5Si_4 or TiSi phase from both interplanar spacing and angular measurements.

In previous studies, no crystalline silicide phases were seen to be present in the a-interlayer of Ti/Si systems after annealing at 450°C for less than 30 min by direct HRTEM imaging [63]. However, with the aid of ACF analysis, small crystallites of Ti_5Si_3, TiSi and Ti_5Si_4 phases can be identified as coexisting in the a-interlayer of Ti/Si systems after annealing at 400–450 °C for 60 s. The results indicate clearly that the silicide crystallites were too small to be observed by direct HRTEM imaging in the initial stage of interfacial reactions. With the aid of ACF analysis, it is verified that the initial stage of intermediate silicide formation was found to

Characterisation of metal silicides

be earlier. As a result, the ACF analysis shall be a powerful technique and is expected to provide new insights in the investigation of the initial stage of silicide formation.

9.6 PHASE FORMATION AND IDENTIFICATION

9.6.1 Silicide formation on patterned (001) and (111)Si wafers

When fabricating electronic devices, silicide reactions are usually restricted to selected and confined areas on the silicon substrate [69, 70]. As the device dimensions scale down to the deep sub-micrometre level, it is of much interest to conduct an in-depth investigation of the effects of lateral confinement on the silicide reactions taking advantages of the new lithographic techniques to define oxide openings as small as 0.1 μm in size [71–73].

Plan-view and cross-sectional SEM and TEM observations are very useful for elucidating the formation mechanisms, such as structural evolution and growth kinetics. In this section, the effects of size and shape of lateral confinement on the formation of $NiSi_2$ and $CoSi_2$ silicides as investigated by SEM and TEM analyses are discussed. With the aid of sheet resistance measurement, estimation of the thickness of reacted silicide was also feasible based on the finding of a single phase silicide.

For Ni (30 nm) on patterned (111)Si samples, uniform coverage of the deposited Ni film inside the contact hole is evident by FE-SEM observation. An example is shown in FIGURE 9.22. For samples after different heat treatment, the morphology of epitaxial $NiSi_2$ phase on (111)Si was found to vary significantly with shape and size of oxide openings as well as annealing temperature. From TEM and electron diffraction pattern analysis, for linear openings of or larger than 0.1, 0.2 and 0.5 μm size, a mixture of type A and type B epitaxy was observed to form on (111)Si in samples annealed at 400, 500 and 600°C, respectively. In addition, the faceted structure was found to be more prone to form in smaller openings and/or at higher temperatures. The epitaxial $NiSi_2$ is hexagonal in shape with faceted edges parallel to ⟨110⟩Si directions viewed along [111]Si direction. An example is shown in FIGURE 9.23. The effects of size and shape of lateral confinement on the epitaxial growth of $NiSi_2$ on (111)Si are correlated with the stress level inside oxide openings.

For Co thin films on the SEG-Si layer on (001)Si samples, from FE-SEM observation, pinhole-free cobalt silicide layers were formed inside linear openings and contact holes. Examples are

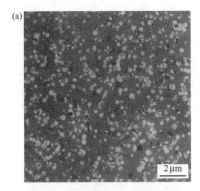
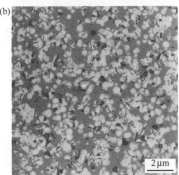

FIGURE 9.18 *In situ* TEM images showing the formation and growth of $CoSi_2$ crystallites in 100 nm thick amorphous $CoSi_2$ thin films after annealing at 140°C for (a) 60 min, (b) 65 min, and (c) 75 min.

Characterisation of metal silicides

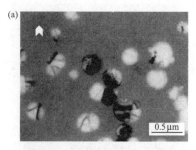

FIGURE 9.19 *In situ* TEM images showing the sequential nucleation and growth of CoSi$_2$ crystallites in samples annealed at 140°C for (a) 60 min, (b) 65 min, and (c) 75 min.

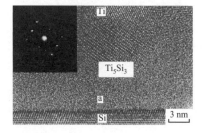

FIGURE 9.20 HRTEM image of a Ti/Si sample annealed at 450°C for 2 h. Inset is the optical diffraction pattern corresponding to [110]Ti$_5$Si$_3$.

shown in FIGURES 9.24(a) and (b). From plan-view TEM examination, epitaxial and polycrystalline CoSi$_2$ were observed to form near the edge and central region of a 1.8 μm linear opening, respectively. Analysis of TEM diffraction patterns of epitaxial CoSi$_2$ films grown inside the linear openings reveals that the silicide grows with an epitaxial orientation: [001]CoSi$_2$//[001]Si and (100)CoSi$_2$//(100)Si. On the other hand, from XTEM observation, the thickness of epitaxial CoSi$_2$ at the oxide edge was much thinner than the polycrystalline CoSi$_2$ in the central region of the 1.8 μm linear opening. An example is shown in FIGURE 9.25.

In order to find the extent of growth of Co silicide at low temperature and its effect on the final structure of the silicide, samples annealed at 480°C for 60 s were examined. From XTEM observation, two distinct regions with partially and completely reacted Co layer near the edge and central region, respectively, of the oxide opening are seen. Analysis of selective area electron diffraction patterns from plan-view TEM images revealed that the polycrystalline CoSi phase was the only silicide phase present. In the partially reacted region, near the oxide sidewall, the thickness of CoSi was increased with the distance from the edge of the oxide sidewall. An example is shown in FIGURE 9.26. A recent work has shown that the substrate regions on the oxide edge are compressively stressed to a significant extent [74]. As a result, the formation of CoSi phase at low temperature was found to be retarded by the local compressive stress near the edge of linear oxide openings. The relative ease in the epitaxial growth of CoSi$_2$ near the oxide edge of linear openings and of 0.7 μm and smaller contact holes is attributed to the thinness of the CoSi layer.

The effect of the oxide opening size on the sheet resistance for the formation of CoSi$_2$ is shown in FIGURE 9.27. From the average thickness determined by TEM, the product of the sheet resistance and average thickness is fairly constant. The result indicates that the resistivity is a constant, as would be expected since the silicide phase was identified to be CoSi$_2$. The sheet resistance versus linewidth data also reflect the fact that the average thickness of silicide is thinner for smaller openings.

9.6.2 Effects of interposing layers on the formation of metal silicides

As the device dimension shrinks to the deep sub-micrometre regime, the extent of the process window was found to decrease with the linewidth [75]. To resolve this problem, a method using a thin interposing layer for enhancing the formation of low-resistivity metal silicides was demonstrated [76–80]. However, the detailed mechanism is still not well understood. As a result,

Characterisation of metal silicides

HRTEM was utilised for studying the interfacial reaction between metal and Si with a thin interposing layer in this section. Grazing incidence X-ray diffractometry (GIXRD) and TEM studies were carried out in parallel for phase identification and structure examination. The incident angle of X-ray is fixed at 0.5°. A four-point probe was used to characterise the trends of sheet resistance variation of the evolving phases. TEM, in conjunction with EDS for elemental identification, was utilised for verifying the chemical composition of local areas in high resolution mode. The probe size can be as small as 4 nm. SIMS was used to trace the existence

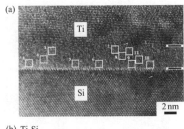

(b) Ti_5Si_3

Ti_5Si_4

TiSi

FIGURE 9.21 (a) HRTEM image of a cross-sectional a-Si/Ti/(001)Si sample after annealing at 400°C for 60 s by RTA. (b) ACF-processed images of the outlined regions shown in (a).

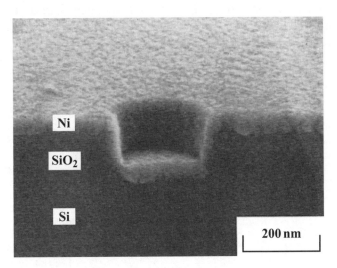

FIGURE 9.22 Cross-sectional FE-SEM image showing as-deposited Ni film on (111)Si inside a contact hole of 0.2 μm in size.

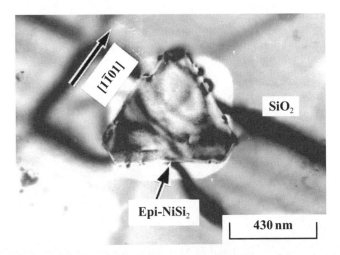

FIGURE 9.23 Plan-view TEM image showing the faceted epitaxial $NiSi_2$ on (111)Si inside a contact hole of 0.6 μm in size, 900°C.

Characterisation of metal silicides

of thin interposing layer via the depth profiling of corresponding elements.

From sheet resistance measurement and GIXRD analysis, the data clearly demonstrate the impact of the interposing layer on the phase transformation. A sharp drop in sheet resistance was found for the Ti/Mo bilayer on (001)Si samples at 650°C. The low-resistivity C54-TiSi$_2$ was identified by GIXRD to be the only silicide phase formed in the Ti/Mo/Si samples annealed at 650°C. The result indicates that with an interposing Mo thin layer, the formation temperature of C54-TiSi$_2$ was lowered by about 100°C compared to what is usually needed for the C49- to C54-TiSi$_2$ transformation. The sheet resistance data and corresponding GIXRD spectra are shown in FIGURES 9.28 and 9.29, respectively. On the other hand, from cross-sectional TEM observation, it is clearly seen that the thicknesses of C54-TiSi$_2$ in

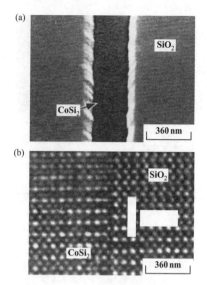

FIGURE 9.24 FE-SEM images of the two-step annealed SEG-Si layer on Si samples inside (a) a 0.25 μm linear opening and (b) a 0.45 μm contact hole.

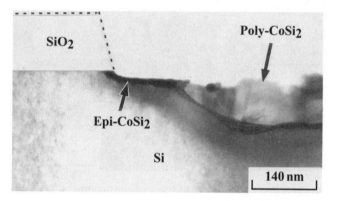

FIGURE 9.25 Cross-sectional TEM images showing the epitaxial and polycrystalline CoSi$_2$ near the oxide edge on (001)Si inside a 1.8 μm linear opening.

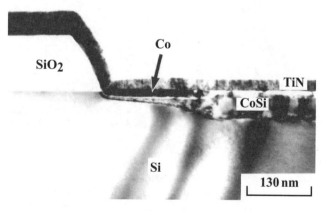

FIGURE 9.26 Cross-sectional TEM images showing the increase in CoSi thickness with the distance from the oxide edge on (001)Si inside a 1.8 μm linear opening after annealing at 480°C for 60 s.

Ti/Mo/Si samples are thinner than those in Ti/Si samples. In addition, it was seen from GIXRD analysis that the ternary (Ti, Mo)Si$_2$ silicide phase was formed in Ti/Mo/Si samples after annealing. These results are correlated with the difference in sheet resistance of C54-TiSi$_2$ between Ti/Mo/Si and Ti/Si samples in FIGURE 9.28. The increases in sheet resistances for samples after annealing at 900°C were correlated to the formation of island structure of silicide thin films.

For Ti/Mo/Si samples annealed at 475°C, the amorphous interlayer with two contrasts was observed by HRTEM. An HRTEM image is shown in FIGURE 9.30. It is thought that Mo atoms are distributed in the lower a-interlayer with dark contrast for the most parts. The upper a-interlayer with light contrast was composed of Si and Ti atoms. In addition, from XTEM and HRTEM analysis, a continuous thin tetragonal Mo$_5$Si$_3$ layer was also found to form at the a-interlayer/Si substrate interface for samples annealed at 500°C. Examples are shown in FIGURE 9.31. As the annealing temperature was increased to 550°C, simultaneous presence of Mo$_5$Si$_3$, Ti$_5$Si$_4$, and C49-TiSi$_2$ phases was found to occur. Furthermore, the Mo$_5$Si$_3$ phase was only occasionally found at the silicide/Si substrate interface. It is thought that Mo$_5$Si$_3$ phase for the most part was decomposed and the Mo atoms were redistributed into the titanium silicides layer.

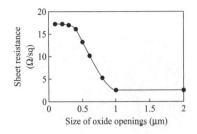

FIGURE 9.27 The sheet resistance versus the size of oxide openings for the CoSi$_2$ films formed by the two steps RTA in the SEG Si layer on (001)Si samples.

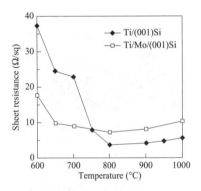

FIGURE 9.28 Sheet resistance of titanium silicides versus annealing temperature curves for Ti/(001)Si and Ti/Mo/Si samples.

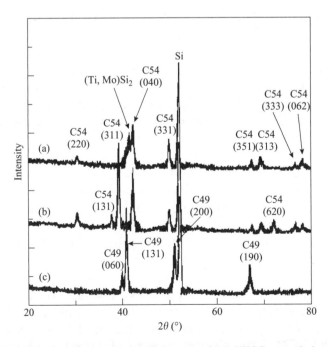

FIGURE 9.29 Glancing angle XRD spectra of (a) 650°C annealed Ti(30 nm)/ Mo(0.5 nm)/(001)Si, (b) 750°C annealed Ti(30 nm)/(001)Si, and (c) 700°C annealed Ti(30 nm)/(001)Si samples.

Characterisation of metal silicides

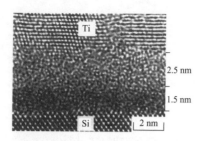

FIGURE 9.30 Cross-sectional HRTEM image of the Ti/Mo/Si sample after annealing at 475°C for 30 s.

In order to analyse the distribution of Mo atoms during silicidation, the energy dispersive X-ray (EDAX) analysis and SIMS measurements were carried out. From EDAX analysis, it is apparent that Mo atoms were not concentrated at the interface of Ti/Si. Instead, Mo atoms were found in $TiSi_2$ and near the surface of $TiSi_2$. From SIMS analysis, it is clearly shown that the peak of Mo was initially concentrated at the Ti/Si interface. As the annealing temperature increased, the Mo atoms diffused from the interface to disperse in $TiSi_2$ films. The SIMS concentration–depth profiles correlate well with the EDAX analysis and HRTEM observation. The combined techniques can also be used to investigate the formation of silicides in other metal/Si systems with a thin interposing layer.

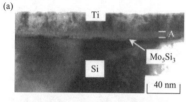
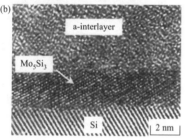

FIGURE 9.31 Cross-sectional (a) TEM image and (b) HRTEM image of the Ti/Mo/Si sample after annealing at 500°C for 30 s.

9.6.3 Silicide formation by metal ion implantation into Si wafers

Mesotaxy, buried silicide layer formation by ion implantation into Si substrate is a technique for providing a Si-rich environment to prepare buried silicide layers, with the advantage that the layers have lower resistivity and greater stability to high temperature annealing than those made by conventional thin film or metal–Si co-deposition methods.

The thin film deposition method is known to form reproducible contacts with clean interfaces and it is compatible with a self-aligned silicide (salicide) technology. Therefore, it has generally been the method of choice in the formation of contacts to shallow junction. However, due to the roughness of the silicide/silicon interface, source and drain junctions must extend at least 50 nm beyond the silicide in order to avoid high leakage currents and silicide shorts to the substrate. Additionally, silicide layers less than 50 nm thick remain difficult to fabricate using this technique. It was reported that high-quality buried single crystalline silicide layers, $CoSi_2$, $TiSi_2$, and $CrSi_2$, with sharp interfaces can be formed consistently through the use of high dose ion implantation and annealing [81–88].

A critical dose, d_c, (corresponding to a threshold peak concentration, ~18.5 at.% Co) is required for the formation of a continuous buried layer under given processing conditions. Below d_c, a discontinuous array of precipitates is obtained. It has been previously reported that the structural perfection of mesotaxial $CoSi_2$ buried layer in Si(111) is better than that in Si(001) as measured by RBS and TEM. Examples are shown in FIGURE 9.32, but electrical measurement showed that $CoSi_2$ layers buried in Si(001) have better characteristics than those in Si(111).

Characterisation of metal silicides

Precipitation of CoSi$_2$ on {111} planes is known to be favourable because {111} planes are lower energy planes than {001} ones. In Si(001), precipitation of CoSi$_2$ occurs equally on each of the four {111} planes. In Si(111), consideration of the implanted Co concentration profile suggests that precipitation of CoSi$_2$ would be enhanced on the (111) plane parallel to the surface rather than on those inclined to the surface. This is because growth along the (111) plane parallel to the surface can be in the region of peak Co concentration. Precipitation on the three inclined {111} planes, however, is limited as the precipitates grow into a region of lower Co atom density (the wings of the implanted profile). Therefore, epitaxial growth of CoSi$_2$ in Si(111), consisting of A type (aligned), B$_0$ type (twinned on the (111) plane parallel to the surface), and B$_{1,2,3}$ types (twinned on one of the three inclined {111} planes), would be expected to be more complex than that in Si(001). FIGURE 9.33 shows XTEM images of implanted CoSi$_2$ precipitates in Si(111). It was subjected to Co$^+$ implantation with a dose of 2.5×10^{16} cm^{-2} and 600 + 1000°C, 1 h annealing. Observations along [011] and [112]Si zone axis enable the direct imaging of CoSi$_2$ precipitates grown on inclined Si(111) planes. However, all five types of precipitates (A, B$_0$, B$_1$, B$_2$, and B$_3$) are only found in the 1000°C annealed sample initially implanted with very low doses (1–3 $\times$ 10^{16} cm^{-2}), where the twinned precipitates are dominant. Imaging with $g = $ [220]Si on at least two poles of [112]B$_0$-CoSi$_2$ will thus reveal all five types of the precipitates. The hexagonal precipitates exhibiting strong contrast are identified to be B$_0$ type twinned on the (111) plane parallel to the surface. The twinned precipitates on two of the three inclined {111} planes, B$_1$, B$_2$, B$_3$ types, are revealed as long, thin strips from the plan-view projection and are present in equal fractions on the inclined {111} planes. Besides the B$_0$ and B$_1$, B$_2$, B$_3$ type precipitates, the remaining

FIGURE 9.32 XTEM micrographs of buried CoSi$_2$ in Si substrates, formed by high energy Co$^+$ ion implantation: (a) CoSi$_2$ in Si(111) substrate, BF on Si[110] pole; (b) CoSi$_2$ in Si(001) substrate, BF on Si[110] pole.

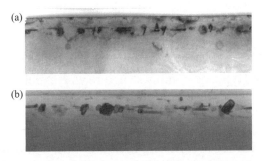

FIGURE 9.33 XTEM/BF micrographs of implanted CoSi$_2$ precipitates in Si(111), where the implant dose was 2.5×10^{16} cm^{-2} and annealing condition was 600 + 1000°C, 1 h: (a) observation along Si[110] zone axis; (b) observation along Si[112] zone axis.

precipitates are mostly identified as A type which show lighter contrast than B_0-type precipitates under a two beam diffraction condition on a pole close to [112] B_0-$CoSi_2$. Only a very small fraction (<2%) of precipitates show epitaxial relationships other than the above-mentioned five types (C-type).

If the buried silicide layer mainly consists of two types (A and B_0) of precipitates, identification of A versus B_0 can be selectively imaged by using Si matrix spots or twinned diffraction spots.

The study of evolution and coalescence of buried silicide layers relied very much on the TEM imaging techniques of selected area diffraction pattern (SADP) and dark field analyses. With the aid of large angle tilting stage, it is possible to image various types of precipitates on different zone axis. It makes ease of microstructural analysis of every single precipitate. As a result, the investigation of areal coverage, thickness variation, and relative fractional variation of A, B_0, and $B_{1,2,3}$ type precipitates was shown to be feasible. Besides, weak beam dark field imaging also helps to indicate the residual defects resulting from ion implantation.

9.7 DEFECT ANALYSIS

Various defects were found to form in silicide thin films on silicon [89–95]. These defects, like cracks, voids, pinholes, vacancies and planar defects, usually play an important role in determining the performances of the devices. Therefore, it is essential to investigate precisely the formation mechanism and the structure of these defects.

For defect analysis, electron microscopes (TEM, SEM) are always the most popular and powerful tool. In this section, various analytical techniques, such as HREM, FIB, electron diffraction pattern, AFM, and image simulation, which are used for investigating these defects in silicide thin films, are discussed.

9.7.1 Cracks (SEM, FIB, PTEM, XTEM, BF, DF, DP)

$CrSi_2$ is of C40 type, hexagonal structure with excellent lattice mismatch between $CrSi_2$ (0001) and Si(111) planes ($a_0/a_0 \sim 0.13\%$ along in-plane directions of $\langle 110 \rangle$). However, epitaxial growth of $CrSi_2$ on Si by e-gun deposition or molecular beam epitaxy co-deposition failed to form single crystal $CrSi_2$. The average grain size of localised epitaxy is less than 5 μm, although it is the first refractory metal silicide grown epitaxially on silicon. It has the smallest effective mismatch and the lowest formation temperature, ~450°C, of all refractory disilicides.

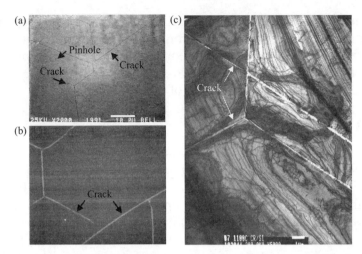

FIGURE 9.34 Single crystalline $CrSi_2$ buried layers in silicon fabricated by Cr^+ metal ion implantation. The samples were subjected to an implant dose of 3×10^{17} cm^{-2}. The cracks induced during buried layer formation were found in samples either subjected to sequential annealings of $600 + 700 + 800 + 900 + 1000 + 1100°C$ or two-step annealed at $600 + 1100°C$. (a) SEM micrograph, (b) FIB micrograph, (c) PTEM micrograph showing the cracks propagating along characteristic [112] or [110]Si orientations.

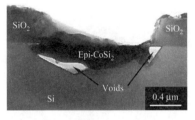

FIGURE 9.35 Cross-sectional TEM image of Co (30 nm)/Ti (10 nm) on (001)Si inside a linear opening of 1 μm size after annealing at 900°C for 60 s by RTA showing the voids.

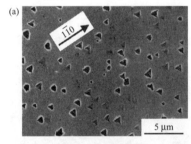

FIGURE 9.36 SEM micrographs of (a) Dy (30 nm) and (b) a-Si (20 nm)/Dy (30 nm) on (111)Si after annealing at 700°C for 10 min in a UHV chamber. The crystallographic directions refer to those in silicon.

As previously illustrated in Section 9.6.3, the mesotaxy technique of metal ion implantation has been used to fabricate $CrSi_2$ buried layers in silicon. In samples subjected to a dose 3×10^{17} cm^{-2}, sequential annealings from 600 to 1000°C, continuous buried layers with large pinholes were obtained after 1000°C annealing. A continuous layer was found only after 1100°C annealing. The epitaxial relationships are identified as: $CrSi_2[0001](10\bar{1}0)//Si[111](2\bar{2}0)$. However, formation of continuous, single crystalline, buried layer is also observed to be associated with cracks and pinholes, as shown in FIGURES 9.34(a)–(c). It was multiply reconfirmed by SEM, FIB and TEM analyses.

In order to understand the formation mechanisms of cracks and pinholes, the samples were examined by XTEM along Si[112] and Si[110] poles perpendicular to the surface normal [111] direction. Abrupt surfaces on the edge of the cracks and an almost buried channel-shaped void beneath the top silicon suggested that the cracks were induced by the internal stress release at the heterostructural interfaces during the cooling process. Since the stress could be originated from the lattice mismatch and the difference between the thermal expansion coefficients of $CrSi_2$ and Si, it can be released by either misfit dislocation generation or by cracks in principle. When viewing the pinholes by XTEM, it was found that the pinholes were filled with Si and the ends of $CrSi_2$ epitaxy

Characterisation of metal silicides

around the pinholes are hemispherical. This structure is apparently an energetically stable configuration resulting from the progression of grain coalescence. Thus, the sequentially annealed samples are closer to equilibrium than the two-step annealed samples, which will thus result in higher density of pinholes [96].

9.7.2 Voids (XTEM)

Epitaxial silicides have received much attention because of their considerable advantages, such as good surface and interface uniformity and high thermal stability over polycrystalline silicides. For the self-aligned formation of epitaxial $CoSi_2$ thin film, Co/Ti bilayer was deposited onto (001)Si inside linear oxide openings. From XTEM observation, for samples annealed at 900°C for 60 s, an epitaxial $CoSi_2$ layer was found to form on the patterned Si substrate. However, defects such as facets, lateral encroachment and voids were clearly seen under the edge of the SiO_2 layers. An example is shown in FIGURE 9.35. Such defects will cause the deterioration of the gate oxide and the shallow junction and yield the device failures [97–99].

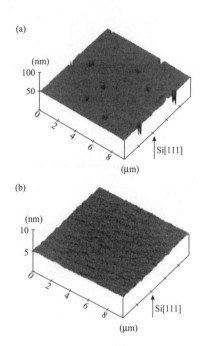

FIGURE 9.37 AFM micrographs of samples with epitaxial $DySi_{1.67}$ film on (111)Si after annealing at 700°C for 10 min in a UHV chamber. (a) Dy (2 nm)/(111)Si and (b) a-Si (1.5 nm)/Dy (2 nm)/(111)Si.

9.7.3 Pinholes (SEM, AFM, PTEM)

The formation of pinholes in metal silicides is a serious issue. [100, 101]. Existence of pinholes may result in short-circuits or direct contact of upper layers to silicon substrate, and thus, may degrade the reliability of devices. From SEM observation, a high density of pinholes was found to form in the epitaxial RE silicide thin films. The fraction of surface coverage and the size of pinholes were found to vary with the thickness of a-Si capping layer. Examples are shown in FIGURE 9.36. AFM examination also shows the presence and absence of pinholes in RE silicides without and with a-Si capping layers, respectively. Examples are shown in FIGURE 9.37. In addition, it can be seen that the pinholes are triangular in shape, reflecting the threefold symmetry of the underlying (111)Si. From TEM analysis, the edges of the triangular pinholes were found to be parallel to the three $\langle 110 \rangle$ directions of the (111)Si surface. An example is shown in FIGURE 9.38. The formation mechanism of these pinholes is attributed to the local depletion of Si atoms [39].

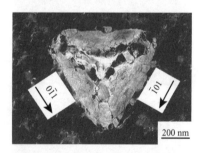

FIGURE 9.38 Plan-view TEM image of $YbSi_{2-x}$/(111)Si samples after annealing at 700°C for 60 s by RTA showing the pinhole.

9.7.4 Planar defects (PTEM, XTEM, 2B-DF, HRTEM)

A high density of planar defects along three $\langle \bar{1}2\bar{1}0 \rangle$ directions and $\langle 0001 \rangle$ direction were also observed by TEM in the epitaxial RE silicides thin films viewed along the [0001] direction. The defects were identified to be stacking faults on $\{10\bar{1}0\}$ planes.

Examples are shown in FIGURES 9.39 and 9.40. In addition, based on the two-beam diffraction analysis, the displacement vectors associated with these defects in (0001) plane were found to be parallel to the enlarged directions of the stacking faults. An example is shown in FIGURE 9.41. Furthermore, from plan-view and cross-sectional HRTEM analysis, the displacement vectors of the stacking faults can be determined. From Burgers' circuit analysis, the displacement vectors were determined to be $1/6\langle\bar{1}2\bar{1}0\rangle$ and $1/2\langle0001\rangle$. As a result, the displacement vectors of the stacking faults are found to be $1/6\langle\bar{1}2\bar{1}3\rangle$ [40, 89, 102]. Examples are shown in FIGURE 9.42.

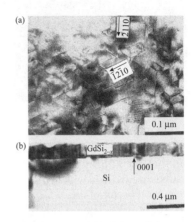

9.7.5 Vacancies (DP, simulation)

In previous reports, the compositions for the stable silicon-rich phase of RE silicides thin films have ranged from $RESi_{1.6}$ to $RESi_2$, with the former having a vacancy distribution in the Si sublattice plane [103–105]. For example, for epitaxial $YbSi_{2-x}/(111)Si$ systems, the orientation relationships were determined to be $[0001]YbSi_{2-x}//[111]Si$ and $(10\bar{1}0)YbSi_{2-x}//(11\bar{2})Si$. An example is shown in FIGURE 9.43. In addition to the diffraction spots of AlB2 type $YbSi_{2-x}$ structure, superlattice spots at $1/3\langle\bar{2}110\rangle$ were also found. The superlattice spots are attributed to $a\sqrt{3} \times a\sqrt{3}R30°$ ordered vacancy mesh [106]. The simulated diffraction pattern shows the same features as those obtained by TEM. In addition, by XTEM, the distribution of superlattice spots along the c axis revealed that the period of ordered vacancy along that axis was $1c$. An example is shown in FIGURE 9.44. The corresponding simulated diffraction pattern depicts the same features. As a result, using the analysis techniques, the unit cell of the vacancy ordering superlattice structure in epitaxial $YbSi_{2-x}/(111)Si$ samples can be found to be $(a\sqrt{3} \times a\sqrt{3}c)$.

The three-dimensional structures of vacancy ordering in other RE silicides were also determined by electron diffraction pattern and simulation analysis [40, 89, 102, 106].

FIGURE 9.39 (a) Plan-view, bright-field TEM and (b) cross-sectional TEM images of a-Si/Gd/(111)Si samples annealed at 600°C for 1 min.

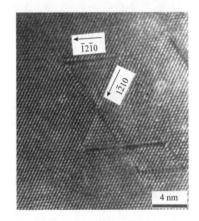

FIGURE 9.40 HRTEM image of stacking faults viewed along [0001] directions of epitaxial $GdSi_{2-x}$ films.

9.8 THERMAL STABILITY (SHEET RESISTANCE, PTEM, XTEM, BF, DF)

As the device dimensions scale down to deep sub-micrometre region, the morphological stability of metal silicide thin films during high temperature annealing is crucial to the development of ULSI technology [108–110]. It is well known that agglomeration in silicide/Si systems starts with grain boundary

Characterisation of metal silicides

FIGURE 9.41 Plan-view, dark-field TEM image under two-beam condition with $[\bar{1}010]$ diffraction vector of the a-Si (20 nm)/Y (30 nm)/(111) Si sample annealed at 800°C for 15 s.

grooving in the silicide films at high temperature annealing, followed by the formation of silicide islands [111–114]. The driving force for the morphological transformation is the reduction of the interface/surface energies. A schematic diagram is shown in FIGURE 9.45. To examine the thermal stability of silicides, sheet resistance measurement as well as TEM analysis will be the most complementary techniques.

For example, FIGURE 9.46 shows the sheet resistance curves of pure Ti, TiN_x, and $TiN_x + Ti$ films on (001)Si samples annealed at different temperatures. The increase in sheet resistance was found for the pure Ti/Si samples at 850°C. However, for the TiN_x and $TiN_x + Ti$ on Si samples, the sheet resistances maintained the

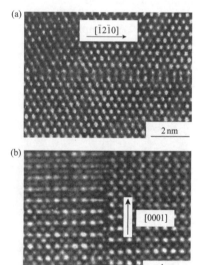

FIGURE 9.42 HRTEM images of stacking faults viewed along (a) [0001] and (b) $[\bar{1}2\bar{1}0]$ directions of epitaxial $DySi_{2-x}$ films.

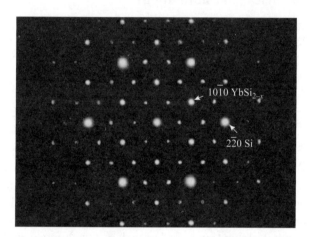

FIGURE 9.43 Diffraction pattern of a plan-view $YbSi_{2-x}/(111)Si$ sample after annealing at 300°C for 30 min.

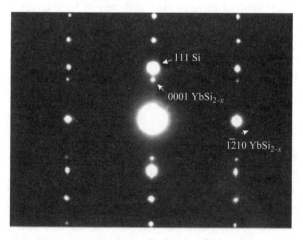

FIGURE 9.44 Diffraction pattern along the $[10\bar{1}0]YbSi_{2-x}//[11\bar{2}]Si$ direction of a cross-sectional $YbSi_{2-x}/(111)Si$ sample after annealing at 300°C for 30 min.

Characterisation of metal silicides

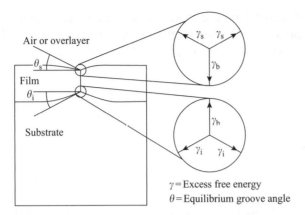

FIGURE 9.45 Energy balances at upper and lower grain boundary grooves. γ_s, γ_i, and γ_b represent the surface, interface and grain boundary energy, respectively.

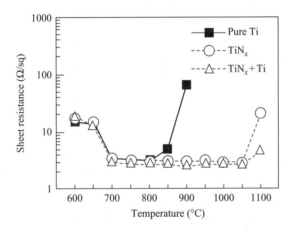

FIGURE 9.46 Sheet resistance curves for pure Ti, TiN_x, and $TiN_x + Ti$ films on (001)Si samples annealed at different temperatures for 30 s.

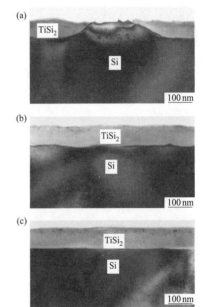

FIGURE 9.47 XTEM images of (a) pure Ti/Si sample annealed at 850°C for 30 s, (b) TiN_x, and (c) $TiN_x + Ti$ on Si samples annealed at 1050°C for 30 s. TiN and unreacted Ti layers of all samples were etched off with a chemical solution.

same low level to a temperature as high as 1050°C. From cross-sectional TEM (XTEM) observation, it is revealed that the increase in sheet resistance at 850°C is correlated to the agglomeration of $TiSi_2$ thin film in the pure Ti/Si samples. In contrast, no morphological degradation of $TiSi_2$ was found for TiN_x, and $TiN_x + Ti$ on Si samples after annealing at 1050°C. Examples are shown in FIGURE 9.47 [115]. Sheet resistance data were found to correlate nicely with the morphological observation. Thus, the thermal stability of C54-$TiSi_2$ can be greatly improved by the addition of N_2 during Ti sputtering.

On the other hand, from plan-view, bright-field and dark-field TEM micrographs analyses, the morphological stability of silicides can be examined clearly. Pinhole structures started to occur in $TiSi_2$ films after annealing at 800°C for 1 h. Agglomeration of silicide films became much more severe at higher annealing temperatures. Island structures of TiSi$_2$ films were found to form in samples annealed at 900°C for 1 h [116].

REFERENCES

[1] P.E.J. Flewitt, R.K. Wild [*Physical Methods for Materials Characterisation* (Institute of Physics Publishing, Bristol and Philadelphia, 1994)]
[2] D.K. Schroder [*Semiconductor Material and Device Characterisation* (Wiley-Interscience, New York, 1990)]
[3] A.H. Landzberg [*Microelectronics Manufacturing Diagnostics Handbook* (Van Nostrand Reinhold, Amsterdam, 1993)]
[4] J.I. Goldstein, D.E. Newbury, P. Echlin, D. Joy, C. Fiori, E. Lifshin [*Scanning Electron Microscopy and X-ray Microanalysis* (Plenum Press, New York, 1984)]
[5] J.W. Edington [*Practical Electron Microscopy in Materials Science* (Philips' Gloeilampenfabrieken, 1976)]
[6] M.H. Loretto, R.E. Smallman [*Defect Analysis in Electron Microscopy* (Chapman and Hall, London, 1975)]
[7] P.B. Hirsch et al. [*Electron Microscopy of Thin Crystals* (R.E. Krieger Pub. Co., 1977)]
[8] J.M. Cowley [*Diffraction Physics* (North-Holland Pub. Co., Amsterdam, 1981)]
[9] P.R. Buseck, J.M. Cowley, L. Eyring [*High Resolution Transmission Electron Microscopy* (Oxford University Press, New York, 1988)]
[10] D.B. Williams, C.B. Carter [*Transmission Electron Microscopy* (Plenum Press, New York and London, 1996)]
[11] K.Y.Y. Doong, J.M. Fu, Y.F. Hsieh [*6th International Symposium on the Physical and Failure Analysis of Integrated Circuit, IPFA*, Singapore, 1997]
[12] K.Y.Y. Doong, J.M. Fu, Y.C. Huang [*23rd International Symposium for Testing and Failure Analysis, ISTFA*, Santa Clara, USA, 1997]
[13] R.H. Livengood, V.R. Rao [*Semiconductor International (USA)* (March 1998) p.111–15]
[14] L. Peto [*Compound Semicond. (USA)* vol.9 no.2 (March 2003) p.25–7]
[15] J.M. Walls (Ed.) [*Methods of Surface Analysis* (Cambridge University Press, Cambridge, 1989)]
[16] A.W. Czanderna (Ed.) [*Methods of Surface Analysis*(Elsevier Scientific Publishing Co., Amsterdam, 1984)]
[17] D.J. O'Connor, B.A. Sexton, R.St.C. Smart (Eds.) [*Surface Analysis Methods in Materials Science* (Springer-Verlag, New York, 1991)]
[18] R.G. Wilson, F.A. Stevie, C.W. Magee [*Secondary Ion Mass Spectroscopy* (John-Wiley & Sons, New York, 1989)]
[19] J.C. Vickerman, Alan Brown, N.M. Reed [*Secondary Ion Mass Spectroscopy* (Clarendon Press, Oxford, 1989)]
[20] R.B. Marcus, T.T. Sheng [*Transmission Electron Microscopy of Silicon VLSI Circuits and Structures* (John-Wiley & Sons, New York, 1983)]
[21] O. Jeandupeux, V. Marsico, A. Acovic, P. Fazan, H. Brune, K. Kern [*Microelectron. Reliabil. (UK)* vol.42 (2002) p.225–31]

[22] T. Kawamura, D. Shinoda, H. Muta [*Appl. Phys. Lett. (USA)* vol.11 (1967) p.101–3]
[23] W.D. Buckley, S.C. Moss [*Solid State Electron. (USA)* vol.15 (1972) p.1331–7]
[24] G.J. van Gurp, C. Langereis [*J. Appl. Phys. (USA)* vol.46 (1975) p.4301–7]
[25] L.J. Chen, J.W. Mayer, K.N. Tu [*Thin Solid Films (Netherlands)* vol.93, (1982) p.135–41]
[26] L.J. Chen, K.N. Tu [*Mater. Res. Rep. (Netherlands)* vol.6 (1991) p.53–140]
[27] R.T. Tung [*Mater. Chem. Phys. (Netherlands)* vol.32 (1992) p.107–33]
[28] H.C. Cheng, I.C. Wu, L.J. Chen [*Appl. Phys. Lett. (USA)* vol.50 (1987) p.174–6]
[29] C.S. Chang, C.W. Nieh, L.J. Chen [*J. Appl. Phys. (USA)* vol.61 (1987) p.2393–5]
[30] J.J. Chu, L.J. Chen, K.N. Tu [*J. Appl. Phys. (USA)* vol.62 (1987) p.461–5]
[31] I.C. Wu, J.J. Chu, L.J. Chen [*J. Appl. Phys. (USA)* vol.62 (1987) p.879–84]
[32] R.T. Tung, J.M. Potate, J.C. Bean, J.M. Gibson, D.C. Jacobson [*Thin Solid Films (Netherlands)* vol.93 (1982) p.77–90]
[33] K.N. Tu, E.I. Alessandriani, W.K. Chu, H. Krautle, J.W. Mayer [*Jpn. J. Appl. Phys. (Japan)* Suppl.2-1 (1974) p.669–72]
[34] G. Molnar, G. Peto, E. Zsoldos [*Appl. Surf. Sci. (Netherlands)* vol.70/71 (1993) p.466–9]
[35] I. Gerocs, E. Bugiel, J. Gyulai, E. Jaroli, G. Molnar, G. Peto, E. Zsoldos [*Appl. Phys. Lett. (USA)* vol.51 (1987) p.2144–5]
[36] Y.K. Lee, N. Fujimura, T. Ito, N. Itoh [*J. Cryst. Growth (Netherlands)* vol.134 (1993) p.247–54]
[37] T.L. Lee, L.J Chen [*J. Appl. Phys. (USA)* vol.73 (1993) p.8258–66]
[38] C.H. Luo, F.R. Chen, L.J. Chen [*J. Appl. Phys. (USA)* vol.76 (1994) p.5744–7]
[39] G.H. Shen, J.C. Chen, C.H. Lou, S.L. Cheng, L.J. Chen [*J. Appl. Phys. (USA)* vol.84 (1998) p.3630–5]
[40] J.C. Chen, G.H. Shen, L.J. Chen [*Appl. Surf. Sci. (Netherlands)* vol.142 (1999) p.291–4]
[41] F.A. d'Avitaya [*Electron. Lett. (UK)* vol.22 (1986) p.699–700]
[42] L. Pahun, Y. Campidelli, F.A. d'Avitaya, P.A. Badoz [*Appl. Phys. Lett. (USA)* vol.60, (1992) p.1166–8]
[43] J.C. Hensel, A.F.J. Levi, R.T. Tung, J.M. Gibson [*Appl. Phys. Lett. (USA)* vol.47 (1985) p.151–3]
[44] J.C. Hensel [*Appl. Phys. Lett. (USA)* vol.49 (1986) p.522–4]
[45] G. Glastre, F.A. Davitaya, G. Vincent, E. Rosencher, M. Pons, C. Puissant, J.C. Pfister [*Appl. Phys. Lett. (USA)* vol.52 (1988) p.898–900]
[46] A. Gruhle, H. Beneking [*IEEE Trans. Electron Devices (USA)* vol.38 (1991) p.1878–82]
[47] J.F. Chen, L.J. Chen [*Mater. Chem. Phys. (Netherlands)* vol.39 (1995) p.229–35]
[48] D. Cherns, D.A. Smith, W. Krakow, P.E. Batson [*Philos. Mag. A (UK)* vol.45 (1982) p.107–25]
[49] H. Foll, P.S. Ho, K.N. Tu [*J. Appl. Phys. (USA)* vol.52 (1981) p.250–5]
[50] K.C.R. Chiu, J.M. Poate, J.E. Howe, T.T. Sheng, A.G. Cullis [*Appl. Phys. Lett. (USA)* vol.38 (1981) p.988–90]
[51] S.P. Murarka, D.B. Fraser, T.F. Retajczyk Jr., T.T. Sheng [*J. Appl. Phys. (USA)* vol.51 (1980) p.5380–5]

[52] S.P. Murarka, D.B. Fraser [*J. Appl. Phys. (USA)* vol.51 (1980) p.1593–8]

[53] S.P. Murarka, M.H. Read, C.C. Chang [*J. Appl. Phys. (USA)* vol.52 (1981) p.7450–2]

[54] K.N. Tu, D.A. Smith, B.Z. Weiss [*Phys. Rev. B (USA)* vol.36 (1987) p.8948–50]

[55] J.M. Liang, L.J. Chen, I. Markov, G.U. Singco, L.T. Shi, C. Farrell, K.N. Tu [*Mater. Chem. Phys. (Netherlands)* vol.38 (1994) p.250–7]

[56] J.R. Abelson, K.B. Kim, D.E. Mercer, C.R. Helms, R. Sinclair, T.W. Sigmon [*J. Appl. Phys. (USA)* vol.63 (1988) p.689–92]

[57] L.J. Chen, I.W. Wu, J.J. Chu, C.W. Nieh [*J. Appl. Phys. (USA)* vol.63 (1988) p.2778–82]

[58] A.E. Morgan, E.K. Boardbent, K.N. Ritz, D.K. Sadana, B.J. Burrow [*J. Appl. Phys. (USA)* vol.64 (1988) p.344–53]

[59] W. Lur, L.J. Chen [*Appl. Phys. Lett. (USA)* vol.54 (1989) p.1217–19]

[60] J.Y. Cheng, L.J. Chen [*J. Appl. Phys. (USA)* vol.68 (1990) p.4002–7]

[61] J.Y. Cheng, L. J. Chen [*J. Appl. Phys. (USA)* vol.69 (1991) p.2161–8]

[62] M.H. Wang, L.J. Chen [*Appl. Phys. Lett. (USA)* vol.58 (1991) p.463–5]

[63] M.H. Wang, L.J. Chen [*J. Appl. Phys. (USA)* vol.71 (1992) p.5918–25]

[64] W.Y. Hsieh, J.H. Lin, L.J. Chen [*Appl. Phys. Lett. (USA)* vol.62 (1993) p.1088–90]

[65] J.H. Lin, W.Y. Hsieh, L.J. Chen [*J. Appl. Phys. (USA)* vol.79 (1996) p.9123–8]

[66] C.H. Luo, L.J. Chen [*Appl. Surf. Sci. (Netherlands)* vol.113/114 (1997) p.556–61]

[67] R. Beyers, R. Sinclair [*J. Appl. Phys. (USA)* vol.57 (1985) p.5240–5]

[68] O.L. Krivanek, P.H. Gaskell, A. Howie [*Nature (UK)* vol.262 (1976) p.454–7]

[69] J.A. Kittl, D.A. Prinslow, P.P. Apte, M.F. Pas [*Appl. Phys. Lett. (USA)* vol.67 (1995) 2308–10]

[70] Z. Ma, L.H. Allen, D.D.J. Allman [*Thin Solid Films (Netherlands)* vol.253 (1994) 451–5]

[71] J.Y. Yew, L.J. Chen, K. Nakamura [*Appl. Phys. Lett. (USA)* vol.69 (1996) 999–1001]

[72] J.Y. Yew, H.C. Tseng, L.J. Chen, Y. Nakamura, C.Y. Chang [*Appl. Phys. Lett. (USA)* vol.69 (1996) p.3692–4]

[73] J.Y. Yew, L.J. Chen, W.F. Wu [*J. Vac. Sci. Technol. B (USA)* vol.17 (1999) p.939–44]

[74] Y.L. Shen, S. Suresh, I.A. Blech [*J. Appl. Phys. (USA)* vol.80 (1996) p.1388–98]

[75] J.B. Lasky, J.S. Nakos, O.J. Chain, P.J. Geiss [*IEEE Trans. Electron Devices (USA)* vol.38 (1991) p.262–9]

[76] A. Mouroux, S.-L. Zhang, W. Kaplan, S. Nygren, M. Östling, C.S. Petersson [*Appl. Phys. Lett. (USA)* vol.69 (1996) p.975–7]

[77] S.L. Cheng, J.J. Jou, L.J. Chen, B.Y. Tsui [*J. Mater. Res. (USA)* vol.14 (1999) p.2061–9]

[78] C. Cabral, L.A. Clevenger, J.M.E. Harper, F.M. Dheurle, R.A. Roy, K.L. Saenger, G.L. Miles, R.U. Mann [*J. Mater. Res. (USA)* vol.12 (1997) p.304–7]

[79] L.W. Cheng, S.L. Cheng, L.J. Chen, H.C. Chien, H.L. Lee, F.M. Pan [*J. Vac. Sci. Technol. A (USA)* vol.18 (2000) p.1176–9]

[80] Y. Kwon, C. Lee [*Mater. Chem. Phys. (Netherlands)* vol.63 (2000) p.202–7]

[81] A.E. White, K.T. Short, R.C. Dynes, J.P. Garno, J.M. Gibson [*Appl. Phys. Lett.* vol.50 (1987) p.95–7]

[82] S.A. Audet, C.S. Rafferery, A.E. White, K.T. Short, Y.F. Hsieh [*Mater. Res. Soc. Symp. Proc.*, Anaheim, CA, USA, 29 April–4 May 1991, Eds. M.L. Green, J.C. Gilpey, J. Wortman, R. Singh, vol.224 (*Mater. Res. Soc.*, USA, 1991) p.109–14]

[83] Y.F. Hsieh, R. Hull, A.E. White, K.T. Short [*Appl. Phys. Lett. (USA)* vol.58 (1991) p.122–4]

[84] R. Hull, Y.F. Hsieh, K.T. Short, A.E. White [*Appl. Phys. Lett. (USA)* vol.59 (1991) p.3467–9]

[85] Y.F. Hsieh, R. Hull, A.E. White, K.T. Short [*J. Appl. Phys. (USA)* vol.70 (1991) p.7354–61]

[86] R. Hull, Y.F. Hsieh, K.T. Short, A.E. White [*Mater. Res. Soc. Symp. Proc.*, San Francisco, CA, USA, 16–21 April 1990, Eds. R. Sinclair, D.J. Smith, U. Dahmen, vol.183 (*Mater. Res. Soc.*, USA, 1990) p.91–96]

[87] Y.F. Hsieh, R. Hull, A.E. White, K.T. Short [*Mater. Res. Soc. Symp. Proc.*, Boston, Massachusetts, USA, 26 Nov.–1 Dec. 1990, Eds. C.V. Thompson, J.Y. Tsao, D.J. Srolovitz, vol.202 (*Mater. Res. Soc.*, USA, 1990) p.665–72]

[88] K. Maex, A.E. White, K.T. Short, Y.F. Hsieh, R. Hull, J.W. Osenbach, H.C. Praefcke [*J. Appl. Phys. (USA)* vol.68 (1990) p.5641–7]

[89] T.L. Lee, L.J. Chen, F.R. Chen [*J. Appl. Phys. (USA)* vol.71 (1992) p.3307–12]

[90] M.P. Siegal, F.H. Kaatz, W.R. Graham, J.J. Santiago, J. Van der Spiegel [*J. Appl. Phys. (USA)* vol.66 (1989) p.2999–3006]

[91] F.H. Kaatz, W.R. Graham, J. Van der Spiegel [*Appl. Phys. Lett. (USA)* vol.62 (1993) p.1748–50]

[92] M.P. Siegal, W.R. Graham, J.J. Santiago-Aviles [*J. Appl. Phys. (USA)* vol.68 (1990) p.574–80]

[93] M. Siegal, L.J. Martinez-Miranda, J.J. Santiago-Aviles, W.R. Graham, M.P. Siegal [*J. Appl. Phys. (USA)* vol.75 (1994) p.1517–20]

[94] N. Frangis, G. Van Tendeloo, J. Van Landuyt, P. Muret, T.T.A. Nguyen [*J. Alloys Compd. (Netherlands)* vol.234 (1996) p.244–50]

[95] K.L. Pey, R. Sundaresan, H. Wong, S.Y. Siah, C.H. Tung [*Mater. Sci. Engng. B (Netherlands)* vol.74 (2000) p.289–95]

[96] Y.F. Hsieh, R. Hull, A.E. White, K.T. Short [*Mater. Res. Soc. Symp. Proc.*, Boston, Massachusetts, USA, 26 Nov.–1 Dec. 1990, Eds. C.V. Thompson, J.Y. Tsao, D.J. Srolovitz, vol.202 (*Mater. Res. Soc.*, USA, 1990)]

[97] J.S. Byun, J.M. Seon, J.W. Park, H. Hwang, J.J. Kim [*Mater. Res. Soc. Symp. Proc.*, San Francisco, CA, USA, 8–12 April 1996, Eds. R.T. Tung, K. Maex, P.W. Pellegrini, L.H. Allen, vol.402 (*Mater. Res. Soc.*, USA, 1990) p.167–72]

[98] R.T. Tung, F. Schrey [*Mater. Res. Soc. Symp. Proc.*, San Francisco, CA, USA, 8–12 April 1996, Eds. R.T. Tung, K. Maex, P.W. Pellegrini, L.H. Allen, vol.402 (*Mater. Res. Soc.*, USA, 1990) p.173–8]

[99] J.S. Byun, J.M. Seon, K.S. Youn, H.S. Hwang, J.W. Park, J.J. Kim [*J. Electrochem. Soc. (USA)* vol.143 (1996) L56–L58]

[100] M.P. Siegal, F.H. Kaatz, W.R. Graham, J.J. Santiago, J. Van der Spiegel [*J. Appl. Phys. (USA)* vol.66 (1989) p.2999–3006]

[101] M.P. Siegal, W.R. Graham, J.J. Santiago-Aviles [*J. Appl. Phys. (USA)* vol.68 (1990) p.574–80]

[102] G.H. Shen, J.C. Chen, L.J. Chen [*Appl. Surf. Sci. (Netherlands)* vol.142 (1999) p.332–5]

[103] J.A. Knapp, S.T. Picraux [*Appl. Phys. Lett. (USA)* vol.48 (1986) p.466–8]

[104] F.A. d'Avitaya, A. Perio, J.-C. Oberlin, Y. Campidelli, J.A. Chroboczak [*Appl. Phys. Lett. (USA)* vol.54 (1989) p.2198–200]

[105] J.E.E. Baglin, F.M. d'Heurle, C.S. Petersson [*J. Appl. Phys. (USA)* vol.52 (1981) p.2841–6]

[106] R. Baptist, S. Ferrer, G. Grenet, H.C. Poon [*Phys. Rev. Lett. (USA)* vol.64 (1990) p.311–14]

[107] C.H. Luo, G.H. Shen, L.J. Chen [*Appl. Surf. Sci. (Netherlands)* vol.113/114 (1997) p.457–61]

[108] P. Revesz, L.S. Hung, J.W. Mayer, L.R. Zheng [*Appl. Phys. Lett. (USA)* vol.48 (1986) p.1591–3]

[109] L.R. Zheng, S.Q. Feng, L.S. Hung, J.W. Mayer, G. Miles, P. Revesz [*Appl. Phys. Lett. (USA)* vol.48 (1986) p.767–9]

[110] C.Y. Wong, P.A. McFarland, C.Y. Ting, L.K. Wang [*J. Appl. Phys. (USA)* vol.60 (1986) p.243–6]

[111] T.P. Nolan. R. Sinclair, R. Beyers [*J. Appl. Phys. (USA)* vol.71 (1992) p.720–4]

[112] C.A. Sukow, R.J. Nemanich [*J. Mater. Res. (USA)* vol.9 (1992) p.1214–27]

[113] Z.G. Xiao, G.A. Rozgonyi, C.A. Canovai, C.M. Osburn [*J. Mater. Res. (USA)* vol.7 (1992) p.269–72]

[114] P. Pevesz, L.R. Zhang, L.S. Hung, J.W. Mayer [*Appl. Phys. Lett. (USA)* vol.48 (1986) p.1591–3]

[115] S.M. Chang, S.L. Cheng, L.J. Chen, C.H. Luo [*J. Appl. Phys. (USA)* vol.90 (2001) p.1779–83]

[116] J.F. Chen, L.J. Chen [*Thin Solid Films (Netherlands)* vol.293 (1997) p.34–9]

APPENDIX: GLOSSARY

Agglomeration
A morphological degradation of microstructure in polycrystalline materials, driven by surface/grain boundary energy reduction, resulting in disintegration of microstructure through grain boundary grooving and islanding.

Amorphous silicide
A metastable titanium silicide phase which can be formed by annealing a Ti/Si (crystalline or amorphous) bilayer at low temperatures. It can be stable up to $\sim 500\,°C$.

Backscattering spectrometry
Measurement of energy of backscattered ions.

Chemical Vapour Deposition (CVD)
A most common thin film deposition method in advanced semiconductor manufacturing. The process involves the introduction of materials in gaseous vapour form into an air-tight deposition chamber. The gas vapours chemically react with an underling substrate (silicon wafer) and form or deposit a solid thin film onto the substrate. It has been widely used to deposit films of semiconductors (crystalline and non-crystalline), insulators as well as metals.

Complementary Metal-Oxide Semiconductor (CMOS)
A key device in state-of-the-art silicon microelectronics, which uses both N-type and P-type transistors to realise logic function. The main advantage of CMOS over NMOS and bipolar technology is the much smaller power consumption and dissipation. Today, CMOS technology is the dominant semiconductor technology for microprocessors, memories and application specific integrated circuits.

Copper Damascene
The name of the copper interconnect formation process for advanced ULSI. The surface of the interlayer dielectric film is patterned by standard photolithography and reactive ion etching, copper is deposited into the recessed trench by electroplating, and excess copper is removed by chemical mechanical polishing. Consequently, low-resistivity inlaid copper metal interconnect lines are formed in ULSI.

Diffusion
Motion under a concentration gradient.

Appendix: Glossary

Dopants
Impurities that provide electrons or holes to Si.

Drain Induced Barrier Lowering (DIBL)
A phenomenon in short-channel devices. When a high drain voltage is applied to a short-channel device, the source and drain fields penetrate deeply into the middle of the channel, which lowers the potential barrier between the source and drain resulting in the decrease of the threshold voltage. It is observed as variations of the measured threshold voltage with reduced gate length and can typically be eliminated by properly scaling the drain and source depths while increasing the substrate doping density.

Enthalpy of formation
Equal to the internal energy of the system plus the work done in moving the particles into the system of which they are comprised. Denoted by $H = U + PV$.

Epitaxial
Layer of atoms aligned structurally with substrate

Equilibrium C54 disilicide
An equilibrium, low resistivity (15 ~ 20 µΩ–cm) disilicide phase with a face-centred orthorhombic structure. It can be formed at elevated temperatures by converting the C49 phase via polymorphic phase transformation.

Eutectic
Minimum freezing point for two components.

Focused ion beam
Implantation system with a selected beam area.

Germanosilicide
An alloy compound, which may be formed through the solid phase reaction between a metal and a Si-Ge alloy. The relative Ge and Si concentration can vary in the compound to achieve thermodynamic equilibrium.

Gibbs free energy
The free energy of a thermodynamic system under constant temperature and pressure. The Gibbs free energy of a system is determined by measuring changes in the free energy with respect to a reference state. Denoted by the equation $\Delta G = \Delta H - T\Delta S$, where ΔH is equal to the enthalpy of formation, ΔS is equal to the entropy of formation, and T is the temperature of the system in question.

Appendix: Glossary

Heterojunction Bipolar Transistor (HBT)
Emitter-base-collector transistor structure formed from the junction of dissimilar materials; characterised by increased efficiency in the injection of charge carriers across the emitter junction. Si/SiGe/Si HBT structures have demonstrated high frequency operation.

Interconnect
Electrical connection between devices.

Ion beam mixing
Use of ion beams to intermix layers.

LDD MOSFET
Lightly doped drain (LDD) MOSFET, is designed with an extended moderately doped source-drain region to relieve high electric fields and related hot-electron effects. LDD structures are prevalent in modern CMOS technology.

Low-Pressure Chemical Vapour Deposition (LPCVD)
Chemical vapour deposition process carried out at reduced pressure. It provides improved conformality of coating and purity of the films as compared to atmospheric pressure CVD.

Metal-Oxide-Semiconductor Field Effect Transistor (MOSFET)
FET with MOS structure as a gate. They come in four different types: enhancement or depletion mode, n-channel or p-channel. The basic structure of an n-channel MOSFET consists of a source and a drain, two highly conducting n-type semiconductor regions which are isolated from the p-type substrate by reversed-biased p-n diode. A metal (or poly-crystalline) gate covers the region between source and drain, but is separated from the semiconductor by the gate oxide. When a sufficiently large bias is applied to the gate so that a surface inversion layer (or channel) is formed between the two n+ regions, the source and drain then connect by a conducting-surface n channel through which a larger current flows. Varying the gate voltage can modulate the conductance of this channel. MOSFET is the most important device for integrated circuits.

Metallisation
Electrical connections between active elements.

Metastable C49 disilicide
A metastable titanium disilicide phase with a base-centred orthorhombic structure. It can be formed through the Ti/Si bilayer

reaction or the crystallisation of co-deposited amorphous $TiSi_2$ alloy. The resistivity of the phase is $60 \sim 100\,\mu\Omega$–cm.

NiSi
Nickel silicide.

Ohmic contact
Metal-semiconductor contact which displays linear I–V characteristics and offers little resistance to the flow of current under biasing conditions.

Oxide
Silicon dioxide.

Pit
A penetrating hollow in Al-Si alloys on Si.

Polymorphic transformation
A phase transformation process where only structural change is involved and the compositions of the initial phase and the final phase remain the same.

polySi
Polycrystalline Si.

Rapid Thermal Anneal (RTA)
A process of raising the temperature of a wafer without applying a lot of thermal energy. This allows implanted dopant atoms to be activated, and temperature critical processes (such as those involved in silicidation) to go to completion without the adverse side effects associated with large amounts of heat.

Recrystallisation
Returning structure to crystal form.

RTP
Rapid thermal processing.

Rutherford Backscattering Spectrometry (RBS)
RBS is based on backscattering of ions incident on a sample and is quantitative without recourse to calibrated standards. It is based on the bombardment of a sample with high energy He ions of 1 to 3 MeV and measuring energy of the backscattered He ions. RBS allows determination of the masses of the elements and depth profiles and the crystalline structure in a nondestructive manner.

Appendix: Glossary

SALICIDE
SALICIDE is an abbreviation of self-aligned silicide formation in source/drain regions and gate electrodes of CMOS transistors. Transition metal is deposited on the surface of CMOS transistors, where the top SiO_2 layer is removed on active source and drain Si regions as well as gate poly-Si electrodes. Metal silicide is formed after rapid thermal annealing only at the interface between metal and Si of the active area. Unreacted metal is removed by selective etching.

Schottky barrier
Potential barrier to the flow of charge carriers from a metal into a semiconductor at a metal-semiconductor contact.

Secondary Ion Mass Spectroscopy (SIMS)
SIMS is the destructive removal of material from the sample by sputtering and the analysis by mass analyser. A primary ion beam impinges on the sample, and the small fraction is ejected as positive and negative ions and detected as a mass spectrum. The technique is element specific and is capable of detecting all elements as well as isotopes and molecular species with detection limits in the 1014 to 1015 cm-3 range.

SEM
Scanning electron microscope.

SOI
Silicon-on-insulator, a structure where a thin silicon layer lies atop of an insulator, such as silicon dioxide, that may be a silicon or other substrate. This allows for isolation of the active silicon layer containing circuit structures from the bulk substrate and each other, permitting improved device performance for CMOS devices, and more flexible integration schemes for electronic, MEMS and MOEMS devices.

Solid state amorphisation
Formation of an amorphous alloy between Ti and Si during low temperature annealing. Its occurrence is kinetically driven due to the inability of forming crystalline phases and the tendency for energy reduction at low temperatures. Two necessary conditions must be met for this to occur. First, the heat of mixing of the two elements must be negative, which provides the thermodynamic driving force. Second, one element must be a sufficiently fast moving species in the other element and must diffuse substantially into the other at temperatures below the crystallisation temperature of the amorphous phase alloy.

Specific contact resistivity, $\rho_c (\Omega - cm^2)$
The effective resistance (normalised to junction area) of the silicide/junction (cm^2) region. The region includes both the actual metallurgical silicide/Si interface as well as the depletion region below the interface.

Spikes
Pit in Al-Si metal layers on Si.

STM
Scanning tunnelling microscope.

Stranski-Krastinov growth mode
Growth mode whereby a thin film with a surface energy less than that of the substrate on which it is grown, is formed. The successive growth of layers induces a build-up of strain in the film, which results in the formation of 3-D islands to accommodate the strain.

Surface reconstruction
The periodic rearrangement of atoms on a crystalline surface, which occurs in order to minimise surface energies through the reduction of unpaired electron bonds and strain relaxation.

Ternary phase diagram
Phase diagram at constant temperature and pressure, depicted as a triangle, illustrating equilibrium between any co-existing phases constituted from three elements. Each apex of the triangle represents 100% pure element.

Thermodynamic stability
Property of a system characterised by a minimum in the total Gibbs free energy of a system.

Time Dependent Dielectric Breakdown (TDDB)
Oxide film integrity is determined by time-dependent measurements. In the constant gate voltage method, a gate voltage near the breakdown voltage is applied and the gate current is measured as a function of time at room temperature. A lower gate voltage is used at an elevated temperature. For constant current measurements, a constant current is forced through the oxide and the gate voltage is measured as a function of time. When the oxide is driven into breakdown, a charge-to-breakdown (which is the charge density flowing through the oxide necessary to break it down) and the time to breakdown are defined.

Appendix: Glossary

Titanium silicides
Alloy compounds which can be formed through the solid-state reaction between titanium and silicon. There are several equilibrium titanium silicide phases in the Ti-Si binary system.

Transmission Electron Microscopy (TEM)
TEM is similar to an optical microscope and contains a series of lenses to magnify the sample. TEM has extremely high resolution approaching 0.15 nm due to the shorter wavelength of 0.004 nm for 100 KV. Transmitted and forward scattered electrons form a diffraction pattern in the back focal plane and a magnified image in the image plane, which is called transmission electron diffraction (TED). With additional lenses either the image or the diffraction pattern is projected on to a fluorescent screen for viewing or for electronic recording.

VLS
Vapour liquid solid.

SUBJECT INDEX

absorption coefficient 170
adhesion layer 37–8, 43–4
adhesion to SiO_2 35
agglomeration 59–61, 64, 66–7, 97, 99, 257–8, 263
   film thickness dependence 131
   NiSi 130–6
Al
   resistivity 16
Al design rules 2
Al metallisation 1–2, 17
Al plug 35–7
Al/Cu alloys 18, 34
Al/Si alloys 16–18
Al/Si/Cu alloys 17–18
amorphisation 211–17
amorphous interlayers 22, 186–8, 191
   silicide formation 243–6
amorphous metal/Si alloy films 242, 245–6, 263
anti-reflection coating 18, 44
atomic diameter
   Co 78
   Si 78
atomic force microscopy 229, 236–7
   defect analysis 254
Au/Si interface 3–4
Auger electron spectroscopy 230–1, 233–5
auto-correlation function analysis 243–5, 247
axiotaxy 95, 128–9

backscattering spectrometry 4, 263
band structure
   $FeSi_2$ 162–5
bandgap 153–4
   $FeSi_2$ 154, 162–5, 170
   metal silicides 154
   $Si_{1-x}Ge_x$ 195

beam-solid interaction 230–2
   electron beam 231
   ion beam 231
   optical beam 231
bond length
   Si/Ge 176
bridging problem 24, 116
bulk modulus
   NiSi 123
buried silicide layers 250–2

C49-to-C54 $TiSi_2$ phase transformation 54–9
   activation energy 55, 58
   enhancement 68–73
   film thickness effects 56–8
   linewidth effects 58–9
C54-$TiSi_2$ phase transformation 54–6, 68–73
   high temperature deposition 68–9
   metal impurity implant 70–2
   pre-amorphisation implant 69–70
carrier mobility
   $FeSi_2$ 155
characterisation 229
characterisation tools 229–37
chemical mechanical polishing 38
chemical stability
   $TiSi_2$ 65
chemical vapour deposition 263
   low-pressure 32–3, 42–3, 265
   metal-organic 43
chemical/elemental analysis 230
CMOS devices 5, 175, 202, 263
   SOI 202–4, 216–17
CMOS transistors 77, 88
   Co salicide 90–1
   I–V characteristics 90–2
Co
   crystal structure 77–8, 81
   density 81

Subject index

Co salicide CMOS process 79, 82, 85, 92–3
   Ti at interface 89–90
   Ti cap 86, 88–9
   TiN cap 85–8
Co salicide structure
   electrical characteristics 87–93
Co silicide formation
   volume changes 82
Co single layer deposition 80–1, 83
   oxidation suppression 80–1
Co/Ti bilayer deposition 89–90
coefficient of thermal expansion
   $FeSi_2$ 160–1
   Ni 102
   Ni/Si phases 102
   NiSi 118–24
   $NiSi_2$ 106
   Si 102, 106, 122
collimated sputtering 40–1
complementary silicide structure 221–2
contact resistance 16–17, 65–6, 221–2, 268
   $CoSi_2$/Si 142
   NiSi/Si 142
   silicide/Si 207–8
   TiN/Si 39
   $TiSi_2$/Si 66–7
contact windows 16
copper damascene 263
corrosion resistance 35
CoSi
   crystal parameters 81
   crystal structure 78
   density 81
   melting point 78
$CoSi_2$ 20, 138, 209
   crystal structure 78, 80–1
   density 81
   layer structure 7–8
   limitations 96–7
   melting point 20, 100
   oxidation 30
   resistivity 5, 8, 20, 104
   sheet resistance 29, 91, 104, 219–20, 246, 249
   Si consumption 104
   thermal stability 20, 84–5, 209–10

$CoSi_2$ formation 79, 82–5
   epitaxial growth 31–2
$CoSi_2$ on Si/Ge 187–90
$CoSi_2$ spherulites 242, 245–6
$CoSi_2$/Si interface roughness 97
Co_2Si
   crystal structure 78, 81
   density 81
   melting point 78
Co_3Si
   crystal parameters 81
cost of ownership 42
Cr-Si/Ge system 193
cracks 252–4
$CrSi_2$ 252–3
crystal structure
   C49-$TiSi_2$ 53–4
   C54-$TiSi_2$ 54
   characterisation 240–6
   Co 77–8, 81
   CoSi 78, 81
   $CoSi_2$ 78, 80–1
   Co_2Si 78, 81
   Co_3Si 81
   $CrSi_2$ 252
   $FeSi_2$ 154
   germanides 182
   Ni 101
   Ni/Si phases 101
   NiSi 101, 119
   $NiSi_2$ 101, 130
   Si 78–9, 101, 130
   silicides 182
Cu interconnects 35
Cu wiring 45
Cu_3Si
   oxidation 30–1

de Broglie wave theory 232
defect analysis 235, 252–6
delay degradation factor 205
density
   Co 81
   CoSi 81
   $CoSi_2$ 81
   Co_2Si 81

Subject index

Ni 101
    Ni/Si phases 101
    Si 101
detection limits 234–5
device characterisation 136–43
device requirements 137–8
diffusion 263
    low temperature 117–18
    Si into Al 2
diffusion barrier layer 19, 39–43
    TiN 39–42
diffusion coefficients
    Ni in Si 111
diffusion-controlled reactions 114–17
dimer-adatom-stacking fault models 178–9
dimers 176–8
dopant segregation 142–3
dopants 26–9, 264
double heterostructure 241
drain-induced-barrier-lowering 201, 264

electrical analysis 230
electroluminescence
    $FeSi_2$ 171–2
electromigration 1, 17–18, 34
    resistance 35
electron spectroscopy for chemical analysis 230
electron wavelength 232
electrostatic force microscopy 229, 237
end-of-range defects 27–9
energy dispersive spectroscopy 230–1, 234–5
energy dispersive X-ray analysis 250
enthalpy of formation 103, 114, 182, 264
    Ni/Si phases 102–3, 115
entropy of mixing 135, 182
epitaxial growth 26–7, 31–2, 240–2, 264
    $CoSi_2$ 31–2
    $FeSi_2$ 161–2
    $NiSi_2$ 31, 130
epitaxial silicides 240–4
equilibrium C54 disilicide 264
$ErSi_2$ 8–9
$ErSi_2$ nanowires 9
eutectic temperature 264
    Al/Si 2
    Au/Si 3, 10–11
    $CoSi_2$ 20
    Ni/Si 108
    NiSi 20
    $TiSi_2$ 20

$FcSi_2$ 153–4
    band structure 162–5
    carrier mobility 155
    coefficient of thermal expansion 160–1
    lattice constants 154, 159, 161–2
    optical properties 169–72
    stress 159–65
    stress measurement 165–7
    α phase 154
    β phase 154
$FeSi_2$ film 161–5
    epitaxial growth mode 161–3
$FeSi_2$ growth 155–9
    ion beam synthesis 155–9, 165–7
    precipitate structure 158–9
    reactive deposition epitaxy 155–7
field emission gun 233
first nucleated phase 21–2, 243, 246
focused ion beam 264
focused ion beam microscopy 229, 231, 233–6
    defect analysis 253
    imaging modes 237–8
force-fill 36–7
formation temperature
    $CoSi_2$ 20
    $CrSi_2$ 252
    NiSi 20
    $TiSi_2$ 20
four-point probe 230

gate dimensions 137–8
gate electrodes 32–4
gate resistance 204–5
Ge^+ pre-amorphisation process 211–18
    doping effects 213–14
germanides 182
germanosilicide contacts 181
germanosilicides 176, 264

Gibbs free energy 182–6, 188, 264
glue layer 43
grain boundary grooving 60–1, 256–7, 263
grain size
 C49-TiSi$_2$ 59, 72
 CrSi$_2$ 252
 TiSi$_2$ 217–18
grazing incidence X-ray diffractometry 247–9
grooving: *see* grain boundary grooving
growth kinetics 22–3

heterojunction bipolar transistor 153, 175, 265
high electron mobility transistor 153

illumination effect 168–72
image resolution 232–4
integration
 TiSi$_2$ 61–6
interconnection limited circuit 16
interconnection lines 16
interconnects 16–17, 34–5, 265
 Al/Cu alloys 34
 Cu 35
 multilevel 15–16
 W 34
interface roughness 97–8, 130
interface thermodynamics 181–4
interfacial oxide layer 25–6
International Technology Roadmap for Semiconductors 15, 137
ion beam synthesis 155–9, 165–7
ion-beam mixing 7–8, 265
 focused 7–8
ionised metal plasma sputtering 41–2
island formation 28–9, 256, 258, 263

junction leakage 140–1

Kirkendall voiding 117
Kissinger analysis 133

laser diodes 153
latch-up 201–2
lateral encroachment 63–4, 67

lattice constants
 Co 77–8, 81
 CoSi 78, 81
 CoSi$_2$ 78, 81
 Co$_2$Si 78, 81
 Co$_3$Si 81
 FeSi$_2$ 154, 159, 161–2
 Ni 101
 Ni/Si phases 101
 NiSi 101, 119–22
 NiSi$_2$ 101, 130
 Si 101, 130, 161–2
lattice mismatch
 CoSi$_2$/Si 130, 210
 CrSi$_2$/Si 252
 ErSi$_2$/Si 8–9
 FeSi$_2$/Si 159
 Ge/Si 176
 NiSi$_2$/Si 130
LDD MOSFET 265
leakage current 89, 92, 140
light emitting diodes 153, 157, 168, 171
light-emitting FeSi$_2$ 153–4
 see also FeSi$_2$
local density approximation 164–5
long throw sputtering 41
low-pressure chemical vapour deposition 32–3, 42–3, 265

markers 23–4
melting point 35
 CoSi 78
 CoSi$_2$ 20, 100
 CoSi$_3$ 78
 Co$_2$Si 78
 Ni 102
 Ni/Si phases 102–3
 NiSi 20, 100, 102
 NiSi$_2$ 132
 Si 102
 TiSi$_2$ 20, 59
mesotaxy 250, 253
metal contacts 16–32
 Al/Si alloys 16–18
 Al/Si/Cu alloys 17–18
 metal silicides 18–32

Subject index

metal induced gap states model 195
metal-organic chemical vapour deposition 43
metallisation 1, 16, 18–19, 265
metastable C49 disilicide 265–6
microstructure
  C49-TiSi$_2$ 57, 69–70
microtwins 242
morphological stability 59–61, 64, 255, 257
  film thickness effects 131–3
  Ni/Si phases 103
  NiSi 130–4
morphology observation 237–42
MOS devices
  series resistance 65
MOSFET devices 29, 32, 176, 265
  LDD 265
  resistance 206
  Si/Ge 176
  silicide process 208–11
  SOI 203, 206, 210, 220–1
multilevel wiring 45
multiphases 21–2

N$^+$ implantation 29
nanodots 168
nanoelectronics 4
nanoscale silicides 4–12
nanotubes 9
nanowires 8–12
  metal contacts 11
  self-assembly 9
  VLS growth method 10
Ni
  coefficient of thermal expansion 102
  crystal structure 101
  density 101
  melting point 102
  resistivity 102
  Young's modulus 102
Ni ambient 29
Ni/Si reactions 106–11
NiSi 20, 95–6, 138–9, 209
  advantages 98
  agglomeration 130–6
  bulk modulus 123
  crystal structure 101, 119–22

limitations 98–9
melting point 20, 100, 102
morphological stability 130–4
resistivity 5–8, 20, 102, 104, 220
Si consumption 104
stresses 103, 121–4
thermal expansion 118–24
thermal stability 20, 209–10
NiSi film
  BF2 implantation 136
  high temperature degradation 129–36
  high temperature stabilisation 134–6
  pole figures 125–8
  Pt addition 135–6
  RBS spectra 5–6
  resistivity 5–8, 141–2
  sheet resistance 5, 7
  substrate effects 5–7
  texture 124–9
  thermal stability 6–7
NiSi formation mechanism 114–17
NiSi on Si/Ge 190–2
NiSi$_2$ 5–6, 99, 141
  coefficient of thermal expansion 106
  crystal structure 101, 130, 106
  epitaxial growth 26–7, 130
  formation at high temperature 129–34
  melting point 132
  oxidation 30
  sheet resistance 29
NMOS transistor 89–90, 92, 143, 205
  source/drain resistance 220–1
  transconductance 204
nucleation 23, 114–16
nucleation theory 97, 135

ohmic contact 3, 194, 266
optical microscopy 232–3
  imaging modes 236–8
  TiSi$_2$ 240
orientation distribution function 124–5
overhang 40
oxidation 29–31
  low temperature 3–4
oxidising ambient 25

Subject index

packing density 16
Pd-Si/Ge system 193–4
phase diagram 268
   Al/Si 17
   Co/Si 78, 81, 100
   Co/Si/Ge 188–9
   Cr/Si/Ge 193
   metal/Si 21–2
   Ni/Si 99–100, 113, 129–30
   Ni/Si/Ge 191
   Si/Ge 175
   Ti/Si 21–2, 50
   Ti/Si/Ge 185
   Ti/Si/O 63
   Zr/Si/Ge 192
phase formation 245–52
   Co/Si 104–6
   dopant effects 26–7, 112–14
   Ni/Si 104–11
   Ti/Si 50–6
phase identification 245–52
phase nucleation 21–2
photoluminescence
   $FeSi_2$ 168, 170–2
physical/structural analysis 229–30
pinholes 253–4, 257
pit formation 1–2, 17, 266
planar defects 254–5
planarisation 36–8
plug 35–9
   Al 35–7
   W 37–9
PMOS transistor 89–90, 92, 143, 205
   source/drain resistance 220–1
   transconductance 204
pole figures 124–8
polycide 32–3, 45, 49
polymorphic transformation 54–5, 266
poly-Si gates 138
positional control 4–5, 8–9
pre-amorphisation implant 69–70
probe size 234–5
Pt-Si/Ge system 193
PtSi 209

quantum dot 161

raised source/drain process 97
Raman spectra
   $FeSi_2$ 166–7
rapid thermal anneal 266
reactive deposition epitaxy 155–6
reactive ion sputtering 40
recessed spacer process 72–3
rectifying contact 194
reflective high energy electron
      diffraction 230
reflow 36
resistivity 35
   Al 16
   $CoSi_2$ 5, 8, 20, 104
   Ni 102
   Ni/Si phases 102, 110
   NiSi 5–8, 20, 102, 104, 220
   Si 102
   $TiSi_2$ 5, 20, 51
reverse linewidth effect 117, 139
ring defects 44
Rutherford backscattering spectroscopy
      230–1, 234–5, 266

SALICIDE 5, 19, 25, 33–4, 45, 267
   Co 77–9, 85
   scaling 61, 72
   scaling limitations 66–8
   Ti 50, 61, 72
scaling 66–8
   Co salicide structure 91–2
scanning capacitance microscopy 229, 237
scanning electron microscopy 229, 231–6
   defect analysis 253–4
   imaging modes 237–8
   silicide on patterned Si 245–8
   $TiSi_2$ 240–1
   WSi_2 238–40
scanning transmission electron microscopy
   229, 237
scanning tunnelling microscopy 230
Schottky barrier height 194–5, 230, 267
   Al on Si/Ge 195–6
   Co on Si/Ge 195–6
   $CoSi_2$ 20, 221

Subject index

ErSi$_2$ 222
NiSi 20
Pt on Si/Ge 195–6
PtSi 222
  strain effects 196
TiSi$_2$ 20, 221
Schottky-Mott model 195
secondary ion mass spectroscopy 230–1, 234–6, 250, 267
segregation 179–81
selective deposition of silicide 222–5
self-aligned silicidation: *see* SALICIDE
self-replication 4–5
shallow junctions 139–41
shallow trench isolation 141
sheet resistance
  Co salicide 87–8
  Co/Si 105
  CoSi$_2$ 29, 91, 104, 219–20, 246, 249
  doping effects 213–14
  Ni/Si 105, 131–2
  NiSi 5, 104
  NiSi$_2$ 29
  SOI 203–4
  Ti 64
  Ti on Si 256–7
  TiN$_x$ 64
  TiN$_x$ on Si 256–7
  TiSi 69
  TiSi$_2$ 28, 69, 214–18, 220, 224
  TiSi$_2$/poly-Si 60
short channel effects 201, 203
Si
  coefficient of thermal expansion 102, 106, 122
  crystal structure 78–9, 101, 106, 130, 161–2
  density 101
  melting point 102
  resistivity 102
  Young's modulus 102–3
Si consumption 104, 137, 139
Si/Ge devices 175–6
  silicide contacts 184–94
Si$_{1-x}$Ge$_x$/Si(001) surfaces 176–8
Si$_{1-x}$Ge$_x$/Si(111) surfaces 178–9

silicide formation 19–25
  dominant diffusing species 23–4
  doping effects 213–14
  epitaxial growth 31–2
  growth kinetics 22
  impurity effects 25–9
  interposing layers 246–50
  mechanisms 24–5
  metal ion implantation into Si 250–2
  on patterned (001) and (111)Si 245–6
silicide lines 6–7
silicides 18–32, 182
  nanodimensions 4–12
  see also entries for individual silicides
SOI 267
SOI devices 201–3
  advanced silicide technology 211–25
  CMOS 202–4, 216–17
  CoSi$_2$ formation 217–20
  fully depleted 202–3, 216–17
  Ge$^+$ pre-amorphisation 211–18
  NiSi formation 220–1
  partially depleted 202–3
  selective deposition of silicide 222–5
  silicide contact formation 208
  silicide design analysis 206–8
  silicide process 208–11
  silicide process flow 216–17
  source/drain series resistance 203–4
  TiSi$_2$ formation 211–19
solid-state amorphisation 53, 267
solubility
  Co in Si 100, 105
  Ni in Si 100, 105
  Si in Al 2, 17
  Ti in CoSi 189
  Ti in CoSi$_2$ 189
source/drain series resistance 203–4, 206–8
space group
  FeSi$_2$ 154
  Ni 101
  Ni/Si phases 101
  Si 101
specific contact resistivity 268
specific heat capacity 35
spiking 2–3, 17, 268

277

Subject index

spreading resistance profiling 230
sputtering 37
  Co 80, 83–5
  collimated 40–1
  ionised metal plasma 41–2
  long throw 41
  reactive ion 40
stacking faults 254–6
stoichiometry
  $Si_{1-x}Ge_x$ surfaces 179–81
strain energy density 123
Stranski-Krastanov growth 176, 268
superconductivity 193
superlattice structure 255
surface electronic structure
  $Si_{1-x}Ge_x/Si$ 176–9
surface morphology
  Ti(Si/Ge) on Si/Ge 184
  $TiSi_2$ 28
surface reconstruction 268
  Ge/Si 177–9
surface roughness
  Ni/Si 110–11
synchrotron radiation-ultraviolet photoemission spectroscopy 230

technology nodes 137
template model 71–2
template-directed synthesis 10–11
ternary phase diagram 268
texture 124–9
thermal conductivity 35
thermal contraction 119
thermal expansion: *see* coefficient of thermal expansion
thermal stability 28, 181–4, 255–8
  $CoSi_2$ 20, 84–5, 209–10
  NiSi 20, 209–10
  Ti-Si/Ge 184–6
  $TiSi_2$ 20, 64–5, 209–10, 225
thermodynamic stability 181, 186, 268
Ti 43
  PECVD 42–3
  sheet resistance 64
Ti cap 86, 88–9
Ti/Co bilayer deposition 88–9

Ti/Si thin film reaction 50–61
  C49-to-C54 $TiSi_2$ 54–9, 68–73
  doping effects 61–3
  metastable amorphous phase 52
  metastable C49-$TiSi_2$ 53–4
Ti(Si/Ge) on Si/Ge 184–7
Ti/Si_3N_4 reaction 63
Ti/SiO_2 reaction 63
time dependent dielectric breakdown 268
TiN 39–44
  LPCVD 42–3
  MOCVD 43
TiN cap 85–8
TiN/Co bilayer deposition 85–8
$TiSi_2$ 20, 209
  chemical stability 65
  crystal structure 53–4
  equilibrium C54 phase 51, 68–73
  grain size 59, 72, 217–18
  melting point 20, 59
  metastable C49 phase 51
  morphological instability 59
  nitrogen implantation effects 29
  oxidation 29–30
  resistivity 5, 20, 51
  sheet resistance 28, 69, 214–18, 220, 224
  surface morphology 28
  thermal stability 20, 64–5, 209–10, 225
$TiSi_2$ formation 49–50, 53–6
  agglomeration 59–61
  C49 phase 53–4
  C49-C54 transformation 54–9
  C54 phase 54–6, 68–73
  on pre-amorphised Si 211–17
total reflection X-ray fluorescence 230–1, 234, 236
transconductance 204
transmission electron diffraction 269
transmission electron microscopy 229, 231–7, 257, 269
  advanced techniques 236
  a-Si/Y/Si 241, 244
  buried silicides 251
  $CoSi_2$ 240–2, 244–6
  defect analysis 253–6
  imaging modes 237–8

initial silicide formation 243–7
Pd$_2$Si 241, 243
silicide on patterned Si 245–8
Ti/Mo/Si 249–50
TiSi$_2$ 240–2
WSi$_2$ 240, 242
triple grain junctions 55–6, 58–9
tungsten etchback 38

ULSI 16, 19–20, 45
Co salicide CMOS 92–3
ultrahigh vacuum deposition 22
unit cell dimensions
NiSi 119–22
NiSi$_2$ 106
Si 106
unit volume
Ni 101
Ni/Si phases 101
Si 101
Urbach tail 170

V/Si/Ge on Si/Ge 193
vacancies 255–6
vapour-liquid-solid growth method 10–11

vertical cavity surface emitting laser 153
vias 16
VLSI technology 181
voids 208–10, 253–4
volume reduction 159–60
FeSi$_2$ 160

W etchback 38
W interconnects 34–5
W plug 37–9
wavelength dispersive spectroscopy 230–1, 234
wetting layer 43
WSi$_2$ 153
WSi$_x$ 32–3

X-ray diffraction 229, 231, 234, 247–9
X-ray photoelectron spectroscopy 230–1, 234

Young's modulus 35
Ni 102
Ni/Si phases 102–3
Si 102–3

Zr(Si$_{1-x}$Ge$_x$)$_2$ on Si/Ge 192